FIELD SINGULARITIES AND WAVE ANALYSIS IN CONTINUUM MECHANICS

Mathematics and its Applications

Series Editor: G. M. BELL, Professor of Mathematics, King's College (KQC), University of London

Mathematics and its applications are now awe-inspiring in their scope, variety and depth. Not only is there rapid growth in pure mathematics and its applications to the traditional fields of the physical sciences, engineering and statistics, but new fields of application are emerging in biology, ecology and social organisation. The user of mathematics must assimilate subtle new techniques and also learn to handle the great power of the computer efficiently and economically.

The need of clear, concise and authoritative texts is thus greater than ever and our series will endeavour to supply this need. It aims to be comprehensive and yet flexible. Works surveying recent research will introduce new areas and up-to-date mathematical methods. Undergraduate texts on established topics will stimulate student interest by including applications relevant at the present day. The series will also include selected volumes of lecture notes which will enable certain important topics to be presented earlier than would otherwise be possible.

In all these ways it is hoped to render a valuable service to those who learn, teach, develop and use mathematics.

For full series list see end of book.

FIELD SINGULARITIES AND WAVE ANALYSIS IN CONTINUUM MECHANICS

W. KOSIŃSKI, M.A., Ph.D.
Associate Professor, Polish Academy of Sciences
Institute of Fundamental Technological Research

Translation Editor:
R. W. OGDEN, B.A., M.A., Ph.D.
George Sinclair Professor of Mathematics
University of Glasgow

ELLIS HORWOOD LIMITED
Publishers · Chichester
Halsted Press: a division of
JOHN WILEY & SONS
New York · Chichester · Brisbane · Toronto
PWN—Polish Scientific Publishers
Warsaw

English Edition first published in 1986 in coedition between
ELLIS HORWOOD LIMITED
Market Cross House, Cooper Street, Chichester, West Sussex, PO19 1EB, England
and
PWN—POLISH SCIENTIFIC PUBLISHERS
Warsaw, Poland.

Revised and enlarged edition of the Polish original *Wstęp do teorii osobliwości pola i analizy fal* published in 1981 by Państwowe Wydawnictwo Naukowe, Warszawa.

Translated by the author (Chapters 6 and 7) and *M. Doliński* (Chapters 1–5 and 8).

The Horwood publisher's colophon is reproduced from James Gillison's drawing of the ancient Market Cross, Chichester.

Distributors:

Australia, New Zealand, South-east Asia:
JACARANDA WILEY LTD.
GPO Box 859, Brisbane, Queensland 4001, Australia

Canada:
JOHN WILEY & SONS CANADA LIMITED
22 Worcester Road, Rexdale, Ontario, Canada

Europe, Africa:
JOHN WILEY & SONS LIMITED
Baffins Lane, Chichester, West Sussex, England

Albania, Bulgaria, Cuba, Czechoslovakia, German Democratic Republic, Hungary, Korean People's Democratic Republic, Mongolia, People's Republic of China, Poland, Rumania, the USSR, Vietnam, Yugoslavia:
ARS POLONA—Foreign Trade Enterprise
Krakowskie Przedmieście 7, 00-068 Warszawa, Poland.

North and South America and the rest of the world:
Halsted Press: a division of
JOHN WILEY & SONS
605 Third Avenue, New York, NY 10158, USA

British Library Cataloguing in Publication Data
Kosiński, W.
Field singularities and wave analysis in
continuum mechanics. — (Ellis Horwood series in maths)
1. Elastic waves — Diffraction
I. Title II. Ogden, R. W. III. Wstęp do teorii
osobliwości pola i analizy fal. *English*
531'.33 QA935

ISBN 0-85312-543-0 (Ellis Horwood Limited)
ISBN 0-470-20278-5 (Halsted Press)

Library of Congress Card No. 85-27031

Printed in Poland

To Ewa

Table of Contents

Preface

Continuum physics is concerned with the description of physical phenomena as observed at the macroscopic level, with no reference to the underlying micro-structure of the matter constituting the medium in which the phenomena occur. The medium itself is regarded as a continuous distribution of matter and is referred to as a *continuous medium* (or simply *continuum*). Physical quantities (such as mass or velocity) are distributed through the medium, and in mathematical terms are treated as *fields*. These fields are subject to a number of physical laws which express general principles common to all forms of matter. Examples of such laws are the laws of balance of mass and energy. The balance laws are formulated as integral equations governing fields defined on regions of space occupied by a material body in motion.

A disturbance in the continuity of a phenomenon or physical field is termed a *singularity*. The aim of this books is to present a unified view of the theory of non-relativistic thermodynamics incorporating phenomena with singularities. These singularities will present as discontinuous functions or their derivatives, and in the form of the discontinuity in respect to the Lebesgue measure of physical quantities. The examples of the first type of singularity are shock and acceleration waves. The second type is usually associated with surface concentrations of physical quantities.

Discontinuities in fields may be caused by discontinuities in material properties or by some discontinuous behaviour of the source which gives rise to the fields. In most problems discontinuities in the source function propagate through the medium, and if the source function is prescribed at the boundary, that is on some initial surface, the carrier of the discontinuity is a moving surface in the medium.

It is mainly with surface singularities that this book is concerned. Accordingly, the first two chapters are devoted to the description of a regular, orientable surface moving in a continuum and to the derivation of compatibility relations for functions suffering jump discontinuities across a surface. Use is made of the classical ideas of Hadamard (the singular surface of a function) and Thomas (the displacement time derivative). Chapters 3 and 4 occupy the central position in the presentation and include the derivation of balance laws for thermodynamic quantities with singularities.

Chapter 5 provides an introductory analysis of first- and second-order waves in a general continuum. The kinematic theory of acceleration waves, which is presented in Chapter 6, may be treated as a formal method of geometrical optics. It generalizes Thomas's theory and extends it to disturbances propagating along curves that are not normal trajectories of waves.

In Chapter 7 in the first two examples a relatively simple material model is used with the intention of providing clear and convincing illustrations of the theory. The surface wave example given subsequently demonstrates the non-trivial influence of surface mass concentration and surface tension on the existence of SH-type waves in a linear elastic medium. At the end of Chapter 7 thermomechanical waves in a non-linear medium are discussed.

The final chapter, which was compiled by Dr. Katarzyna Wołoszyńska-Saxton, includes a detailed analysis of the propagation of coupled thermomechanical waves in an elastic-viscoplastic material in addition to a synthetic description of the general theory of one-dimensional waves.

Most sections in the book are followed by surveys of the relevant literature by way of reference to original sources or to thematically related papers which may provide further examples or interpretations. Limitations placed on the length of the book preclude the inclusion of more examples in the text.

The list of references at the end of the book is given in alphabetical order. The symbol □ signifies the end of the proof of a theorem, corollary, proposition or example. When it occurs immediately after the statement it means that the proof is elementary.

I wish to thank my teachers and colleagues who have inspired my investigations of phenomena in non-linear continua. I am much indebted to colleagues from the Division of the Theory of Inelastic Media in the Institute of Fundamental Technological Research (IPPT-PAN) at the Polish Academy of Sciences, and particularly to Professor Piotr Perzyna, Head of the Division, for his valuable help.

Witold Kosiński

Warsaw, April 1983

Chapter 1

Moving surfaces in a continuum

In this book we take a wave to mean either a member of a class of solutions or a moving surface of discontinuity, that is a singular surface whose velocity depends on the non-linear properties of the medium through which it propagates. We shall be concerned mainly with problems directly connected with the latter of these meanings.

The idea of modelling wave phenomena in terms of moving surfaces is due to Hugoniot (1887–1889). According to his interpretation any disturbance which is strictly confined to a surface and is of arbitrary magnitude may be regarded as a wave. This approach enables a whole range of problems in wave analysis to be encompassed within a single strictly mathematical theory, free from any linearizations and simplifying assumptions.

The one-parameter family of surfaces introduced at the beginning of this chapter is a mathematical model of a surface moving in a three-dimensional medium. In view of the needs of subsequent chapters one section of the present chapter is devoted to the time description of the intrinsic geometry of a moving surface. Some of the formulae for the displacement derivatives of geometrical invariants presented in that section have not appeared previously in the literature.

From the analytical point of view both types of singularity mentioned in the introduction may be modelled by three-dimensional hypersurfaces (or by parts of such surfaces) in four-dimensional space-time.

1 SURFACES IN A CONTINUUM

In order to describe three-dimensional problems in the theory of field singularities we shall require certain information about the geometry of surfaces. The basic object is a one-parameter family of surfaces $\{\mathscr{S}(t)\}_{t \in I}$ in three-dimensional Euclidean space $\mathscr{E}^3$, the parameter t being identified with time. Equivalently, instead of describing a family of surfaces we may consider a hypersurface $\mathscr{S}$ in four-dimensional space-time. If we assume that t varies in a certain open interval I of the real line $\boldsymbol{R}$ then the hypersurface $\mathscr{S}$ in $I \times \mathscr{E}^3$ and the family of surfaces $\mathscr{S}(t)$ in $\mathscr{E}^3$ are related by

$$\mathscr{S} = \bigcup_{t \in I} \{t\} \times \mathscr{S}(t).$$

In most applications to mechanics of continua $\mathscr{S}(t)$ need not be a material surface; it is an imaginary surface on which field functions possess certain properties.

Definition 1.1

A one-parameter family of surfaces $\{\mathscr{S}(t)\}_{t \in I}$ will be called a **surface moving in a three-dimensional continuum** if there is an open subset $\mathscr{U} \subset \mathscr{E}^2$ and a one-one (injective) vector function $\boldsymbol{\varphi}(t, l^1, l^2)$ of class C^2 on $I \times \mathscr{U}$ such that for every t (that is, for every instant t) in the interval I the relation

$$\mathscr{S}(t)\colon \mathbf{x} = \boldsymbol{\varphi}(t, l^1, l^2) \tag{1.1}$$

is a parametric representation of a regular and orientable surface $\mathscr{S}(t)$ of class C^2 in $\mathscr{E}^3$.

We recall that a surface is regular if for every point of it there is a parametrization (1.1) such that the matrix

$$\left[\frac{\partial \varphi^i}{\partial l^\alpha}\right], \qquad i = 1, 2, 3;\ \alpha = 1, 2$$

has rank two; a surface is orientable if a normal vector of fixed orientation (see (1.3) below) is a continuous function on the surface. In what follows we gather together the basic relations for a surface. We use the convention that lower case Roman indices (such as $i, j, k, \ldots$) take values 1, 2, 3 and Greek indices ($\alpha, \beta, \ldots$) take values 1, 2. The usual Einstein summation convention is also adopted.

For a regular surface $\mathscr{S}(t)$ we introduce the tangent vectors

$$\boldsymbol{\varphi}_{,\alpha} = \frac{\partial \boldsymbol{\varphi}}{\partial l^\alpha} \tag{1.2}$$

and the unit normal vector

$$\mathbf{n} = \|\boldsymbol{\varphi}_{,1} \times \boldsymbol{\varphi}_{,2}\|_3^{-1} \boldsymbol{\varphi}_{,1} \times \boldsymbol{\varphi}_{,2}, \tag{1.3}$$

where $\|\cdot\|_3$ denotes the length of a vector in $\mathscr{E}^3$. We have

$$\mathbf{n} \cdot \mathbf{n} = 1, \qquad \mathbf{n} \cdot \boldsymbol{\varphi}_{,\alpha} = 0, \tag{1.4}$$

where the dot denotes the scalar product in $\mathscr{E}^3$.

The first fundamental form (the metric tensor of the surface) is defined by its components

$$g_{\alpha\beta} = \boldsymbol{\varphi}_{,\alpha} \cdot \boldsymbol{\varphi}_{,\beta}. \tag{1.5}$$

The reciprocal terms are given by the relation

$$g_{\alpha\beta}g^{\beta\gamma} = \delta^{\gamma}_{\alpha},$$

where δ^{γ}_{α} is the Kronecker symbol. The components $b_{\alpha\beta}$ of the second fundamental form satisfy

$$b_{\alpha\beta} = \boldsymbol{\varphi}_{,\alpha\beta} \cdot \mathbf{n} = -\boldsymbol{\varphi}_{,\alpha} \cdot \mathbf{n}_{,\beta}. \tag{1.6}$$

In what follows we shall use the Gauss and Weingarten formulae

$$\boldsymbol{\varphi}_{,\alpha\beta} = \Gamma^{\gamma}_{\alpha\beta}\boldsymbol{\varphi}_{,\gamma} + b_{\alpha\beta}\mathbf{n}, \qquad \mathbf{n}_{,\alpha} = -b^{\beta}_{\alpha}\boldsymbol{\varphi}_{,\beta}, \tag{1.7}$$

where $\Gamma^{\gamma}_{\alpha\beta}$ are Christoffel symbols of the second kind* and $b^{\beta}_{\alpha} = g^{\beta\lambda}b_{\lambda\alpha}$. The **Gaussian curvature** K_g is the second invariant of the second fundamental form, that is

$$K_g = \frac{\det[b_{\alpha\beta}]}{\det[g_{\alpha\beta}]} = \det[b^{\alpha}_{\beta}], \tag{1.8}$$

and the **mean curvature** K_m is half the first invariant:

$$K_m = \tfrac{1}{2}b_{\alpha\beta}g^{\alpha\beta} = \tfrac{1}{2}b^{\alpha}_{\alpha}. \tag{1.9}$$

From the Cayley–Hamilton theorem for 2×2 tensors we then have

$$b^{\alpha}_{\sigma}b^{\sigma}_{\beta} - 2K_m b^{\alpha}_{\beta} + K_g\delta^{\alpha}_{\beta} = 0. \tag{1.10}$$

Before describing the time changes of the quantities characterizing a surface we present one more useful relation, obtained by introducing the basis $\{\boldsymbol{\varphi}^1, \boldsymbol{\varphi}^2, \mathbf{n}\}$ dual to the basis $\{\boldsymbol{\varphi}_{,1}, \boldsymbol{\varphi}_{,2}, \mathbf{n}\}$ through

$$\boldsymbol{\varphi}^{\alpha} = g^{\alpha\beta}\boldsymbol{\varphi}_{,\beta}.$$

We observe that

$$\begin{aligned} &\boldsymbol{\varphi}^{\alpha} \cdot \boldsymbol{\varphi}_{,\beta} = \delta^{\alpha}_{\beta}, \\ &\boldsymbol{\varphi}^{\alpha} \cdot \mathbf{n} = 0, \qquad \boldsymbol{\varphi}_{,\alpha} \cdot \mathbf{n} = 0, \qquad \mathbf{n} \cdot \mathbf{n} = 1. \end{aligned} \tag{1.11}$$

It follows that the tensor product of the basis $\{\boldsymbol{\varphi}_{,1}, \boldsymbol{\varphi}_{,2}, \mathbf{n}\}$ and the basis $\{\boldsymbol{\varphi}^1, \boldsymbol{\varphi}^2, \mathbf{n}\}$ yields the unit tensor, that is

$$\boldsymbol{\varphi}^1 \otimes \boldsymbol{\varphi}_{,1} + \boldsymbol{\varphi}^2 \otimes \boldsymbol{\varphi}_{,2} + \mathbf{n} \otimes \mathbf{n} = \mathbf{1}. \tag{1.12}$$

* The Christoffel symbols $\Gamma^1_{\alpha\beta}$, $\Gamma^2_{\alpha\beta}$ of the second kind are by definition the coefficients of the decomposition of the vector $\boldsymbol{\varphi}_{,\alpha\beta}$ onto tangent directions $\boldsymbol{\varphi}_{,1}$ and $\boldsymbol{\varphi}_{,2}$, while the components $b_{\alpha\beta}$ are the corresponding coefficients in the normal direction.

In index notation this assumes the form

$$g^{\alpha\beta}\varphi^i_{,\alpha}\varphi^j_{,\beta}+n^in^j = G^{ij},$$

where G^{ij} are the components of the metric tensor of a local coordinate system in $\mathscr{E}^3$. In the case of a rectangular Cartesian system $G^{ij} = \delta_{ij}$.

We now introduce the **displacement velocity c of points of the surface** $\mathscr{S}(t)$, defined by

$$\mathbf{c} = \frac{\partial \boldsymbol{\varphi}}{\partial t}. \tag{1.13}$$

Note that $\mathbf{c}$ depends on the choice of parametrization (l^1, l^2). Indeed, if we assume that $(\bar{l}^1, \bar{l}^2)$ is a second system of local coordinates on the surface, related at each instant t to (l^1, l^2) by means of the equations

$$l^\alpha = \Theta^\alpha(t, \bar{l}^1, \bar{l}^2), \tag{1.14}$$

then the velocity of displacement $\bar{\mathbf{c}}$ of the surface $\mathscr{S}(t)$ measured in the parametrization $(\bar{l}^\alpha)$ differs from $\mathbf{c}$ by a linear combination of vectors tangential to the surface at time t. This follows simply from

$$\bar{\mathbf{c}}-\mathbf{c} = \boldsymbol{\varphi}_{,\alpha}\frac{\partial \Theta^\alpha}{\partial t}. \tag{1.15}$$

Equation (1.15) implies that the velocities $\bar{\mathbf{c}}$ and $\mathbf{c}$ have the same normal component, $\mathbf{u}$ say, where

$$\mathbf{u} = u_n\mathbf{n}, \qquad u_n = \mathbf{c}\cdot\mathbf{n} = \bar{\mathbf{c}}\cdot\mathbf{n}. \tag{1.16}$$

Thus the quantities $\mathbf{u}$ and u_n are independent of the choice of parametrization of a moving surface and are called the **normal velocity** and the **normal component of velocity** (or **intrinsic velocity**) **of the surface** respectively.

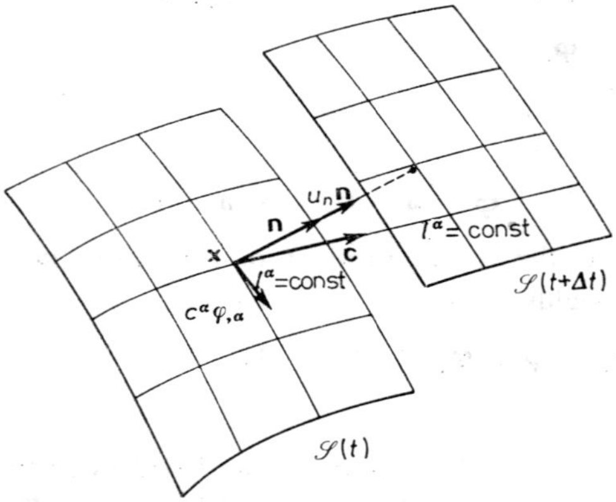

Fig. 1.1

The displacement velocity $\mathbf{c}$ may be decomposed into the normal and tangential directions. Thus

$$\mathbf{c} = c^{\alpha}\boldsymbol{\varphi}_{,\alpha}+\mathbf{u}, \tag{1.17}$$

where the tangential components c^{α} are given by the formula

$$c^{\alpha} = \mathbf{c}\cdot\boldsymbol{\varphi}^{\alpha} = \frac{\partial\boldsymbol{\varphi}}{\partial t}\cdot\boldsymbol{\varphi}_{,\beta}g^{\beta\alpha}. \tag{1.18}$$

From (1.17) we may infer that the vector $c^{\alpha}\boldsymbol{\varphi}_{,\alpha}$, called the tangential velocity of the parametrization (l^{α}) by Truesdell and Toupin (1960, p. 499), is the velocity of displacement of a point (l^1, l^2) = const on the surface $\mathscr{S}(t)$ in a direction tangential to the surface. The velocity vector $\mathbf{u}$ determines the normal trajectory of the surface. By a normal trajectory we mean a space curve which is tangent to the normal velocity $\mathbf{u}$ at every instant $t \in I$ and begins at a certain point of the initial surface $\mathscr{S}(t_0)$. Each point of $\mathscr{S}(t_0)$ is the starting point for a certain normal trajectory; moreover, different points of $\mathscr{S}(t_0)$ determine different normal trajectories. In Section 2.4 conditions will be given under which normal trajectories may intersect.

Clearly, the condition that the tangential components c^{α} vanish is necessary and sufficient for the displacement velocity $\mathbf{c}$ of points of $\mathscr{S}(t)$, measured in the parametrization (l^1, l^2), to be equal to the normal velocity $\mathbf{u}$ of $\mathscr{S}(t)$, that is

$$\mathbf{c} = \mathbf{u}. \tag{1.19}$$

The parametrization for which (1.19) holds will be called the **convected parametrization** following Bowen and Wang (1971). In the convected parametrization the trace of a surface point with coordinates (l^1, l^2) = const for different times is the normal trajectory of the surface. In the convected surface coordinate system (l^{α}) the normal trajectory is the geometrical locus of the surface point (l^{α}). Such a coordinate system exists locally, that is for a sufficiently short time interval including a given instant.

Example 1.1

Let $\{\mathscr{S}(t)\}_{t\in I}$ be a family of moving surfaces of revolution, given by

$$\mathscr{S}(t)\colon\ \mathbf{x} = \boldsymbol{\varphi}(t, l^1, l^2) \equiv \big(r(t, l^2)\cos l^1,\ r(t, l^2)\sin l^1,\ l^2\big) = (x^i),$$

where $(x^i) = \mathbf{x}$ denotes the Cartesian coordinates of a point in space, while the parameters (l^1, l^2) run through the intervals $l^1 \in (0, 2\pi)$, $l^2 \in (a, b)$. Then

$$\boldsymbol{\varphi}_{,1} = (-r\sin l^1, r\cos l^2, 0), \qquad \boldsymbol{\varphi}_{,2} = (r_{,2}\cos l^1,\ r_{,2}\sin l^1,\ 1),$$

$$\mathbf{n} = (\cos l^1, \sin l^1, -r_{,2})/\sqrt{1+(r_{,2})^2},$$

$$\boldsymbol{\varphi}^1 = \boldsymbol{\varphi}_{,1}/r^2, \quad \boldsymbol{\varphi}^2 = \boldsymbol{\varphi}_{,2}/\left(1+(r_{,2})^2\right),$$

$$\mathbf{c} = (r_{,t}\cos l^1, r_{,t}\sin l^1, 0),$$

$$u_n = r_{,t}/\sqrt{1+(r_{,2})^2},$$

$$c^1 \equiv \mathbf{c}\cdot\boldsymbol{\varphi}^1 = 0, \quad c^2 \equiv \mathbf{c}\cdot\boldsymbol{\varphi}^2 = \frac{r_{,t}r_{,2}}{\sqrt{1+(r_{,2})^2}},$$

where

$$r_{,2} \equiv \frac{\partial r}{\partial l^2}(t, l^2), \quad r_{,t} \equiv \frac{\partial r}{\partial t}(t, l^2).$$

In this example the displacement velocity of the surface is not the same as the normal velocity. This means that (l^α) is a convected parametrization if $r_{,2} = 0$.

We observe that the regularity of $\mathscr{S}(t)$ prescribed in Definition 1.1 allows us to eliminate the parameters l^α in (1.1) and represent the surface in implicit form, namely

$$\mathscr{S}(t)\colon\ g(t, \mathbf{x}) = 0, \tag{1.20}$$

where the function g is of class C^2 on its domain. (In practice it is often difficult, if not impossible, to find the explicit form of $\boldsymbol{\varphi}$ when g is given.)

Differentiating (1.20) with respect to t we obtain

$$\frac{\partial g}{\partial t} + \mathbf{c}\cdot\mathbf{grad}\, g = 0. \tag{1.21}$$

The unit vector $\mathbf{n}$ normal to the surface is then given by

$$\mathbf{n} = \mathbf{grad}\, g/|\mathbf{grad}\, g|_3.$$

Hence

$$u_n = \mathbf{c}\cdot\mathbf{n} = \mathbf{c}\cdot\frac{\mathbf{grad}\, g}{|\mathbf{grad}\, g|_3} = -\frac{\partial g/\partial t}{|\mathbf{grad}\, g|_3}.$$

Finally, we note that if $\partial g/\partial t \neq 0$ then, without loss of generality, we may assume that the moving surface $\{\mathscr{S}(t)\}_{t\in I}$ may be represented in the form

$$\mathscr{S}(t)\colon\ t = \tau(\mathbf{x}). \tag{1.22}$$

It then follows that

$$\mathbf{n} = \pm\frac{\mathbf{grad}\,\tau}{|\mathbf{grad}\,\tau|_3}, \quad u_n = \pm\frac{1}{|\mathbf{grad}\,\tau|_3} \tag{1.23}$$

and

$$\mathbf{grad}\,\tau = \frac{\mathbf{n}}{u_n}. \tag{1.24}$$

The choice of either the plus or minus sign on the right-hand sides in (1.23) depends on the way in which the representation of $\mathscr{S}(t)$ by (1.22) is rewritten as $g(t, \mathbf{x}) = 0$. In particular, if the vector $\mathbf{n}$ at time t_0 is directed into the region described by the inequality $t_0 < \tau(\mathbf{x})$ then we must take the plus sign.

The vector $\mathbf{n}/u_n$ appearing in (1.24) is called the **slowness vector.**

The expediency of selecting the representation (1.22) will become apparent when we introduce the expression for the displacement derivative and discuss the ray method (also called the method of bicharacteristics) in connection with acceleration waves.

2 THE DISPLACEMENT DERIVATIVE

In this section we shall determine the first partial derivatives of functions defined on a surface moving in a three-dimensional continuum.

Let f be a scalar-, vector- or tensor-valued function defined on a Cartesian product $I \times \mathscr{D}$, where $\mathscr{D}$ is an open subset in $\mathscr{E}^3$ containing a moving surface $\{\mathscr{S}(t)\}_{t \in I}$ or having a non-empty intersection with the surface. Since our results will only refer to that part of $\{\mathscr{S}(t)\}_{t \in I}$ contained in $\mathscr{D}$, we shall replace the notation $I \times \mathscr{D}$ by $\mathscr{S}$.

We assume that f is of class C^1 on $I \times \mathscr{D}$. Then we define the functions $\tilde{f}$ and $\bar{f}$ on $\mathscr{S}$ by

$$\begin{aligned} &\tilde{f}(t, l^1, l^2) \equiv f(t, \boldsymbol{\varphi}(t, l^1, l^2)), \\ &\bar{f}(\mathbf{x}) \equiv f(\tau(\mathbf{x}), \mathbf{x}). \end{aligned} \tag{2.1}$$

For the derivatives of $\tilde{f}$ and $\bar{f}$ we have the formulae

$$\begin{aligned} &\frac{\partial \tilde{f}}{\partial t} = \frac{\partial f}{\partial t} + \mathbf{c} \cdot \operatorname{grad} f, \quad \tilde{f}_{,\alpha} \equiv \frac{\partial \tilde{f}}{\partial l^\alpha} = \boldsymbol{\varphi}_{,\alpha} \cdot \operatorname{grad} f, \\ &\operatorname{grad} \bar{f} = \frac{\partial f}{\partial t} \operatorname{grad} \tau + \operatorname{grad} f \quad \text{if } u_n \neq 0. \end{aligned} \tag{2.2}$$

Let $\partial f/\partial n$ and $d\bar{f}/dn$ denote the derivatives of f and $\bar{f}$ respectively in the direction of the unit normal vector $\mathbf{n}$ to $\mathscr{S}(t)$, where n is the distance measured from $\mathscr{S}(t)$ in the direction $\mathbf{n}$. We then obtain

$$\frac{\partial f}{\partial n} \equiv \mathbf{n} \cdot \operatorname{grad} f, \qquad \frac{d\bar{f}}{dn} \equiv \mathbf{n} \cdot \operatorname{grad} \bar{f}. \tag{2.3}$$

On multiplying the second equation in (2.2) by $g^{\alpha\sigma}\boldsymbol{\varphi}_{,\sigma} \equiv \boldsymbol{\varphi}^\alpha$ we obtain

$$\tilde{f}_{,\alpha}\boldsymbol{\varphi}^\alpha = (\boldsymbol{\varphi}_{,\alpha} \cdot \operatorname{grad} f)\boldsymbol{\varphi}^\alpha$$

and, with the help of (1.24), the third equation in (2.2) coupled with (2.3) leads to

$$\frac{\mathrm{d}\bar{f}}{\mathrm{d}n} = \frac{1}{u_n}\frac{\partial f}{\partial t} + \frac{\partial f}{\partial n}.$$

Use of (1.12) and (2.3) then yields the required first derivative, which we write as

$$\operatorname{grad} f = \tilde{f}_{,\alpha}\boldsymbol{\varphi}^{\alpha} + \frac{\partial f}{\partial n}\mathbf{n} = \operatorname{grad}\bar{f} - \frac{\partial f}{\partial t}\operatorname{grad}\tau. \tag{2.4}$$

Elimination of $\operatorname{grad} f$ between the first equations in (2.2) and (2.4) then leads to

$$\frac{\partial f}{\partial t} = \frac{\partial \tilde{f}}{\partial t} - \tilde{f}_{,\alpha}\boldsymbol{\varphi}^{\alpha}\cdot\mathbf{c} - u_n\frac{\partial f}{\partial n} = u_n\frac{\mathrm{d}\bar{f}}{\mathrm{d}n} - u_n\frac{\partial f}{\partial n}$$

or

$$\frac{\partial f}{\partial t} + u_n\frac{\partial f}{\partial n} = \frac{\partial \tilde{f}}{\partial t} - \tilde{f}_{,\alpha}\boldsymbol{\varphi}^{\alpha}\cdot\mathbf{c} = u_n\frac{\mathrm{d}\bar{f}}{\mathrm{d}n}. \tag{2.5}$$

Following Thomas (1957) we introduce the **displacement derivative** (or δ-time derivative) for an arbitrary time-dependent field f in $\mathscr{D}$ confined to the surface $\mathscr{S}(t)$. It is defined as the time derivative of f along the normal trajectory. Since, in a convected parametrization (k^{α}), the geometrical locus of the surface point $(k^{\alpha}) = \text{const}$ is the normal trajectory of the surface, the displacement derivative $\delta\tilde{f}/\delta t$ is defined as the time derivative of $\tilde{f}$ if the moving surface $\{\mathscr{S}(t)\}_{t\in I}$ is given in the convected parametrization $\mathbf{x} = \boldsymbol{\varphi}(t, k^1, k^2)$, that is in the convected surface coordinate system. Thus

$$\frac{\delta\tilde{f}}{\delta t} := \frac{\partial_k \tilde{f}}{\partial t}, \tag{2.6}$$

where we have used the index k to stress that the time derivative of $\tilde{f}$ calculated for the right-hand side requires that the parametric representation of the surface is convected.

We now extend the definition of the displacement derivative to apply in any (not necessarily convected) surface coordinate system. Let $l^{\alpha} = \Theta^{\alpha}(t, k^1, k^2)$ describe the connections between an arbitrary surface coordinate system and the convected coordinate system. Then, since $\mathbf{c}(t, k^1, k^2) = \mathbf{u}$, we obtain

$$\frac{\partial\Theta^{\alpha}}{\partial t} = -c^{\alpha}(t, l^1, l^2),$$

use having been made of (1.15) with $\bar{\mathbf{c}} = \mathbf{c}(t, k^1, k^2)$. It follows that the form

$$\frac{\delta \tilde{f}}{\delta t} = \frac{\partial \tilde{f}}{\partial t} - c^\alpha \tilde{f}_{,\alpha} \tag{2.7}$$

of the displacement derivative is valid in any surface parametrization (l^α). From equations (2.5) and (2.6) we obtain expressions for the displacement derivatives of f and $\bar{f}$. These are

$$\frac{\delta f}{\delta t} = \frac{\partial f}{\partial t} + u_n \frac{\partial f}{\partial n}, \qquad \frac{\delta \bar{f}}{\delta t} = u_n \frac{\mathrm{d}\bar{f}}{\mathrm{d}n}. \tag{2.8}$$

We therefore have the connections

$$\frac{\partial f}{\partial t} = \frac{\partial \tilde{f}}{\partial t} - u_n \frac{\partial f}{\partial n} = u_n \left(\frac{\mathrm{d}\bar{f}}{\mathrm{d}n} - \frac{\partial f}{\partial n} \right). \tag{2.9}$$

Equations (2.4) and (2.9) determine the derivatives $\partial f/\partial t$ and grad$f$ in terms of the interior derivatives $\partial \tilde{f}/\partial t$ and $\tilde{f}_{,\alpha}$, the quantities defining the geometry and kinematics of the moving surface $\{\mathscr{S}(t)\}_{t \in I}$ and the normal derivative $\partial f/\partial n$. From this it is evident that in spite of knowing the function f on the surface not all the first partial derivatives of f can be determined. The normal derivative on $\mathscr{S}$ or an additional first-order partial differential equation for f is needed. We shall return to this problem when discussing the propagation of waves in particular material media.

The displacement derivative was originally introduced by Thomas (1957) in the form

$$\frac{\delta}{\delta t} \equiv \frac{\partial}{\partial t} + u_n \frac{\partial}{\partial n}. \tag{2.10}$$

Unfortunately, a definition of this kind is not **universal** (that is, invariant in form under changes of surface parametrization), particularly for fields which are functions of time and of both spatial and surface coordinate systems. In order to improve this definition Truesdell and Toupin (1960) suggested a generalization of Thomas's derivative to two-point tensor fields. Their derivative reduces to (2.10) in the case of one-point fields.

However, as pointed out by Bowen and Wang (1971), the generalization of the displacement derivative given by Truesdell and Toupin contains an error: the value of the Truesdell–Toupin derivative of a given geometrical object depends on the basis in which that object is represented, that is whether it is the spatial basis in $\mathscr{E}^3$ or the basis of surface vectors. In order to rectify this error Bowen and Wang introduced what they called the **total displacement derivative** of a function, this being the partial time derivative of the function defined on a surface given in convected parametrization, as in (2.6).

With the notation $\delta_d/\delta t$ for the displacement derivative of Truesdell and Toupin, Bowen and Wang obtain the following connection* between the full displacement derivative and $\delta_d/\delta t$:

$$\frac{\delta}{\delta t}\boldsymbol{\psi} = \frac{\delta_d \boldsymbol{\psi}}{\delta t} + \psi^\alpha (u_{n,\alpha}\mathbf{n} - b_{\alpha\beta} u_n \boldsymbol{\varphi}^\beta),$$

where $\boldsymbol{\psi}$ is an arbitrary vector function given on the surface and represented in the form

$$\boldsymbol{\psi}(t, l^1, l^2) = \psi_n(t, l^1, l^2)\mathbf{n} + \psi^\alpha(t, l^1, l^2)\boldsymbol{\varphi}_{,\alpha}.$$

To complete the picture we note that the displacement derivative of an arbitrary function g of time, coordinates $\mathbf{x}$ and curvilinear coordinates (l^α) on a surface is given by

$$\frac{\delta}{\delta t} g(t, \mathbf{x}, l^1, l^2) = \frac{\partial g}{\partial t} + u_n \frac{\partial g}{\partial n} - c^\alpha g_{,\alpha} \tag{2.11}$$

(Bowen and Wang, 1971).

2.1 Time Description of Surface Geometry

We now derive formulae for the displacement derivatives of the quantities which characterize a moving surface $\mathscr{S}$. We begin with the tangent vectors $\boldsymbol{\varphi}_{,\alpha}$. From definition (2.7) we have

$$\frac{\delta}{\delta t}\boldsymbol{\varphi}_{,\alpha} = \frac{\partial}{\partial t}\boldsymbol{\varphi}_{,\alpha} - c^\gamma \frac{\partial}{\partial l^\gamma}\boldsymbol{\varphi}_{,\alpha}.$$

Commutativity of the differentiation with respect to t and l^α yields

$$\frac{\partial}{\partial t}\boldsymbol{\varphi}_{,\alpha} = \frac{\partial}{\partial l^\alpha}\mathbf{c} = \frac{\partial}{\partial l^\alpha}(c^\gamma \boldsymbol{\varphi}_{,\gamma}) + \frac{\partial \mathbf{u}}{\partial l^\alpha},$$

use having been made of (1.17).

Carrying out the necessary differentiation we obtain

$$\frac{\delta}{\delta t}\boldsymbol{\varphi}_{,\alpha} = \frac{\partial u_n}{\partial l^\alpha}\mathbf{n} - u_n b_\alpha^\beta \boldsymbol{\varphi}_{,\beta} + \frac{\partial c^\gamma}{\partial l^\alpha}\boldsymbol{\varphi}_{,\gamma} + c^\gamma \boldsymbol{\varphi}_{,\gamma\alpha} - c^\gamma \boldsymbol{\varphi}_{,\alpha\gamma}$$

and hence

$$\frac{\delta}{\delta t}\boldsymbol{\varphi}_{,\alpha} = u_{n,\alpha}\mathbf{n} - u_n b_\alpha^\beta \boldsymbol{\varphi}_{,\beta} + c^\beta_{,\alpha}\boldsymbol{\varphi}_{,\beta}, \tag{2.12}$$

where, α denotes $\partial/\partial l^\alpha$.

---

* Compare with Braun (1974).

In order to derive the formula for the derivative of a normal vector we first differentiate (1.4) to give

$$\frac{\delta \mathbf{n}}{\delta t} \cdot \mathbf{n} = 0, \qquad \frac{\delta \mathbf{n}}{\delta t} \cdot \boldsymbol{\varphi}_{,\alpha} = -\mathbf{n} \cdot \frac{\delta}{\delta t} \boldsymbol{\varphi}_{,\alpha}. \tag{2.13}$$

Since $\{\boldsymbol{\varphi}_{,1}, \boldsymbol{\varphi}_{,2}, \mathbf{n}\}$ is a basis, we obtain

$$\frac{\delta \mathbf{n}}{\delta t} = -\mathbf{n} \cdot \frac{\delta \boldsymbol{\varphi}_{,\alpha}}{\delta t} \boldsymbol{\varphi}^{\alpha}.$$

From (2.12),

$$\mathbf{n} \cdot \frac{\delta}{\delta t} \boldsymbol{\varphi}_{,\alpha} = u_{n,\alpha} \tag{2.14}$$

and hence

$$\frac{\delta \mathbf{n}}{\delta t} = -u_{n,\alpha} \boldsymbol{\varphi}^{\alpha} \tag{2.15}$$

or, by use of (1.12),

$$\frac{\delta \mathbf{n}}{\delta t} = -u_{n,\alpha} g^{\alpha\beta} \boldsymbol{\varphi}_{,\beta} = (\mathbf{n} \otimes \mathbf{n} - \mathbf{1}) \operatorname{grad} u_n. \tag{2.16}$$

On differentiation of (1.11) we obtain

$$\frac{\delta \boldsymbol{\varphi}^{\alpha}}{\delta t} \cdot \boldsymbol{\varphi}_{,\beta} = -\boldsymbol{\varphi}^{\alpha} \cdot \frac{\delta}{\delta t} \boldsymbol{\varphi}_{,\beta}, \qquad \frac{\delta \boldsymbol{\varphi}^{\alpha}}{\delta t} \cdot \mathbf{n} = -\boldsymbol{\varphi}^{\alpha} \cdot \frac{\delta \mathbf{n}}{\delta t}. \tag{2.17}$$

Use of (2.12) and (2.15) then yields

$$\frac{\delta \boldsymbol{\varphi}^{\alpha}}{\delta t} = u_{n,\beta} g^{\beta\alpha} \mathbf{n} - c^{\alpha}_{\cdot\beta} \boldsymbol{\varphi}^{\beta} + u_n b^{\alpha}_{\beta} \boldsymbol{\varphi}^{\beta} \tag{2.18}$$

when decomposed on the basis $\{\boldsymbol{\varphi}^1, \boldsymbol{\varphi}^2, \mathbf{n}\}$, or, on the basis $\{\boldsymbol{\varphi}_{,1}, \boldsymbol{\varphi}_{,2}, \mathbf{n}\}$

$$\frac{\delta \boldsymbol{\varphi}^{\alpha}}{\delta t} = u_{n,\beta} g^{\beta\alpha} \mathbf{n} - c^{\alpha}_{\cdot\gamma} g^{\gamma\beta} \boldsymbol{\varphi}_{,\beta} + u_n b^{\alpha\beta} \boldsymbol{\varphi}_{,\beta}. \tag{2.19}$$

The derivatives of the components of the metric tensor are computed by differentiating (1.5) and making use of the formulae for the derivatives of the tangent vectors. Thus,

$$\begin{aligned}\frac{\delta}{\delta t} g_{\alpha\beta} &= \frac{\delta}{\delta t} \boldsymbol{\varphi}_{,\alpha} \cdot \boldsymbol{\varphi}_{,\beta} + \boldsymbol{\varphi}_{,\alpha} \cdot \frac{\delta}{\delta t} \boldsymbol{\varphi}_{,\beta} \\ &= c^{\gamma}_{\cdot\alpha} g_{\gamma\beta} + c^{\gamma}_{\cdot\beta} g_{\gamma\alpha} - u_n (b^{\gamma}_{\alpha} g_{\gamma\beta} + b^{\gamma}_{\beta} g_{\gamma\alpha}).\end{aligned}$$

Hence

$$\frac{\delta g_{\alpha\beta}}{\delta t} = c^{\gamma}_{\cdot\alpha} g_{\gamma\beta} + c^{\gamma}_{\cdot\beta} g_{\gamma\alpha} - 2u_n b_{\alpha\beta}. \tag{2.20}$$

By use of the formula

$$\frac{\partial}{\partial g_{\alpha\beta}}\det[g_{\alpha\beta}] = \det[g_{\alpha\beta}]g^{\alpha\beta}$$

and the definition (1.9) of the mean curvature K_m, we arrive at

$$\frac{\delta}{\delta t}\det[g_{\alpha\beta}] = 2\det[g_{\alpha\beta}]c^{\beta}_{,\beta} - 4\det[g_{\alpha\beta}]u_n K_m. \tag{2.21}$$

It follows from (2.21) that if the surface $\mathscr{S}(t)$ is minimal then the derivative of the determinant vanishes (in a convected coordinate system).

In order to determine $\delta b_{\alpha\beta}/\delta t$ we first differentiate the Gaussian formula* (1.7). This gives

$$\frac{\delta}{\delta t}\boldsymbol{\varphi}_{,\alpha\beta} = \frac{\delta}{\delta t}\Gamma^{\gamma}_{\alpha\beta}\boldsymbol{\varphi}_{,\gamma} + \Gamma^{\gamma}_{\alpha\beta}\frac{\delta}{\delta t}\boldsymbol{\varphi}_{,\gamma} + \frac{\delta}{\delta t}b_{\alpha\beta}\mathbf{n} + b_{\alpha\beta}\frac{\delta \mathbf{n}}{\delta t}.$$

Multiplying by $\mathbf{n}$ and making use of (1.4) and (2.13) we then obtain

$$\frac{\delta b_{\alpha\beta}}{\delta t} = \frac{\delta}{\delta t}\boldsymbol{\varphi}_{,\alpha\beta}\cdot\mathbf{n} - \Gamma^{\gamma}_{\alpha\beta}\frac{\delta\boldsymbol{\varphi}_{,\gamma}}{\delta t}\cdot\mathbf{n}. \tag{2.22}$$

The first term on the right-hand side of (2.22) may be written as

$$\frac{\delta\boldsymbol{\varphi}_{,\alpha\beta}}{\delta t}\cdot\mathbf{n} = \frac{\partial\boldsymbol{\varphi}_{,\alpha\beta}}{\partial t}\cdot\mathbf{n} - c^{\gamma}\boldsymbol{\varphi}_{,\alpha\beta\gamma}\cdot\mathbf{n} = \mathbf{c}_{,\alpha\beta}\cdot\mathbf{n} - c^{\gamma}\boldsymbol{\varphi}_{,\alpha\beta\gamma}\cdot\mathbf{n}$$

since

$$\frac{\partial\boldsymbol{\varphi}_{,\alpha\beta}}{\partial t} = \left(\frac{\partial\boldsymbol{\varphi}}{\partial t}\right)_{,\alpha\beta}.$$

Further transformations and use of the equation $\mathbf{n}_{,\alpha}\cdot\mathbf{n} = 0$ lead to

$$\frac{\delta\boldsymbol{\varphi}_{,\alpha\beta}}{\delta t}\cdot\mathbf{n} = \mathbf{u}_{,\alpha\beta}\cdot\mathbf{n} + c^{\gamma}_{,\alpha}b_{\gamma\beta} + c^{\gamma}_{,\beta}b_{\gamma\alpha}$$
$$= u_{n,\alpha\beta} + \mathbf{u}\cdot\mathbf{n}_{,\alpha\beta} + c^{\gamma}_{,\alpha}b_{\gamma\beta} + c^{\gamma}_{,\beta}b_{\gamma\beta}.$$

With the help of (1.5) and (1.7) this becomes

$$\frac{\delta\boldsymbol{\varphi}_{,\alpha\beta}}{\delta t}\cdot\mathbf{n} = u_{n,\alpha\beta} - u_n b^{\gamma}_{\alpha}b_{\gamma\beta} + c^{\gamma}_{,\alpha}b_{\gamma\beta} + c^{\gamma}_{,\beta}b_{\gamma\beta}. \tag{2.23}$$

On the use of (2.23) and (2.14) in (2.22) the latter becomes

$$\frac{\delta b_{\alpha\beta}}{\delta t} = u_{n,\alpha\beta} - u_{n,\gamma}\Gamma^{\gamma}_{\alpha\beta} + c^{\gamma}_{,\alpha}b_{\gamma\beta} + c^{\gamma}_{,\beta}b_{\gamma\alpha} - u_n b^{\gamma}_{\alpha}b_{\gamma\beta}, \tag{2.24}$$

---

* In deriving this formula it has to be assumed that $\boldsymbol{\varphi}$ is of class C^3.

and finally, by means of (1.10),

$$\frac{\delta b_{\alpha\beta}}{\delta t} = u_{n,\alpha\beta} - u_{n,\gamma}\Gamma^{\gamma}_{\alpha\beta} + c^{\gamma}_{,\alpha}b_{\gamma\beta} + c^{\gamma}_{,\beta}b_{\gamma\alpha} + u_n K_g g_{\alpha\beta} - 2u_n K_m b_{\alpha\beta}. \tag{2.25}$$

To end this section we discuss the usefulness of the formulae derived above in respect of a convected parametrization of the moving surface. In accordance with the definition given in Section 1 a parametrization (k^1, k^2) is convected when the tangential components c^α of the velocity $\mathbf{c}$ of a surface measured in that parametrization vanish. Then (2.7) assumes the simplified form

$$\frac{\delta \tilde{f}}{\delta t} = \frac{\partial \tilde{f}}{\partial t} \tag{2.26}$$

for an arbitrary function on $\mathscr{S}$ and given in a convected parametrization.

Let

$$\mathscr{S}(t):\ \mathbf{x} = {}_k\boldsymbol{\varphi}(t, k^1, k^2) \tag{2.27}$$

be a convected parametric representation of the surface and let

$$l^\alpha = \Theta^\alpha(t, k^1, k^2) \tag{2.28}$$

define an arbitrary parametrization. Since

$${}_kc^\alpha = c^\alpha(t, k^1, k^2) \equiv 0 \tag{2.29}$$

equations (2.12), (2.15), (2.20), (2.21), (2.24) and (2.19) are replaced by

$$\begin{aligned}
\frac{\delta}{\delta t}{}_k\boldsymbol{\varphi}_{,\alpha} &= \frac{\partial u_n}{\partial k^\alpha}\mathbf{n} - u_{n\,k}b^{\beta}_{\alpha\,k}\boldsymbol{\varphi}_{,\beta},\\
\frac{\delta \mathbf{n}}{\delta t} &= -\frac{\partial u_n}{\partial k^\alpha}{}_k\boldsymbol{\varphi}^\alpha = -u_{n,\alpha\,k}g^{\alpha\beta}{}_k\boldsymbol{\varphi}_{,\beta},\\
\frac{\delta}{\delta t}{}_kg_{\alpha\beta} &= -2u_{n\,k}b_{\alpha\beta},\\
\frac{\delta}{\delta t}\det[{}_kg_{\alpha\beta}] &= -4\det[{}_kg_{\alpha\beta}]u_n K_m,\\
\frac{\delta}{\delta t}{}_kb_{\alpha\beta} &= \frac{\partial^2 u_n}{\partial k^\alpha \partial k^\beta} - {}_k\Gamma^{\gamma}_{\alpha\beta}\frac{\partial u_n}{\partial k^\gamma} - u_{n\,k}b^{\gamma}_{\alpha\,k}b_{\gamma\beta},\\
\frac{\delta}{\delta t}{}_k\boldsymbol{\varphi}^\alpha &= \frac{\partial u_n}{\partial k^\beta}{}_kg^{\alpha\beta}\mathbf{n} + u_{n\,k}b^{\alpha\beta}{}_k\boldsymbol{\varphi}_{,\beta}
\end{aligned} \tag{2.30}$$

respectively. In the above the pre-subscript k signifies that a convected parametrization is used.

2.2 Basic Results in Index Notation

In index notation the fundamental results in Section 2 may be expressed as follows, but with f restricted to being a scalar field.

Equation (2.4) becomes

$$\frac{\partial f}{\partial x^i} = G_{ik}\left\{\tilde{f}_{,\alpha}\varphi^{k\alpha} - \frac{\partial f}{\partial n}n^k\right\} = G_{ik}\left\{g^{\alpha\beta}\tilde{f}_{,\alpha}\varphi^k_{.\beta} + \frac{\partial f}{\partial n}n^k\right\}, \tag{2.31}$$

where $\partial f/\partial n$ is given by (2.3), that is

$$\frac{\partial f}{\partial n} \equiv n^p \frac{\partial f}{\partial x^p}. \tag{2.32}$$

From (2.7) we obtain

$$\frac{\delta \tilde{f}}{\delta t} = \frac{\partial \tilde{f}}{\partial t} - G_{ik}c^i\varphi^{k\alpha}\tilde{f}_{,\alpha} = \frac{\partial \tilde{f}}{\partial t} - G_{ik}g^{\alpha\beta}c^i\varphi^k_{.\alpha}\tilde{f}_{,\beta}, \tag{2.33}$$

in which we have used (1.18) written in the form

$$c^\alpha = G_{ik}c^i\varphi^{k\alpha} = G_{ik}c^i\varphi^k_{.\beta}g^{\alpha\beta}. \tag{2.34}$$

The symbols G_{ik} denote the components of the metric tensor of a (generally non-Cartesian) system of coordinates in $\mathscr{E}^3$.

Equation (2.8) specializes to

$$\frac{\delta f}{\delta t} = \frac{\partial f}{\partial t} + G_{ik}c^i n^k \frac{\partial f}{\partial n}, \tag{2.35}$$

where (1.16), written as

$$u_n = G_{ik}c^i n^k, \tag{2.36}$$

has been used, and (2.12) to

$$\frac{\delta \varphi^i_{,\alpha}}{\delta t} = u_{n,\alpha}n^i - u_n b^\beta_\alpha \varphi^i_{.\beta} + c^\beta_{,\alpha}\varphi^i_{.\beta}. \tag{2.37}$$

Equations (2.15) and (2.16) become

$$\frac{\delta n^i}{\delta t} = -u_{n,\alpha}\varphi^{i\alpha} = -g^{\alpha\beta}u_{n,\alpha}\varphi^i_{.\beta} = (n^i n^j - G^{ij})u_{n,j}. \tag{2.38}$$

Finally, (2.18) and (2.19) yield

$$\begin{aligned} \frac{\delta \varphi^{i\alpha}}{\delta t} &= u_{n,\beta}n^i g^{\alpha\beta} - \varphi^{i\beta}c^\alpha_{.\beta} + u_n\varphi^{i\beta}b^\alpha_\beta, \\ \frac{\delta \varphi^{i\alpha}}{\delta t} &= u_{n,\beta}n^i g^{\alpha\beta} - g^{\gamma\beta}\varphi^i_{.\beta}c^\alpha_{.\gamma} + u_n\varphi^i_{.\beta}b^{\alpha\beta}. \end{aligned} \tag{2.39}$$

Equations (2.20), (2.21), (2.24) and (2.25) are already given in the index notation of the surface coordinate system.

2.3 The Covariant Derivative

We now introduce the covariant derivative for the geometrical quantities defined on a surface $\mathscr{S}$. Covariant differentiation is signified by a semi-colon; thus, for a scalar field c we have

$$c_{;\alpha} = \frac{\partial c}{\partial l^{\alpha}}, \tag{2.40}$$

while, if c_{γ} are the covariant components of a vector field (that is the components of a covector),

$$c_{\gamma;\alpha} \equiv \frac{\partial c_{\gamma}}{\partial l^{\alpha}} - \Gamma^{\delta}_{\gamma\alpha} c_{\delta}. \tag{2.41}$$

For the contravariant components c^{γ} of a vector field (that is the components of a vector),

$$c^{\gamma}_{;\alpha} \equiv \frac{\partial c^{\gamma}}{\partial l^{\alpha}} + \Gamma^{\gamma}_{\alpha\delta} c^{\delta}. \tag{2.42}$$

It should be noted that the components of a tensor (or vector) field defined on the Euclidean space in which the surface is situated are treated as two-point tensors as far as covariant differentiation is concerned. For example, if

$$\mathbf{\Psi} = \psi^{i}\mathbf{e}_{i}, \tag{2.43}$$

where $\mathbf{e}_i$ are constant basis vectors in $\mathscr{E}^3$, then

$$\mathbf{\Psi}_{;\alpha} \equiv \frac{\partial \psi^{i}}{\partial l^{\alpha}}\mathbf{e}_{i} = \mathbf{\Psi}_{,\alpha}. \tag{2.44}$$

However, for the field of tangent vectors $\boldsymbol{\varphi}_{,\alpha}$, we have from (2.41)

$$\boldsymbol{\varphi}_{,\alpha;\beta} = \left\{\frac{\partial^2 \varphi^{i}}{\partial l^{\alpha}\partial l^{\beta}} - \Gamma^{\lambda}_{\alpha\beta}\frac{\partial \varphi^{i}}{\partial l^{\lambda}}\right\}\mathbf{e}_{i} = \boldsymbol{\varphi}_{,\alpha\beta} - \Gamma^{\lambda}_{\alpha\beta}\boldsymbol{\varphi}_{,\lambda} \equiv \boldsymbol{\varphi}_{;\alpha\beta}. \tag{2.45}$$

For the derivatives of u_n we have

$$u_{n;\alpha} = u_{n,\alpha} \tag{2.46}$$

and

$$u_{n;\alpha\beta} = u_{n,\alpha\beta} - \Gamma^{\gamma}_{\alpha\beta} u_{n,\gamma}. \tag{2.47}$$

In terms of the covariant derivative the Gaussian formula (1.7) takes the form

$$\boldsymbol{\varphi}_{;\alpha\beta} = b_{\alpha\beta}\mathbf{n}, \quad \varphi^{i}{}_{;\alpha\beta} = b_{\alpha\beta}n^{i} \tag{2.48}$$

which may be used to obtain the useful identities

$$g_{\alpha\beta;\delta} = 0, \quad g^{\alpha\beta}{}_{;\delta} = 0.$$

Substitution of the above formulae into (2.24) and $(2.30)_5$ yields

$$\begin{aligned}\frac{\delta b_{\alpha\beta}}{\delta t} &= u_{n;\alpha\beta}+c^{\gamma}{}_{,\alpha}b_{\gamma\beta}+c^{\gamma}{}_{,\beta}b_{\gamma\alpha}-u_n b^{\gamma}_{\alpha}b_{\gamma\beta},\\ \frac{\delta}{\delta t}{}_k b_{\alpha\beta} &= u_{n;\alpha\beta}-u_{n\,k}b^{\gamma}_{\alpha\,k}b_{\gamma\beta}.\end{aligned} \tag{2.49}$$

A derivation of (2.49) may be found in Truesdell and Toupin (1960) or Moeckel (1974).

Example 2.1

We determine the derivative $\partial g_{\alpha\beta}/\partial t$. Since

$$\frac{\partial g_{\alpha\beta}}{\partial t} = \frac{\partial \boldsymbol{\varphi}_{,\alpha}}{\partial t}\cdot\boldsymbol{\varphi}_{,\beta}+\boldsymbol{\varphi}_{,\alpha}\cdot\frac{\partial \boldsymbol{\varphi}_{,\beta}}{\partial t}$$

and

$$\frac{\partial \boldsymbol{\varphi}_{,\alpha}}{\partial t} = \frac{\partial \mathbf{c}}{\partial l^{\alpha}}$$

we must calculate $\partial\mathbf{c}/\partial l^{\alpha}$. Applying successive transformations we may write

$$\begin{aligned}\frac{\partial \mathbf{c}}{\partial l^{\alpha}} &= (c^{\delta}\boldsymbol{\varphi}_{,\delta}+\mathbf{u})_{,\alpha} = c^{\delta}{}_{,\alpha}\boldsymbol{\varphi}_{,\delta}+c^{\delta}(\Gamma^{\gamma}_{\delta\alpha}\boldsymbol{\varphi}_{,\gamma}+b_{\alpha\beta}\mathbf{n})+\mathbf{u}_{,\alpha},\\ \mathbf{c}_{,\alpha}\cdot\boldsymbol{\varphi}_{,\beta} &= (c^{\delta}_{,\alpha}+c^{\gamma}\Gamma^{\delta}_{\gamma\alpha})g_{\delta\beta}+\mathbf{u}_{,\alpha}\cdot\boldsymbol{\varphi}_{,\beta}.\end{aligned}$$

In a similar manner we calculate $\mathbf{c}_{,\beta}\cdot\boldsymbol{\varphi}_{,\alpha}$ to obtain the required result

$$\begin{aligned}\frac{\partial g_{\alpha\beta}}{\partial t} &= c^{\delta}_{;\alpha}g_{\delta\beta}+c^{\delta}_{;\beta}g_{\delta\alpha}-u_n b^{\gamma}_{\beta}g_{\gamma\alpha}-u_n b^{\gamma}_{\alpha}g_{\gamma\beta}\\ &= c_{\beta;\alpha}+c_{\alpha;\beta}-2u_n b_{\alpha\beta}.\ \square\end{aligned}$$

Example 2.2

Here we determine the derivative $\delta b^{\alpha}_{\beta}/\delta t$. From the identity

$$g^{\gamma\beta}g_{\alpha\beta} = \delta^{\gamma}_{\alpha}$$

we obtain

$$\frac{\delta g^{\alpha\beta}}{\delta t}g_{\beta\gamma} = -g^{\alpha\beta}\frac{\delta}{\delta t}g_{\beta\gamma},$$

and hence, from (2.20),

$$\frac{\delta}{\delta t}g^{\alpha\beta} = -g^{\alpha\delta}\frac{\delta}{\delta t}g_{\delta\gamma}g^{\gamma\beta} = 2u_n b^{\alpha\beta}-c^{\alpha}_{,\gamma}g^{\gamma\beta}-c^{\beta}_{,\gamma}g^{\gamma\alpha}. \tag{2.50}$$

On use of (2.24) we see that

$$\begin{aligned}\frac{\delta b_\alpha^\beta}{\delta t} &= \frac{\delta b_{\alpha\gamma}}{\delta t} g^{\gamma\beta} + b_{\alpha\gamma}\frac{\delta g^{\gamma\beta}}{\delta t}\\ &= u_{n;\alpha\gamma} g^{\gamma\beta} + c^\delta_{,\alpha} b_\delta^\beta - c^\beta_{,\delta} b_\alpha^\delta + u_n b_\alpha^\delta b_\delta^\beta. \qquad (2.51)\end{aligned}$$

For a convected parametrization this reduces to

$$\frac{\delta}{\delta t}{}_k b_\alpha^\beta = u_{n;\alpha\gamma\,k} g^{\gamma\beta} + u_{n\,k} b_{\alpha\,k}^\delta b_\delta^\beta. \quad \square$$

2.4 Families of Parallel Surfaces

We return briefly to the expression (2.15) for the displacement derivative of the normal vector, namely

$$\frac{\delta \mathbf{n}}{\delta t} = -u_{n,\alpha} \boldsymbol{\varphi}^\alpha.$$

If $u_{n,\alpha} = 0$, that is u_n is constant on a moving surface $\mathscr{S}(t)$, then the normal vector is constant along the normal trajectory. This means that normal trajectories are straight lines, and a one-parameter family of surfaces $\{\mathscr{S}(t)\}_{t\in I}$ forms a family of parallel surfaces. The fact that u_n is independent of the curvilinear parameters of an individual member $\mathscr{S}(t_0)$ of the family means that after a time Δt the position of the surface $\mathscr{S}(t_0+\Delta t)$ is determined by measuring off equal segments n from the surface $\mathscr{S}(t_0)$ along the normal straight lines in the direction of the unit normal $\mathbf{n}$ (see Figure 2.1).

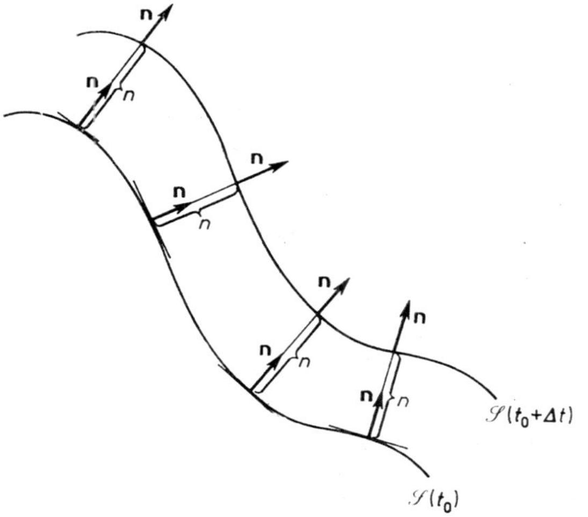

Fig. 2.1

In this case it is convenient to make use of a convected parametrization in which a surface point $(l^1, l^2) = \text{const}$ moves in time along the normal trajectory. By (2.26) the displacement derivative of the position function $\boldsymbol{\varphi}$ of a surface in space becomes a partial derivative with respect to time, and its value coincides with the normal speed, that is

$$\frac{\delta\boldsymbol{\varphi}}{\delta t} = \frac{\partial\boldsymbol{\varphi}}{\partial t} = u_n\mathbf{n}.$$

This leads to the following representation of the surface at instant t relative to that at t_0:

$$\boldsymbol{\varphi}(t, l^1, l^2) = \boldsymbol{\varphi}(t_0, l^1, l^2) + \mathbf{n}\int_{t_0}^{t} u_n(\tau)\,\mathrm{d}\tau. \tag{2.52}$$

The integral on the right-hand side may be replaced by $(t-t_0)u_n$ if u_n is independent of time.

Parallel surfaces play an important role in the analysis of acceleration waves propagating through a medium in a homogeneous state.

The tangent vectors to a surface may be expressed in the form

$$\boldsymbol{\varphi}_{,\alpha} = \boldsymbol{\varphi}^0_{,\alpha} + n\mathbf{n}_{,\alpha},$$

where

$$n(t) = \int_{t_0}^{t} u_n(\tau)\,\mathrm{d}\tau, \qquad t > t_0, \tag{2.53}$$

and $\boldsymbol{\varphi}^0$ denotes the function $\boldsymbol{\varphi}(t_0, l^1, l^2)$. By use of Weingarten's formulae

$$\mathbf{n}_{,\alpha} = -\mathbf{b}\boldsymbol{\varphi}_{,\alpha} = -\mathbf{b}^0\boldsymbol{\varphi}^0_{,\alpha} \tag{2.54}$$

we obtain finally

$$\boldsymbol{\varphi}_{,\alpha} = (\delta^\beta_\alpha - nb^{0\beta}_\alpha)\boldsymbol{\varphi}^0_{,\beta} = (\mathbf{1}_{\mathscr{S}} - n\mathbf{b}^0)\boldsymbol{\varphi}^0_{,\alpha}. \tag{2.55}$$

In (2.54) we have denoted by $\mathbf{b}$ and $\mathbf{b}^0$ the second metric tensors (curvature tensors) of the surfaces $\mathscr{S}(t)$ and $\mathscr{S}(t_0)$ respectively, that is

$$\mathbf{b} = b^\alpha_\beta\boldsymbol{\varphi}_{,\alpha}\otimes\boldsymbol{\varphi}^\beta = -\mathbf{n}_{,\alpha}\otimes\boldsymbol{\varphi}^\alpha,$$
$$\mathbf{b}^0 = b^{0\alpha}_\beta\boldsymbol{\varphi}^0_{,\alpha}\otimes\boldsymbol{\varphi}^{0\beta} = -\mathbf{n}_{,\alpha}\otimes\boldsymbol{\varphi}^{0\alpha}.$$

On the right-hand side of (2.55) the symbol $\mathbf{1}_{\mathscr{S}}$ denotes the surface identity (metric) tensor.

Equations (2.54) and (2.55) allow us to write

$$\mathbf{b}(\mathbf{1}_{\mathscr{S}} - n\mathbf{b}^0)\boldsymbol{\varphi}^0_{,\alpha} = \mathbf{b}^0\boldsymbol{\varphi}^0_{,\alpha}, \qquad \alpha = 1, 2.$$

Hence we obtain the relationship

$$\mathbf{b} = \mathbf{b}^0(\mathbf{1}_{\mathscr{S}} - n\mathbf{b}^0)^{-1} \tag{2.56}$$

between the curvature tensors of $\mathscr{S}(t)$ and $\mathscr{S}(t_0)$, where $(\mathbf{1}_{\mathscr{S}}-n\mathbf{b}^0)^{-1}$ is an inverse surface tensor so that

$$(\mathbf{1}_{\mathscr{S}}-n\mathbf{b}^0)^{-1}(\mathbf{1}_{\mathscr{S}}-n\mathbf{b}^0) = \mathbf{1}_{\mathscr{S}}.$$

Observe that by differentiating (2.56) with respect to time we obtain

$$\frac{\partial \mathbf{b}}{\partial t} = u_n \mathbf{b}^0(\mathbf{1}_{\mathscr{S}}-n\mathbf{b}^0)^{-1}\mathbf{b}^0(\mathbf{1}_{\mathscr{S}}-n\mathbf{b}^0)^{-1} = u_n \mathbf{b}^2,$$

which corresponds to the equation derived in Example 2.2 for the derivative $\delta b_\alpha^\beta/\delta t$ in a convected parametrization under the assumption that u_n is constant on the surface.

Application of the Cayley–Hamilton theorem to $(\mathbf{1}_{\mathscr{S}}-n\mathbf{b}^0)^{-1}$ leads to

$$(\mathbf{1}_{\mathscr{S}}-n\mathbf{b}^0)^{-1} = \frac{\mathbf{1}_{\mathscr{S}}-nK_g^0(\mathbf{b}^0)^{-1}}{\det_{\mathscr{S}}(\mathbf{1}_{\mathscr{S}}-n\mathbf{b}^0)},$$

where

$$\det{}_{\mathscr{S}}(\mathbf{1}_{\mathscr{S}}-n\mathbf{b}^0) = 1-2K_m^0 n+K_g^0 n^2.$$

Hence

$$\mathbf{b} = \frac{\mathbf{b}^0-K_g^0 n\mathbf{1}_{\mathscr{S}}}{\det_{\mathscr{S}}(\mathbf{1}_{\mathscr{S}}-n\mathbf{b}^0)} = \frac{\mathbf{b}^0-K_g^0 n\mathbf{1}_{\mathscr{S}}}{1-2K_m^0 n+K_g^0 n^2}. \tag{2.57}$$

The invariants of $\mathbf{b}$, namely the mean curvature K_m and the Gaussian curvature K_g, may be expressed in terms of the corresponding invariants of $\mathbf{b}^0$ through (2.57). Thus

$$\begin{aligned} K_m &= \frac{K_m^0-K_g^0 n}{\det_{\mathscr{S}}(\mathbf{1}_{\mathscr{S}}-n\mathbf{b}^0)} = \frac{K_m^0-K_b^0 n}{1-2K_m^0 n+K_g^0 n^2}, \\ K_g &= \frac{K_b^0}{\det_{\mathscr{S}}(\mathbf{1}_{\mathscr{S}}-n\mathbf{b}^0)} = \frac{K_g^0}{1-2K_m^0 n+K_g^0 n^2}. \end{aligned} \tag{2.58}$$

To find the expression for the covariant components $\boldsymbol{\varphi}_{,\alpha}\cdot\boldsymbol{\varphi}_{,\beta}$ of the metric tensor we apply (2.55) to give

$$\boldsymbol{\varphi}_{,\alpha}\cdot\boldsymbol{\varphi}_{,\beta} = (\mathbf{1}_{\mathscr{S}}-n\mathbf{b}^0)\boldsymbol{\varphi}^0_{,\alpha}\cdot(\mathbf{1}_{\mathscr{S}}-n\mathbf{b}^0)\boldsymbol{\varphi}^0_{,\beta}.$$

On use of the symmetry of $\mathbf{b}^0$ and the Cayley–Hamilton theorem we deduce that

$$\boldsymbol{\varphi}_{,\alpha}\cdot\boldsymbol{\varphi}_{,\beta} \equiv g_{\alpha\beta} = (1-n^2K_g^0)g^0_{\alpha\beta}-2n(1-nK_m^0)b^0_{\alpha\beta}. \tag{2.59}$$

From the equality

$$\boldsymbol{\varphi}_{,\alpha}\cdot\mathbf{b}\boldsymbol{\varphi}_{,\beta} = \boldsymbol{\varphi}_{,\alpha}\cdot\mathbf{b}^0(\mathbf{1}_{\mathscr{S}}-n\mathbf{b}^0)^{-1}\boldsymbol{\varphi}_{,\beta}$$

we determine the covariant components

$$b_{\alpha\beta} \equiv \boldsymbol{\varphi}_{,\alpha}\cdot\mathbf{b}\boldsymbol{\varphi}_{,\beta} = (\mathbf{1}_{\mathscr{S}}-n\mathbf{b}^0)\boldsymbol{\varphi}^0_{,\alpha}\cdot\mathbf{b}^0\boldsymbol{\varphi}^0_{,\beta} = b_\alpha^{0\delta}g^0_{\delta\beta}-n\,b_\beta^{0\gamma}b_\alpha^{0\delta}g^0_{\delta\gamma},$$

and finally, from (1.10),

$$b_{\alpha\beta} = (1-2K_m^0 n)b_{\alpha\beta}^0 + nK_g^0 g_{\alpha\beta}^0. \tag{2.60}$$

It is easy to prove from the above that

$$\begin{aligned} \det[g_{\alpha\beta}] &= \det_{\mathscr{S}}(\mathbf{1}_{\mathscr{S}} - n\mathbf{b}^0)\det[g_{\alpha\beta}^0], \\ \det[b_{\alpha\beta}] &= \det_{\mathscr{S}}(\mathbf{1}_{\mathscr{S}} - n\mathbf{b}^0)\det[b_{\alpha\beta}^0]. \end{aligned} \tag{2.61}$$

We observe that in certain situations the set $\{\mathscr{S}(t)\}_{t\in I}$ violates the conditions of Definition 1.1. In particular, equations (2.57), (2.58) and (2.61) involve the term $\det_{\mathscr{S}}(\mathbf{1}_{\mathscr{S}} - n\mathbf{b}^0)$ and when this vanishes certain singularities of the surface $\mathscr{S}(t)$, $t > t_0$, arise, as detailed below:

(a) loss of regularity of the representation of the surface if

$$\boldsymbol{\varphi}_{,\alpha} = (\mathbf{1}_{\mathscr{S}} - n\mathbf{b}^0)\boldsymbol{\varphi}_{,\alpha}^0 = \mathbf{0},$$

(b) unboundedness of one of the surface curvatures,

(c) the vanishing of an elementary field of the surface if $\det[g_{\alpha\beta}] = 0$.

The zeros of the determinant are the singular points of the surface or its parametrization. The singular points may each be represented by a surface cusp, a focus in the form of the vertex of a cone, or, it they are infinite in number and if they assume the shapes of certain curves (caustics), they may form edges (in the sense of Goetz, 1970) along which the surface is refracted.

The vanishing of $\det_{\mathscr{S}}(\mathbf{1}_{\mathscr{S}} - n\mathbf{b}^0)$ implies that n^{-1} is an eigenvalue of $\mathbf{b}^0$ as well as being one of the principal curvatures of the surface $\mathscr{S}(t_0)$.

In accordance with (2.53) the function $n(t)$, $t > t_0$, is positive under the assumption that u_n is positive.

We have proved

Lemma 2.1

If both principal curvatures of the surface $\mathscr{S}(t_0)$ are non-positive then neither of the singularities (a)–(c) can occur at any instant $t > t_0$. □

The condition for non-positive principal curvatures of $\mathscr{S}(t_0)$ is expressible as

$$K_m^0 \leqslant 0, \qquad K_g^0 \geqslant 0, \tag{2.62}$$

but it should be remembered that the mean curvature changes sign with the change in orientation of the surface.

The above proposition is obvious if we write

$$\det_{\mathscr{S}}(\mathbf{1}_{\mathscr{S}} - n\mathbf{b}^0) = 1 - 2K_m^0 n + K_g^0 n^2$$

since $n > 0$.

In the language of surface geometry the inequalities (2.62) express the fact that the surface $\mathscr{S}(t_0)$ has no hyperbolic points (this condition being only necessary for the non-appearance of the singularities (a)–(c)). Examples of such surfaces are the spherical and ellipsoidal surfaces.

In order to reject the possibility of occurrence of singular points of the parametrization of $\mathscr{S}(t)$, $t > t_0$, $t \in I$, it is necessary to assume that the tangent vectors $\boldsymbol{\varphi}_{,1}$, $\boldsymbol{\varphi}_{,2}$ to $\mathscr{S}(t_0)$ have no principal directions, that is that the equation

$$\mathbf{b}^0 \boldsymbol{\varphi}^0_{,\alpha} = k^0 \boldsymbol{\varphi}^0_{,\alpha}$$

has no solution with $k^0 > 0$ for either $\alpha = 1$ or 2 globally on $\mathscr{S}(t_0)$.

We recall that a point of $\mathscr{S}(t_0)$ with mean curvature K_m is a spherical point (umbilical point in the terminology of Goetz, 1970) if

$$\mathbf{b}^0 = K^0_m \mathbf{1}_{\mathscr{S}}.$$

At a spherical point the second metric tensor is proportional to the first metric tensor and the two principal curvatures are equal. We therefore conclude that if $\mathscr{S}(t_0)$ has a spherical point with positive curvature K_m then for the value $n(t) = (K^0_m)^{-1}$ a cusp is formed since at the point in question both tangent vectors to $\mathscr{S}(t)$ vanish.

The singularities (a)–(c) arise in configurations of $\mathscr{S}(t_0)$ for which one of the following conditions holds:

(i) one of the principal curvatures is positive,

(ii) the surface is developable (that is $K^0_g = 0$) with one principal curvature positive,

(iii) the surface $\mathscr{S}(t_0)$ is minimal but not a plane.

We now interpret these possibilities in terms of normal trajectories.

Let us regard the function

$$\mathbf{x} = \boldsymbol{\varphi}(t, l^1, l^2), \qquad (t, l^1, l^2) \in I \times \mathscr{U} \tag{2.63}$$

as a mapping of the open region $I \times \mathscr{U}$ onto a region $\mathscr{V}$ in $\mathscr{E}^3$. According to our assumptions $\boldsymbol{\varphi}$ is differentiable and its first derivatives are

$$\frac{\partial \boldsymbol{\varphi}}{\partial t} = u_n \mathbf{n}, \qquad \frac{\partial \boldsymbol{\varphi}}{\partial l^1} = \boldsymbol{\varphi}_{,1}, \qquad \frac{\partial \boldsymbol{\varphi}}{\partial l^2} = \boldsymbol{\varphi}_{,2}. \tag{2.64}$$

The derivatives of the inverse transformation to (2.63) may be found with the help of (2.64) and

$$\frac{\partial l^\alpha}{\partial \mathbf{x}} \cdot \frac{\partial \boldsymbol{\varphi}}{\partial l^\alpha} = 1, \qquad \frac{\partial t}{\partial \mathbf{x}} \cdot \frac{\partial \boldsymbol{\varphi}}{\partial t} = 1, \qquad \frac{\partial l^\alpha}{\partial l^\beta} = \delta^\alpha_\beta, \qquad \frac{\partial t}{\partial l^\alpha} = 0.$$

Thus

$$\frac{\partial t}{\partial \mathbf{x}} = \frac{\mathbf{n}}{u_n}, \qquad \frac{\partial l^1}{\partial \mathbf{x}} = \boldsymbol{\varphi}^1, \qquad \frac{\partial l^2}{\partial \mathbf{x}} = \boldsymbol{\varphi}^2. \tag{2.65}$$

The relations (2.65) are well defined if $\tilde{j}$, the Jacobian of (2.63), is non-vanishing. In other words, if the mapping of $I \times \mathscr{U}$ onto $\mathscr{V}$ is one-to-one then to each triplet (t, l^1, l^2) corresponds one and only one point $\mathbf{x}$, and conversely. Since the point $(l^1, l^2) = \text{const}$ moves in t along the normal trajectory (straight line), a second point $(\bar{l}^1, \bar{l}^2) = \text{const}$ cannot, at the same time, occupy the same place $\mathbf{x} = \boldsymbol{\varphi}(t, l^1, l^2)$.

If it happens that $\boldsymbol{\varphi}(t, \bar{l}^1, \bar{l}^2) = \boldsymbol{\varphi}(t, l^1, l^2)$ for two distinct pairs $(\bar{l}^1, \bar{l}^2)$ and (l^1, l^2) then the normal trajectories will intersect.

The geometrically obvious fact that the loss of the unambigous character of $\boldsymbol{\varphi}$ occurs simultaneously with the vanishing of $\tilde{j}$ is supported by the following reasoning. We have

$$\tilde{j} = \det\left[\frac{\partial \boldsymbol{\varphi}}{\partial(t, l^1, l^2)}\right] = (\boldsymbol{\varphi}_{,1} \times \boldsymbol{\varphi}_{,2}) \cdot u_n \mathbf{n} = \det[g_{\alpha\beta}] u_n, \tag{2.66}$$

and it follows that the fundamental singularities (a)–(c) occur if the normal trajectories intersect.

The determinant $\det[g_{\alpha\beta}]$ may be interpreted as the surface area of the cross-section of an elementary "tube" bounded by trajectories (straight lines) which are normal to the surface $\mathscr{S}(t)$. Since $u_n \neq 0$ the vanishing of $\tilde{j}$ implies that the surface area is equal to zero.

As is easily seen from Figure 2.1, if the surface $\mathscr{S}(t_0)$ is not convex it should always be expected that the normal straight lines will intersect for a finite value of n.

The above case of a family of parallel surfaces together with the analysis of possible singularities serves as an introduction to the general situation of a family of moving, not necessarily parallel, surfaces, for which there exist curves (rays), not necessarily rectilinear normal trajectories, with properties similar to those discussed above. In Chapter 6, which is devoted to the kinematic theory of waves (the theory of rays), we shall show how to generalize the results derived here.

2.5 Bibliographical Notes

The displacement derivative in the form $(2.8)_1$ was introduced by Thomas (1957). For geometrical quantities defined on a moving surface a similar suggestion for the time derivative can be found in Hayes (1957). The first derivations of the derivatives of geometrical quantities defined on a surface were given by Thomas (1957) and repeated in his book (1961).

The formulae (2.11) and (2.18) were presented by Wang and Truesdell

(1973), and the derivation of equation (2.14) may be found in Chadwick and Powdrill (1965).

Equation (2.16) was first formulated by Hayes (1957) and derivations were given by Thomas (1957), Truesdell and Toupin (1960), Chadwick and Powdrill (1965) and Bowen and Wang (1971). The formulae (2.20), (2.21) and (2.23)–(2.25), which are valid for an arbitrary parametrization, have not been given previously. For a convected parametrization the derivations of $(2.30)_{1-4}$ may be found in Moeckel (1974) and Truesdell and Toupin (1960).

The geometrical properties of a family of parallel surfaces are discussed in Thomas (1965). Coordinate systems in the form of parallel surfaces (the so-called normal coordinates) are discussed extensively by Naghdi (1972) and Napolitano (1982).

3 THE SECOND PARTIAL DERIVATIVES ON A SURFACE

Here we determine the second derivatives of a function f which is assumed to be of class C^2 on $I \times \mathscr{D}$. On $\mathscr{S}$, f is defined as in (2.1).

On replacing f by grad$f$ in (2.4) we obtain

$$\operatorname{grad}\operatorname{grad} f = (\operatorname{grad} f)^{\sim}_{,\alpha} \otimes \boldsymbol{\varphi}^{\alpha} + \frac{\partial}{\partial n}(\operatorname{grad} f) \otimes \mathbf{n}, \tag{3.1}$$

where $(\operatorname{grad} f)^{\sim}$ is defined on the surface by

$$(\operatorname{grad} f)^{\sim}(t, l^1, l^2) = \operatorname{grad} f(t, \boldsymbol{\varphi}(t, l^1, l^2)). \tag{3.2}$$

The symbol $\otimes$ denotes the tensor product.

Multiplying both sides of (3.1) by $\mathbf{n}$, we obtain

$$\frac{\partial}{\partial n}(\operatorname{grad} f) = \left((\operatorname{grad} f)^{\sim}_{,\alpha} \cdot \mathbf{n}\right)\boldsymbol{\varphi}^{\alpha} + \frac{\partial^2 f}{\partial n^2}\mathbf{n}, \tag{3.3}$$

in which we have used (2.3) and the notation

$$\frac{\partial^2 f}{\partial n^2} = \frac{\partial}{\partial n}(\operatorname{grad} f) \cdot \mathbf{n} = (\operatorname{grad}\operatorname{grad} f) \cdot (\mathbf{n} \otimes \mathbf{n}).$$

Next, inserting (3.3) into (3.1), we obtain

$$\operatorname{grad}\operatorname{grad} f = (\operatorname{grad} f)^{\sim}_{,\alpha} \otimes \boldsymbol{\varphi}^{\alpha} + \left((\operatorname{grad} f)^{\sim}_{,\alpha} \cdot \mathbf{n}\right)\boldsymbol{\varphi}^{\alpha} \otimes \mathbf{n} + \frac{\partial^2 f}{\partial n^2}\mathbf{n} \otimes \mathbf{n}. \tag{3.4}$$

With the help of (2.4) we have

$$(\operatorname{grad} f)^{\sim}_{,\alpha} = \tilde{f}_{,\alpha\gamma}\boldsymbol{\varphi}^{\gamma} + \tilde{f}_{,\gamma}\boldsymbol{\varphi}^{\gamma}_{,\alpha} + \left(\frac{\partial f}{\partial n}\right)^{\sim}_{,\alpha}\mathbf{n} + \frac{\partial f}{\partial n}\mathbf{n}_{,\alpha},$$

where the function $(\partial f/\partial n)^{\sim}$ is defined on the surface in a similar way to $(\mathrm{grad} f)^{\sim}$ in (3.2).

The two terms needed for the right-hand side of (3.4) are thus

$$
\begin{aligned}
(\mathrm{grad} f)^{\sim}_{,\alpha}\cdot\mathbf{n} &= \tilde{f}_{,\gamma}\boldsymbol{\varphi}^{\gamma}_{,\alpha}\cdot\mathbf{n}+\left(\frac{\partial f}{\partial n}\right)^{\sim}_{,\alpha},\\
(\mathrm{grad} f)^{\sim}_{,\alpha}\otimes\boldsymbol{\varphi}^{\alpha} &= \tilde{f}_{,\gamma\alpha}\boldsymbol{\varphi}^{\gamma}\otimes\boldsymbol{\varphi}^{\alpha}+\tilde{f}_{,\gamma}\boldsymbol{\varphi}^{\gamma}_{,\alpha}\otimes\boldsymbol{\varphi}^{\alpha}\\
&\quad+\left(\frac{\partial f}{\partial n}\right)^{\sim}_{,\alpha}\mathbf{n}\otimes\boldsymbol{\varphi}^{\alpha}+\frac{\partial f}{\partial n}\mathbf{n}_{,\alpha}\otimes\boldsymbol{\varphi}^{\alpha}.
\end{aligned}
\tag{3.5}
$$

We now require the derivatives $\boldsymbol{\varphi}^{\gamma}_{,\alpha}$. Since $\boldsymbol{\varphi}^{\gamma} = g^{\gamma\delta}\boldsymbol{\varphi}_{,\delta}$, we obtain

$$
\boldsymbol{\varphi}^{\gamma}_{,\alpha} = g^{\gamma\delta}_{,\alpha}\boldsymbol{\varphi}_{,\delta}+g^{\gamma\delta}\boldsymbol{\varphi}_{,\delta\alpha}. \tag{3.6}
$$

The derivatives of the covariant components of the metric tensor are given in terms of Christoffel symbols of the first kind. Thus

$$
g_{\delta\nu,\alpha} = \Gamma_{\delta\alpha\nu}+\Gamma_{\nu\alpha\delta}.
$$

On differentiation of the identity $g^{\gamma\delta}g_{\delta\nu} = \delta^{\gamma}_{\nu}$ we obtain

$$
g^{\gamma\delta}_{,\alpha}g_{\delta\nu} = -g^{\gamma\delta}(\Gamma_{\delta\alpha\nu}+\Gamma_{\nu\alpha\delta}). \tag{3.7}
$$

Equations (3.6) and (3.7), together with (1.7), are then used to write

$$
\begin{aligned}
\boldsymbol{\varphi}^{\gamma}_{,\alpha}\otimes\boldsymbol{\varphi}^{\alpha} &= -g^{\gamma\delta}(\Gamma_{\delta\alpha\nu}+\Gamma_{\nu\alpha\delta})\boldsymbol{\varphi}^{\nu}\otimes\boldsymbol{\varphi}^{\alpha}+g^{\gamma\delta}(\Gamma^{\nu}_{\delta\alpha}\boldsymbol{\varphi}_{,\nu}\otimes\boldsymbol{\varphi}^{\alpha}\\
&\quad+b_{\delta\alpha}\mathbf{n}\otimes\boldsymbol{\varphi}^{\alpha}),\\
\boldsymbol{\varphi}^{\gamma}_{,\alpha}\cdot\mathbf{n} &= g^{\gamma\delta}\mathbf{b}_{\delta\alpha}.
\end{aligned}
\tag{3.8}
$$

Inserting the expressions (3.8) into (3.5) and then (3.5) into (3.4) we obtain

$$
\begin{aligned}
\mathrm{grad\,grad} f = {} & \tilde{f}_{,\gamma\alpha}\boldsymbol{\varphi}^{\alpha}\otimes\boldsymbol{\varphi}^{\gamma}+\left\{\left(\frac{\partial f}{\partial n}\right)^{\sim}_{,\alpha}+\tilde{f}_{,\gamma}b^{\gamma}_{\alpha}\right\}(\mathbf{n}\otimes\boldsymbol{\varphi}^{\alpha}+\boldsymbol{\varphi}^{\alpha}\otimes\mathbf{n})\\
&+\tilde{f}_{,\gamma}g^{\gamma\delta}\{\Gamma^{\nu}_{\delta\alpha}\boldsymbol{\varphi}_{,\nu}\otimes\boldsymbol{\varphi}^{\alpha}-(\Gamma_{\delta\alpha\nu}+\Gamma_{\nu\alpha\delta})\boldsymbol{\varphi}^{\nu}\otimes\boldsymbol{\varphi}^{\alpha}\}\\
&+\frac{\partial^2 f}{\partial n^2}\mathbf{n}\otimes\mathbf{n}-\frac{\partial f}{\partial n}b^{\beta}_{\alpha}\boldsymbol{\varphi}_{,\beta}\otimes\boldsymbol{\varphi}^{\alpha}.
\end{aligned}
$$

Since $\Gamma^{\nu}_{\delta\alpha}\boldsymbol{\varphi}_{,\nu}\otimes\boldsymbol{\varphi}^{\alpha} = \Gamma^{\beta}_{\delta\alpha}g_{\beta\nu}\boldsymbol{\varphi}^{\nu}\otimes\boldsymbol{\varphi}^{\alpha}$ and $\Gamma^{\beta}_{\delta\alpha}g_{\beta\nu} = \Gamma_{\delta\alpha\nu}$ we obtain finally

$$
\begin{aligned}
\mathrm{grad\,grad} f = {} & \tilde{f}_{,\gamma\alpha}\boldsymbol{\varphi}^{\gamma}\otimes\boldsymbol{\varphi}^{\alpha}+\left\{\left(\frac{\partial f}{\partial n}\right)^{\sim}_{,\alpha}+\tilde{f}_{,\gamma}b^{\gamma}_{\alpha}\right\}(\mathbf{n}\otimes\boldsymbol{\varphi}^{\alpha}+\boldsymbol{\varphi}^{\alpha}\otimes\mathbf{n})\\
&-\tilde{f}_{,\gamma}\Gamma^{\gamma}_{\nu\alpha}\boldsymbol{\varphi}^{\nu}\otimes\boldsymbol{\varphi}^{\alpha}-\frac{\partial f}{\partial n}b^{\beta}_{\alpha}g_{\beta\nu}\boldsymbol{\varphi}^{\nu}\otimes\boldsymbol{\varphi}^{\alpha}+\frac{\partial^2 f}{\partial n^2}\mathbf{n}\otimes\mathbf{n}.
\end{aligned}
\tag{3.9}
$$

This formula is one of the fundamental relations in the theory of the motion of surfaces in deformable media. In addition to this formula we now derive two others, one for the mixed derivative and one for the time derivative. First, however, we note that by contraction of (3.9)

$$\mathrm{grad}\cdot(\mathrm{grad}f) = (\tilde{f}_{,\nu\alpha}-\tilde{f}_{,\gamma}\Gamma^{\gamma}_{\nu\alpha})g^{\nu\alpha}-2K_m\frac{\partial f}{\partial n}+\frac{\partial^2 f}{\partial n^2}. \tag{3.10}$$

With f replaced by $\partial f/\partial t$ in (2.4), we have

$$\mathrm{grad}\left(\frac{\partial f}{\partial t}\right) = \left(\frac{\partial f}{\partial t}\right)^{\sim}_{,\alpha}\boldsymbol{\varphi}^{\alpha}+\frac{\partial}{\partial n}\left(\frac{\partial f}{\partial t}\right)\mathbf{n}, \tag{3.11}$$

where the function $(\partial f/\partial t)^{\sim}$ is defined similarly to (3.2). Because of the assumed smoothness of f, we also have

$$\mathrm{grad}\left(\frac{\partial f}{\partial t}\right) = \frac{\partial}{\partial t}(\mathrm{grad}f).$$

It follows from (2.3), (2.4), (2.10), (2.15) and (3.3) that

$$\begin{aligned}\frac{\delta}{\delta t}\left(\frac{\partial f}{\partial n}\right) &= \frac{\delta}{\delta t}(\mathrm{grad}f)\cdot\mathbf{n}+\mathrm{grad}f\cdot\frac{\delta\mathbf{n}}{\delta t}\\ &= \frac{\partial}{\partial n}\left(\frac{\partial f}{\partial t}\right)+u_n\frac{\partial^2 f}{\partial n^2}-u_{n,\gamma}\tilde{f}_{,\alpha}g^{\gamma\alpha}.\end{aligned} \tag{3.12}$$

Computing $\partial(\partial f/\partial t)/\partial n$ from (3.12) and inserting the result into (3.11) we see that

$$\mathrm{grad}\left(\frac{\partial f}{\partial t}\right) = \left(\frac{\partial f}{\partial t}\right)^{\sim}_{,\alpha}\boldsymbol{\varphi}^{\alpha}+\left\{\frac{\delta}{\delta t}\left(\frac{\partial f}{\partial n}\right)+u_{n,\gamma}\tilde{f}_{,\alpha}g^{\gamma\alpha}-u_n\frac{\partial^2 f}{\partial n^2}\right\}\mathbf{n}. \tag{3.13}$$

After use of (2.9) the required mixed derivative is expressed as

$$\begin{aligned}\mathrm{grad}\left(\frac{\partial f}{\partial t}\right) &= \left\{\frac{\delta\tilde{f}}{\delta t}-u_n\left(\frac{\partial f}{\partial n}\right)^{\sim}\right\}_{,\alpha}\boldsymbol{\varphi}^{\alpha}\\ &\quad+\left\{\frac{\delta}{\delta t}\left(\frac{\partial f}{\partial n}\right)+u_{n,\gamma}\tilde{f}_{,\alpha}g^{\alpha\gamma}-u_n\frac{\partial^2 f}{\partial n^2}\right\}\mathbf{n}.\end{aligned} \tag{3.14}$$

The missing relation for the second time derivative is determined with the use of (2.9) applied to $\partial f/\partial t$ instead of f. Elimination of $\partial(\partial f/\partial t)/\partial n$, use of (3.12) and a further application of (2.9) then yields

$$\begin{aligned}\frac{\partial^2 f}{\partial t^2} &= \frac{\delta}{\delta t}\left\{\frac{\delta\tilde{f}}{\delta t}-u_n\left(\frac{\partial f}{\partial n}\right)\right\}\\ &\quad-u_n\left\{\frac{\delta}{\delta t}\left(\frac{\partial f}{\partial n}\right)-u_n\frac{\partial^2 f}{\partial n^2}+u_{n,\gamma}\tilde{f}_{,\alpha}g^{\alpha\gamma}\right\}.\end{aligned} \tag{3.15}$$

The formulae (3.9), (3.14) and (3.15) were first given in index notation by Chadwick and Powdrill (1965), and are called the compatibility conditions for the second partial derivatives of a function f on the surface $\mathscr{S}$. They require knowledge, not only of f and its normal derivative as functions of t, l^1, l^2 on $\mathscr{S}$, but also the second normal derivative. Thus, the determination of the second derivatives of f depends on additional information about the behaviour of f or its derivatives on $\mathscr{S}$. This information, coupled with the compatibility conditions, may be given in the form of a partial differential equation governing f.

The equation for $\operatorname{grad}\operatorname{grad} f$ may also be expressed in terms of the covariant derivative since, by (2.47),

$$\tilde{f}_{;\nu\alpha} = \tilde{f}_{,\nu\alpha} - \Gamma^{\gamma}_{\nu\alpha}\tilde{f}_{,\gamma}. \tag{3.16}$$

Thus, (3.9) becomes

$$\begin{aligned}\operatorname{grad}\operatorname{grad} f = \tilde{f}_{;\nu\alpha}\boldsymbol{\varphi}^{\nu}\otimes\boldsymbol{\varphi}^{\alpha} + \left\{\left(\frac{\partial f}{\partial n}\right)^{\sim}_{;\alpha} + \tilde{f}_{;\gamma}b^{\gamma}_{\alpha}\right\}(\mathbf{n}\otimes\boldsymbol{\varphi}^{\alpha}+\boldsymbol{\varphi}^{\alpha}\otimes\mathbf{n})\\ - \frac{\partial f}{\partial n}b^{\beta}_{\alpha}g_{\beta\nu}\boldsymbol{\varphi}^{\nu}\otimes\boldsymbol{\varphi}^{\alpha} + \frac{\partial^2 f}{\partial n^2}\mathbf{n}\otimes\mathbf{n},\end{aligned} \tag{3.17}$$

which makes use of

$$\left(\frac{\partial f}{\partial n}\right)^{\sim}_{;\alpha} = \left(\frac{\partial f}{\partial n}\right)^{\sim}_{,\alpha}.$$

Observe that in (3.17), as in (3.9), the right-hand side is symmetric with respect to the indices ν and α. This indicates that the equations may also be written as follows:

$$\begin{aligned}\operatorname{grad}\operatorname{grad} f = \left\{\tilde{f}_{,(\nu\alpha)} - \Gamma^{\gamma}_{(\nu\alpha)}\tilde{f}_{,\gamma} - \frac{\partial f}{\partial n}b_{\nu\alpha}\right\}\boldsymbol{\varphi}^{(\nu}\otimes\boldsymbol{\varphi}^{\alpha)}\\ + \left\{\left(\frac{\partial f}{\partial n}\right)^{\sim}_{,\alpha} + \tilde{f}_{,\gamma}b^{\gamma}_{\alpha}\right\}(\mathbf{n}\otimes\boldsymbol{\varphi}^{\alpha}+\boldsymbol{\varphi}^{\alpha}\otimes\mathbf{n}) + \frac{\partial^2 f}{\partial n^2}\mathbf{n}\otimes\mathbf{n},\end{aligned} \tag{3.9a}$$

$$\begin{aligned}\operatorname{grad}\operatorname{grad} f = \left\{\tilde{f}_{;(\nu\alpha)} - \frac{\partial f}{\partial n}b_{\nu\alpha}\right\}\boldsymbol{\varphi}^{(\nu}\otimes\boldsymbol{\varphi}^{\alpha)}\\ + \left\{\left(\frac{\partial f}{\partial n}\right)^{\sim}_{;\alpha} + \tilde{f}_{;\gamma}b^{\gamma}_{\alpha}\right\}(\mathbf{n}\otimes\boldsymbol{\varphi}^{\alpha}+\boldsymbol{\varphi}^{\alpha}\otimes\mathbf{n}) + \frac{\partial^2 f}{\partial n^2}\mathbf{n}\otimes\mathbf{n},\end{aligned} \tag{3.17a}$$

where the bracketing $(\nu\alpha)$ signifies symmetrization with respect to the indices ν and α.

3.1 Basic Results in Index Notation

In index notation the basic results of Section 3 may be written as follows. Equations (3.9) and (3.17) become

$$
\begin{aligned}
f_{;kl} &= G_{ki}G_{lj}\Big\{(\tilde{f}_{,\nu\alpha}-\tilde{f}_{,\gamma}\Gamma^{\gamma}_{\nu\alpha})\varphi^{i\nu}\varphi^{j\alpha}-\frac{\partial f}{\partial n}b_{\nu\alpha}\varphi^{i\nu}\varphi^{j\alpha} \\
&\quad +\Big[\Big(\frac{\partial f}{\partial n}\Big)^{\sim}_{,\alpha}+\tilde{f}_{,\gamma}b^{\gamma}_{\alpha}\Big](n^{i}\varphi^{j\alpha}+n^{j}\varphi^{i\alpha})+\frac{\partial^{2}f}{\partial n^{2}}n^{i}n^{j}\Big\} \\
&= G_{ki}G_{lj}\Big\{\Big(f_{;\nu\alpha}-\frac{\partial f}{\partial n}b'_{\nu\alpha}\Big)\varphi^{i\nu}\varphi^{j\alpha}+\Big[\Big(\frac{\partial f}{\partial n}\Big)^{\sim}_{;\alpha}+\tilde{f}_{;\gamma}b^{\gamma}_{\alpha}\Big](n^{i}\varphi^{j\alpha} \\
&\quad +n^{j}\varphi^{i\alpha})+\frac{\partial^{2}f}{\partial n^{2}}n^{i}n^{j}\Big\},
\end{aligned}
\tag{3.18}
$$

where $\partial^2 f/\partial n^2 \equiv f_{;ij}n^i n^j$, the semi-colon indicating covariant differentiation in the coordinate system of the space $\mathscr{E}^3$.

Similarly, from (3.9a) and (3.17a) we have

$$
\begin{aligned}
f_{;kl} &= G_{ki}G_{lj}\Big\{\Big(\tilde{f}_{,(\nu\alpha)}-\tilde{f}_{,\gamma}\Gamma^{\gamma}_{(\nu\alpha)}-\frac{\partial f}{\partial n}b_{\nu\alpha}\Big)\varphi^{i(\nu}\varphi^{j\alpha)} \\
&\quad +2\Big[\Big(\frac{\partial f}{\partial n}\Big)^{\sim}_{,\alpha}+\tilde{f}_{,\gamma}b^{\gamma}_{\alpha}\Big]n^{(i}\varphi^{j)\alpha}+\frac{\partial^{2}f}{\partial n^{2}}n^{i}n^{j}\Big\} \\
&= G_{ki}G_{lj}\Big\{\Big[\tilde{f}_{;(\nu\alpha)}-\frac{\partial f}{\partial n}b_{\nu\alpha}\Big]\varphi^{i(\nu}\varphi^{j\alpha)} \\
&\quad +2\Big[\Big(\frac{\partial f}{\partial n}\Big)^{\sim}_{;\alpha}+\tilde{f}_{;\gamma}b^{\gamma}_{\alpha}\Big]n^{(i}\varphi^{j)\alpha}+\frac{\partial^{2}f}{\partial n^{2}}n^{i}n^{j}\Big\},
\end{aligned}
\tag{3.19}
$$

and from (3.14)

$$
\begin{aligned}
\frac{\partial}{\partial t}f_{;k} &= G_{ki}\Big\{\Big(\frac{\delta\tilde{f}}{\delta t}-u_n\Big(\frac{\partial f}{\partial n}\Big)^{\sim}\Big)_{,\alpha}\varphi^{i\alpha} \\
&\quad +\Big[\frac{\delta}{\delta t}\Big(\frac{\partial f}{\partial n}\Big)+u_{n,\gamma}\tilde{f}_{,\alpha}g^{\gamma\alpha}-u_n\frac{\partial^{2}f}{\partial n^{2}}\Big]n^{i}\Big\}.
\end{aligned}
\tag{3.20}
$$

The derivations of (3.15), (3.18) and (3.20) for the Cartesian coordinate system may also be found in Chadwick and Powdrill (1965).

3.2 The Invariant Derivative

The representations (3.15), $(3.18)_1$, (3.20) and also (2.31)–(2.33) for tensor fields in index notation require further comment. We examine the simplest case of the evolution in time of a vector field $\mathbf{f}=f^i\mathbf{e}_i$. If the coordinate system in $\mathscr{E}^3$ is not rectangular Cartesian then the basis vectors $\mathbf{e}_i$ change from point to point. With Γ^k_{ij} denoting the Christoffel symbols of this system we obtain

$$\operatorname{grad}\mathbf{f} = \frac{\partial}{\partial x^j}(f^i\mathbf{e}_i)\otimes\mathbf{e}^j = (f^i_{,j}+\Gamma^i_{jl}f^l)\,\mathbf{e}_i\otimes\mathbf{e}^j = f^i_{;j}\,\mathbf{e}_i\otimes\mathbf{e}^j. \tag{3.21}$$

If we now introduce the vector function $\tilde{\mathbf{f}}$ on a moving surface $\mathscr{S}$ in accordance with (2.1), then

$$\frac{\partial\tilde{\mathbf{f}}}{\partial l^\alpha} = \left(\frac{\partial\tilde{f}^i}{\partial l^\alpha}+\Gamma^i_{jl}\tilde{f}^l\frac{\partial\varphi^j}{\partial l^\alpha}\right)\mathbf{e}_i. \tag{3.22}$$

The displacement derivative of the vector field $\mathbf{f}$ may be written

$$\frac{\delta\mathbf{f}}{\delta t} = \frac{\delta f^i}{\delta t}\mathbf{e}_i+f^i\frac{\delta\mathbf{e}_i}{\delta t}.$$

Since the basis vectors are time independent and

$$\frac{\delta\boldsymbol{\varphi}}{\delta t} = \frac{\partial\boldsymbol{\varphi}}{\partial t}-c^\alpha\frac{\partial\boldsymbol{\varphi}}{\partial l^\alpha} = u_n\mathbf{n} = u_n n^l\mathbf{e}_l \tag{3.23}$$

we obtain

$$\frac{\delta\mathbf{e}_i}{\delta t} = \operatorname{grad}\mathbf{e}_i\frac{\delta\boldsymbol{\varphi}}{\delta t} = u_n\Gamma^k_{il}n^l\mathbf{e}_k, \tag{3.24}$$

and hence

$$\frac{\delta\mathbf{f}}{\delta t} = \left(\frac{\delta f^i}{\delta t}+\Gamma^i_{jk}f^j u_n n^k\right)\mathbf{e}_i. \tag{3.25}$$

Adopting the notation*

$$\frac{\mathrm{d}f^i}{\delta t} := \frac{\delta f^i}{\delta t}+u_n\Gamma^i_{jk}f^j n^k \tag{3.26}$$

and remembering that

$$\frac{\delta f^i}{\delta t} = \frac{\partial f^i}{\partial t}+u_n f^i_{,j}n^j$$

we obtain

$$\frac{\mathrm{d}f^i}{\delta t} = \frac{\partial f^i}{\partial t}+u_n(f^i_{,k}+\Gamma^i_{jk}f^j)n^k. \tag{3.27}$$

We shall now deal with the displacement derivative of the function $\tilde{\mathbf{f}}$. In view of (3.22) we may write

---

* Thomas (1961, p. 53) calls $\mathrm{d}f^i/\delta t$ the **invariant derivative.** See also Braun (1974).

$$\frac{\delta\tilde{\mathbf{f}}}{\delta t} = \frac{\delta\tilde{f}^i}{\delta t}\mathbf{e}_i + \tilde{f}^i\frac{\delta\mathbf{e}_i}{\delta t}.$$

Hence, by (3.23) and (2.7), we obtain

$$\frac{\delta\tilde{\mathbf{f}}}{\delta t} = \left(\frac{\delta\tilde{f}^i}{\delta t} + \Gamma^i_{jk}\tilde{f}^j u_n n^k\right)\mathbf{e}_i \tag{3.28}$$

or

$$\frac{\delta\tilde{\mathbf{f}}}{\delta t} = \left(\frac{\partial\tilde{f}^i}{\partial t} - c^\alpha\frac{\partial\tilde{f}^i}{\partial l^\alpha} + u_n\Gamma^i_{jk}\tilde{f}^j n^k\right)\mathbf{e}_i. \tag{3.29}$$

In the present case the derivative $\mathrm{d}/\delta t$ acting on a function defined on a surface $\mathscr{S}$ is of the form

$$\frac{\mathrm{d}\tilde{f}^i}{\delta t} = \frac{\partial\tilde{f}^i}{\partial t} - c^\alpha\tilde{f}^i_{,\alpha} + u_n\Gamma^i_{jk}\tilde{f}^j n^k. \tag{3.30}$$

For the vector function $\mathbf{g} = g^i\mathbf{e}_i$ of the variables t, $\mathbf{x}$ and l^α the derived relations (3.22) and (3.30) together with (2.11) lead to a general form, namely

$$\frac{\delta}{\delta t}\mathbf{g}(t, \mathbf{x}, l^1, l^2) = \left\{\frac{\partial g^i}{\partial t} + u_n(g^i_{,k} + \Gamma^i_{jk}g^j)n^k - c^\alpha g^i_{,\alpha}\right\}\mathbf{e}_i. \tag{3.31}$$

It is relatively easy to derive the appropriate relations for tensor quantities defined not only on the basis $\mathbf{e}_i$ but also on the dual basis $\mathbf{e}^j$ of the space $\mathscr{E}^3$. Let $\mathbf{T} = T^i_j\,\mathbf{e}_i\otimes\mathbf{e}^j$ be a time-dependent tensor field in $\mathscr{E}^3$. Then

$$\begin{aligned}\frac{\delta\mathbf{T}}{\delta t} &= \left\{\frac{\delta T^i_j}{\delta t} + u_n(\Gamma^i_{lk}T^l_j - \Gamma^m_{jk}T^i_m)n^k\right\}\mathbf{e}_i\otimes\mathbf{e}^j \\ &= \left\{\frac{\partial T^i_j}{\partial t} + u_n(T^i_{j,k} + \Gamma^i_{lk}T^l_j - \Gamma^m_{jk}T^i_m)n^k\right\}\mathbf{e}_i\otimes\mathbf{e}^j.\end{aligned} \tag{3.32}$$

Since

$$T^i_{j;k} \equiv T^i_{j,k} + \Gamma^i_{kl}T^l_j - \Gamma^m_{jk}T^i_m$$

we deduce from (3.32) that

$$\frac{\delta\mathbf{T}}{\delta t} = \left(\frac{\partial T^i_j}{\partial t} + u_n T^i_{j;k}n^k\right)\mathbf{e}_i\otimes\mathbf{e}^j. \tag{3.33}$$

The relations derived so far enable us to define the **invariant displacement derivative** $\mathrm{d}/\delta t$ according to

$$\frac{\mathrm{d}f^{ij\cdots}_{kl\cdots}}{\delta t} \equiv \frac{\partial f^{ij\cdots}_{kl\cdots}}{\partial t} + u_n f^{ij\cdots}_{kl\cdots;p}n^p \tag{3.34}$$

for the components $f^{ij\cdots}_{kl\cdots}$ of an arbitrary tensor function $\mathbf{f}(t, \mathbf{x})$. With the help of this function the displacement derivative $\delta/\delta t$ may be written in index notation as follows:

$$\frac{\delta}{\delta t}\left(f^{ij\cdots}_{kl\cdots}\mathbf{e}_i\otimes\mathbf{e}_j\otimes\ \ldots\ \otimes\mathbf{e}^k\otimes\mathbf{e}^l\otimes\ \ldots\right)$$

$$= \frac{\mathrm{d}f^{ij\cdots}_{kl\cdots}}{\delta t}\,\mathbf{e}_i\otimes\mathbf{e}_j\otimes\ldots\ \otimes\mathbf{e}^k\otimes\mathbf{e}^l\otimes\ \ldots \tag{3.35}$$

Chapter 2

Functions with surface singularities

In the preceding chapter we examined the behaviour of a continuously differentiable function on a moving surface $\mathscr{S}(t)$. We now consider certain functions which lose their regularity on $\mathscr{S}(t)$. The choice of the class of field functions we shall investigate in this chapter is determined by the possibility of their occurrence in problems of wave propagation. As indicated in the introduction to Chapter 1, the model of wave phenomena adopted in the present book is due to Hugoniot, and it characterizes waves as disturbances inherently associated with a moving surface. This provides the key to our choice of field functions. In particular, we shall deal with functions which lose continuity or differentiability across a surface. For such functions a moving surface $\mathscr{S}(t)$ is a surface on which singularities occur, and is referred to briefly as a **singular surface** (see Section 5 for details).

Our aim is to find the conditions which must be satisfied by the so-called **jump discontinuities** of a function and/or its derivatives across $\mathscr{S}(t)$. To find these conditions we shall draw on the results of the previous two sections. Firstly, however, we introduce the notion of a smooth extension of a function defined on a non-open set, and, for a function defined on a set intersected by a singular surface $\{\mathscr{S}(t)\}_{t \in I}$, we define "left-sided" and "right-sided" smooth extensions, labelled respectively with minus and plus signs. The required conditions are obtained by applying the formulae for the first and second partial derivatives of a smooth function on $\{\mathscr{S}(t)\}_{t \in I}$ to the smooth extensions.

The method used to derive the compatibility conditions is radically different from the classical method based on the well-known Hadamard lemma.

4 SMOOTH EXTENSIONS OF FUNCTIONS

In Chapter 1 we assumed that a certain function f, for which we derived formulae for the first and second partial derivatives on a moving surface $\mathscr{S}$, was continuously differentiable on an open set $\mathscr{D}$. Openness of the set is a natural requirement since in mathematical analysis continuity and differentiability of a function are defined on open sets.

In the problems which will be discussed later in the book the classical definitions of continuity and differentiability are not sufficient. Indeed, the functions we shall employ are defined on non-open sets and therefore require new definitions of continuity and differentiability. To this end we use the following definition (see Maurin, 1976, p. 222; Nickerson, Spencer and Steenrod, 1951, p. 411, or Chadwick and Powdrill, 1965).

Definition 4.1

A function f defined on a (generally non-open) subset $\mathscr{V}$ of a normed space $\mathscr{X}$ with values in a normed (Banach) space $\mathscr{Y}$ is said to be **p-times continuously differentiable*** on $\mathscr{V}$ if there is an open set $\mathscr{U}$ containing $\mathscr{V}$ and a function f_1 of class C^p on $\mathscr{U}$ (that is p-times continuously differentiable) which coincides with f on $\mathscr{V}$, that is

$$f_1|_{\mathscr{V}} = f. \tag{4.1}$$

By the p-th derivative of the function f we understand the p-th derivative of f_1. The function f_1 is called a **smooth extension** of class C^p of the function f, while f is called an **extendible function.**

Clearly, if f is extendible then its extension is not uniquely determined, and the choice of f_1 and $\mathscr{U}$ is fairly arbitrary. For example, if $\mathscr{V} \subset \mathscr{U}_1 \subset \mathscr{U}$, where $\mathscr{U}_1$ is an open set, then the restriction of f_1 to $\mathscr{U}_1$, that is the function $f_2 \equiv f_1|_{\mathscr{U}_1}$, is also an extension of f.

The question of whether the derivatives up to the p-th order of f are uniquely determined on a non-open set $\mathscr{V}$ is of vital importance for further considerations. The following lemma provides the answer.

Lemma 4.1

Let a function f defined on a non-empty set $\mathscr{V}$ be extendible and p-times continuously differentiable on $\mathscr{V}$, that is, let it satisfy Definition 4.1. If the set $\mathscr{V}$ is contained in the closure of its interior, then all the derivatives of f up to the p-th order are uniquely determined on $\mathscr{V}$, that is they do not depend on the choice of the extension f_1.

---

* In particular, if $p = 0$, the function is said to be continuous.

Proof

Let f_1 and f_2 be two smooth extensions of f, to the open sets $\mathcal{U}_1$ and $\mathcal{U}_2$ respectively. Then, in view of (4.1), we have

$$f_1|_{\mathcal{V}} = f_2|_{\mathcal{V}} = f. \tag{4.2}$$

Let $\operatorname{Int}\mathcal{V}$ denote the interior of the set $\mathcal{V}$, that is the largest open set included in $\mathcal{V}$. By assumption, $\operatorname{Int}\mathcal{V} \neq \varnothing$, where $\varnothing$ denotes the empty set. Since f_1 is of class C^p on $\mathcal{U}_1$ and f_2 is of class C^p on $\mathcal{U}_2$, and both $\mathcal{V} \subset \mathcal{U}_1$ and $\mathcal{V} \subset \mathcal{U}_2$, both functions are of class C^p on $\mathcal{U} = \mathcal{U}_1 \cap \mathcal{U}_2$, which is an open set. Moreover,

$$\operatorname{Int}\mathcal{V} \subset \mathcal{V} \subset \mathcal{U}. \tag{4.3}$$

If the k-th derivatives (where $k \leqslant p$) of f_1 and f_2 are denoted by $\mathrm{D}_k f_1$ and $\mathrm{D}_k f_2$ respectively then, by (4.2) and (4.3), we have

$$\mathrm{D}_k f_1|_{\operatorname{Int}\mathcal{V}} = \mathrm{D}_k f_2|_{\operatorname{Int}\mathcal{V}}, \tag{4.4}$$

both functions being continuous on $\mathcal{U}$.

Let y denote an arbitrary point of $\mathcal{V}$. We have assumed that

$$\mathcal{V} \subset \overline{\operatorname{Int}\mathcal{V}}, \tag{4.5}$$

where the horizontal bar denotes the closure of the set $\operatorname{Int}\mathcal{V}$. Each element a of the set $\overline{\operatorname{Int}\mathcal{V}}$ admits the property* that there exists a sequence composed of elements of the set $\operatorname{Int}\mathcal{V}$ convergent to a. This property, together with (4.5), indicates that there exists a sequence $\{y_n\}_0^\infty \subset \operatorname{Int}\mathcal{V}$ convergent to y, that is $y_n \to y$.

Because of (4.4) we have

$$\mathrm{D}_k f_1(y_n) = \mathrm{D}_k f_2(y_n)$$

for each n, and since $\mathrm{D}_k f_1$ and $\mathrm{D}_k f_2$ are continuous, it follows that

$$\mathrm{D}_k f_1(y_n) \to \mathrm{D}_k f_1(y) \quad \text{and} \quad \mathrm{D}_k f_2(y_n) \to \mathrm{D}_k f_2(y).$$

This means that

$$\mathrm{D}_k f_1(y) = \mathrm{D}_k f_2(y). \tag{4.6}$$

Since y is arbitrary we deduce from (4.4) and (4.6) the identity

$$\mathrm{D}_k f_1|_{\mathcal{V}} = \mathrm{D}_k f_2|_{\mathcal{V}}. \tag{4.7}$$

Since k is an arbitrary natural number not greater than p, this completes the proof. □

---

* For a topological space this property is formulated in the language of generalized sequences, called nets (see Engelking, 1979, p. 54).

In commenting on the above lemma we note that the essential assumption about the set $\mathscr{V}$ is expressed by (4.5). Without this assumption the derivatives are determined uniquely only for the interior of $\mathscr{V}$ (Int $\mathscr{V}$ is not the empty set* since $\mathscr{V}$ satisfying (4.5) is not empty). The assumption (4.5) guarantees that for an arbitrary point of $\mathscr{V}$ there exists a sequence of points of Int $\mathscr{V}$ convergent to it, and *ipso facto* the possibility of a unique extension of the derivative over the whole set $\mathscr{V}$ with continuity preserved.

The method of proof used here suggests the possibility of determining a continuously differentiable extension of a function defined on a non-open set. To this end we consider the following example:

Example 4.1 (Sikorski, 1969, p. 110)

Let f be a continuous mapping of a subset $\mathscr{A}$ of a metric space $\mathscr{X}$ into a metric space $\mathscr{Y}$. Let $\mathscr{F}$ be the set of points $y \in \mathscr{X} - \mathscr{A}$ in which a limit of f exists. The mapping g: $\mathscr{A} \cup \mathscr{F} \to \mathscr{Y}$ defined by the formula

$$g(y) = \begin{cases} f(y) & \text{for } y \in \mathscr{A}, \\ \lim\limits_{\substack{q \to y \\ q \in \mathscr{A}}} f(q) & \text{for } y \in \mathscr{F} \end{cases} \tag{4.8}$$

is continuous. The proof makes direct use of the definition of continuity. The function g is continuous on the set $\mathscr{A}$ by (4.8). Continuity at points of $\mathscr{F}$ is obvious since the existence of the limit value of f at $y \in \mathscr{F}$ implies that, for every sequence of points $\{y_n\}_0^\infty$ which satisfies the condition

$$\lim_{n \to \infty} y_n = y,$$

$\lim\limits_{n \to \infty} f(y_n)$ exists and does not depend on the choice of the sequence $\{y_n\}_0^\infty$. This condition follows from the definition of continuity of a function at a point since the value of g at y_n coincides with $f(y_n)$ while $g(y) = \lim\limits_{n \to \infty} f(y_n)$. □

This example and Lemma 4.1 point to the proof of the following:

Lemma 4.2

Let $\mathscr{V}$ and $\mathscr{U}$ be subsets of a normed space $\mathscr{X}$, the subset $\mathscr{V}$ having the property

$$\mathscr{V} \subset \overline{\operatorname{Int}\mathscr{U}}, \quad \operatorname{Int}\mathscr{U} \subset \mathscr{V}. \tag{4.9}$$

Let f be a function (mapping) of class C^p defined on $\mathscr{A} := \operatorname{Int}\mathscr{U}$ with values in $\mathscr{Y}$ and having the property that for each $0 \leqslant k \leqslant p$ and a point

* If the interior of the set $\mathscr{V}$ is empty then $\mathscr{V}$ is a **boundary set**. In $\mathscr{E}^3$ surfaces and curves are boundary sets.

$y \in \mathscr{V}$ there exists a limit of the k-th derivative of f at y. The function g and its k-th derivatives, defined on $\mathscr{V}$ by means of (4.8) and

$$\mathrm{D}_k g(q) = \begin{cases} \mathrm{D}_k f(q) & \text{for } q \in \mathscr{A}, \\ \lim\limits_{\substack{p \to q \\ p \in \mathscr{A}}} \mathrm{D}_k f(p) & \text{for } q \in \mathscr{V} - \mathscr{A}, \end{cases} \tag{4.10}$$

are continuous, and all derivatives of g up to order p are uniquely determined on $\mathscr{V}$ and continuous. □

We note that if $\mathscr{F} \subset \partial \mathscr{U}$ then the set $\mathscr{V} = \mathscr{F} \cup \mathrm{Int}\, \mathscr{U}$ satisfies (4.9). (Sometimes $\partial \mathscr{U}$ is denoted by $\mathrm{Fr}\, \mathscr{U}$; then $\mathrm{Fr}\, \mathscr{U} = \overline{\mathscr{U}} - \mathrm{Int}\, \mathscr{U}$.)

Definition 4.2

The function g determined* in the above way will be called a smooth **continuation** (of class C^p) of the function $f\colon \mathscr{A} \to \mathscr{Y}$ to the non-open set $\mathscr{V}$.

In later parts of the book we shall use a smooth continuation of a function when the function in question is defined not on the whole set $\mathscr{V}$ (as in Definition 4.1) but only on the subset $\mathrm{Int}\, \mathscr{V}$. Observe, however, that the set $\mathscr{V}$ appearing in Lemma 4.2 has the property (4.5). Indeed, from (4.9) we have

$$\mathrm{Int}\, \mathscr{V} = \mathrm{Int}\, \mathscr{U} \subset \mathscr{V}$$

and

$$\mathscr{V} \subset \overline{\mathrm{Int}\, \mathscr{U}} \subset \overline{\mathrm{Int}\, \mathscr{V}}.$$

The reason for introducing a smooth continuation in addition to a smooth extension is the following: a typical function f, defined on a set $\mathscr{N}$ intersected by a hypersurface $\mathscr{S}$ and appearing in descriptions of surface phenomena, may have values defined on $\mathscr{S}$ independently of the limits (traces) f^- and f^+ which are the limit values of f if $\mathscr{S}$ is approached from those parts of the region labelled minus and plus respectively. That is, for any $y \in \mathscr{S} \cap \mathscr{N}$ and $t \in I$

$$f^-(y, t) = \lim_{\substack{z \to y \\ \overrightarrow{yz} \cdot \mathbf{m} < 0}} f(z, t), \qquad f^+(y, t) = \lim_{\substack{z \to y \\ \overrightarrow{yz} \cdot \mathbf{m} > 0}} f(z, t),$$

where $\mathbf{m}$ is a normal vector to the orientable hypersurface $\mathscr{S}$ at y.

It follows that the restrictions $f|_{\mathscr{N}^+ \cup \mathscr{S}}$ and $f|_{\mathscr{N}^- \cup \mathscr{S}}$ are not likely to be continuous in the sense of Definition 4.1. Here we choose $\mathbf{m}$ to be the outward normal vector to $\mathscr{N}^-$. Hence the restrictions $f|_{\mathscr{N}^+}$ and $f|_{\mathscr{N}^-}$ have to be continued separately to the non-open sets $\mathscr{N}^+ \cup \mathscr{S}$ and $\mathscr{N}^- \cup \mathscr{S}$ respectively whenever $f|_{\mathscr{N}^+}$ and $f|_{\mathscr{N}^-}$ are continuous. The precise meaning of this operation is given in Definition 4.2.

* In the book by Sikorski (1969, pp. 338–341) a mapping of class C^p on a **closed** set is defined.

5 SINGULAR SURFACES

In Section 2 we considered a function f defined on the Cartesian product $I \times \mathscr{D}$, where I is an open interval of the real axis and $\mathscr{D}$ is an open subset of $\mathscr{E}^3$. Let f be such a function and let $\mathscr{N}$ denote the set

$$\mathscr{N} \equiv \bigcup_{t \in I} \{t\} \times \mathscr{D}_t, \tag{5.1}$$

where $\{\mathscr{D}_t\}_{t \in I}$ is a one-parameter family of open sets.

Clearly, $\mathscr{N}$ is an open subset of $\mathscr{E}^4$. Suppose that $\mathscr{S}$ is a moving surface such that if

$$\mathscr{S} = \bigcup_{t \in I} \{t\} \times \mathscr{S}(t) \tag{5.2}$$

then

$$\mathscr{S}(t) \subset \mathscr{D}_t, \qquad t \in I. \tag{5.3}$$

We also suppose that for every $t \in I$ the surface $\mathscr{S}(t)$ divides the region $\mathscr{D}_t$ into two disjoint open sets. More precisely, if, for an orientable surface $\mathscr{S}(t)$, we choose one of the orientations then we can determine two open and non-intersecting sets $\mathscr{D}_t^+$ and $\mathscr{D}_t^-$ in such a way that their set-theoretic sum is the set $\mathscr{D}_t - \mathscr{S}(t)$, where "$-$" denotes the set-theoretic difference, that is

$$\mathscr{D}_t^+ \cup \mathscr{D}_t^- \cup \mathscr{S}(t) = \mathscr{D}_t, \tag{5.4}$$

$$\mathscr{D}_t^+ \cap \mathscr{D}_t^- = \emptyset, \tag{5.5}$$

and

$$\mathscr{S}(t) \subset (\mathscr{D}_t^+)^{\mathrm{d}}, \qquad \mathscr{S}(t) \subset (\mathscr{D}_t^-)^{\mathrm{d}}, \tag{5.6}$$

where $(\mathscr{D}_t^+)^{\mathrm{d}}$ and $(\mathscr{D}_t^-)^{\mathrm{d}}$ denote the derivatives of $\mathscr{D}_t^+$ and $\mathscr{D}_t^-$ respectively. We recall that by the derivative of a set $\mathscr{A}$ we mean the set of all accumulation (limit) points* of $\mathscr{A}$ (Engelking, 1968, p. 38). The set $\mathscr{D}_t^+$ denotes the subset of $\mathscr{D}_t$ towards which the vector $\mathbf{n}$ normal to $\mathscr{S}(t)$ is directed (or, equivalently, $\mathbf{n}$ is the outward normal to $\mathscr{D}_t^-$).

With the help of $\mathscr{D}_t^+$ and $\mathscr{D}_t^-$ we may define the decomposition of $\mathscr{N}$ into two open, non-intersecting subsets $\mathscr{N}^+$ and $\mathscr{N}^-$ and the surface $\mathscr{S}$. Thus

$$\mathscr{N} = \mathscr{N}^+ \cup \mathscr{N}^- \cup \mathscr{S}, \tag{5.7}$$

where

$$\mathscr{N}^+ \equiv \bigcup_{t \in I} \{t\} \times \mathscr{D}_t^+, \qquad \mathscr{N}^- \equiv \bigcup_{t \in I} \{t\} \times \mathscr{D}_t^-. \tag{5.8}$$

---

* If $p \in \overline{\mathscr{A} - \{p\}}$ then p is called an **accumulation** (or **limit**) **point** of $\mathscr{A}$.

It follows from (5.6) that

$$\mathscr{S} \subset (\mathscr{N}^+)^{\mathrm{d}}, \quad \mathscr{S} \subset (\mathscr{N}^-)^{\mathrm{d}}. \tag{5.9}$$

Instead of introducing the topological conditions (5.9) (and (5.6)) it is enough to assume that for every point y of $\mathscr{S}$ there exist sequences of points in each of $\mathscr{N}^+$ and $\mathscr{N}^-$ converging to y.

The assumption (5.6) expresses the simple fact, illustrated by Figure 5.1, that the surface $\mathscr{S}(t)$ "precisely" divides the region $\mathscr{D}_t$ into two parts

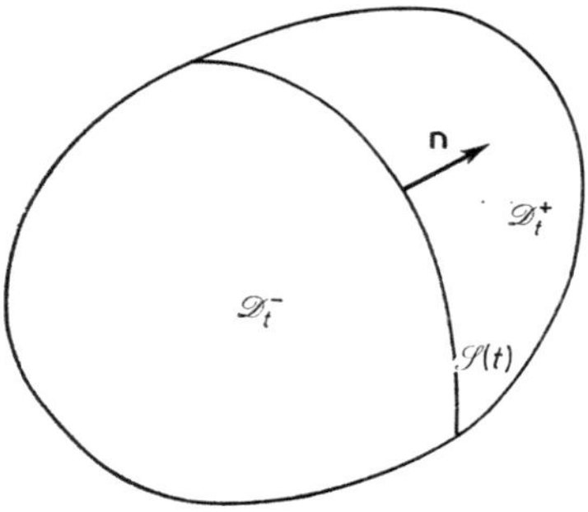

Fig. 5.1

$\mathscr{D}_t^+$ and $\mathscr{D}_t^-$ without any "free space" between them except the common "boundary" $\mathscr{S}(t)$. The same can be said of the decomposition of $\mathscr{N}$ into $\mathscr{N}^+$, $\mathscr{N}^-$ and $\mathscr{S}$.

We observe that the decomposition (5.7) with the properties (5.8) and (5.9) has one further essential characteristic feature, which is determined as follows. The fact that the sets $\mathscr{N}^+$ and $\mathscr{N}^-$ are open and the surface $\mathscr{S}$ is a boundary set in $\mathscr{E}^4$, that is

$$\operatorname{Int}\mathscr{S} = \varnothing, \tag{5.10}$$

leads to

$$\operatorname{Int}(\mathscr{N}^+ \cup \mathscr{S}) = \mathscr{N}^+, \quad \operatorname{Int}(\mathscr{N}^- \cup \mathscr{S}) = \mathscr{N}^-. \tag{5.11}$$

Since the operation of the derivative of a set has the property

$$\overline{\mathscr{A}} = \mathscr{A} \cup \mathscr{A}^{\mathrm{d}}$$

for an arbitrary set $\mathscr{A}$ (Engelking, 1968, p. 38), it follows that

$$\overline{\operatorname{Int}(\mathscr{N}^+ \cup \mathscr{S})} = \overline{\mathscr{N}^+} = \mathscr{N}^+ \cup (\mathscr{N}^+)^{\mathrm{d}} \supset \mathscr{N}^+ \cup \mathscr{S} \tag{5.12}$$

and similarly

$$\overline{\operatorname{Int}(\mathscr{N}^- \cup \mathscr{S})} = \overline{\mathscr{N}^-} = \mathscr{N}^- \cup (\mathscr{N}^-)^{\mathrm{d}} \supset \mathscr{N}^- \cup \mathscr{S}. \tag{5.13}$$

This means that the non-open sets* $\mathscr{N}^+ \cup \mathscr{S}$ and $\mathscr{N}^- \cup \mathscr{S}$ satisfy (4.5).

* It is assumed that $\mathscr{S}$ is non-empty.

The function f defined on the whole set $\mathcal{N}$ may be discontinuous at certain points and even on the whole surface $\mathcal{S}$. It may happen that at such points f has jump discontinuities; each is characterized by the existence of a limit of the function not coinciding with the value of the function at the point in question. In the theory of real functions such points are called the discontinuity points of the first order, and a function with such points is called a **regulated function.** To generalize this notion to functions determined on regions we use the following definition:

Definition 5.1

Let f be a function defined on the region $\mathcal{N}$ which is intersected by the surface $\mathcal{S}$. By the function $f\|_{\mathcal{N}^+\cup\mathcal{S}}$ we mean a continuation (that is, a smooth continuation of class C^0) of the function $f|_{\mathcal{N}^+}$ to the set $\mathcal{N}^+\cup\mathcal{S}$, and by $f\|_{\mathcal{N}^-\cup\mathcal{S}}$ a continuation of the function $f|_{\mathcal{N}^-}$ to the set $\mathcal{N}^-\cup\mathcal{S}$, in the sense of Definition 4.2, provided $f|_{\mathcal{N}^+}$ and $f|_{\mathcal{N}^-}$ are continuous.

We observe that the equalities $f|_{\mathcal{N}^+\cup\mathcal{S}} = f\|_{\mathcal{N}^+\cup\mathcal{S}}$, $f|_{\mathcal{N}^-\cup\mathcal{S}} = f\|_{\mathcal{N}^-\cup\mathcal{S}}$ hold only if f is continuous on $\mathcal{N}$. Moreover, if f is not continuously differentiable over $\mathcal{N}$ as a whole, the functions $f\|_{\mathcal{N}^+\cup\mathcal{S}}$ and $f\|_{\mathcal{N}^-\cup\mathcal{S}}$ can be continuously differentiable in the sense of Definition 4.1.

We now introduce two basic definitions.

Definition 5.2

A surface $\mathcal{S}$ is said to be a **discontinuity surface* of a function** f or a **singular surface of zero order** if f is not continuous on $\mathcal{N}$ but is continuous on each of the sets $\mathcal{N}^+$ and $\mathcal{N}^-$ and there exist continuations $f\|_{\mathcal{N}^+\cup\mathcal{S}}$ and $f\|_{\mathcal{N}^-\cup\mathcal{S}}$ which are at least continuous in the sense of Definition 4.1.

For most wave problems $f\|_{\mathcal{N}^-\cup\mathcal{S}}$ will coincide with the restriction $f|_{\mathcal{N}^-\cup\mathcal{S}}$, but in general $f\|_{\mathcal{N}^+\cup\mathcal{S}} \neq f|_{\mathcal{N}^+\cup\mathcal{S}}$.

Definition 5.3

A surface $\mathcal{S}$ is said to be a **singular surface of order** $p\,(\geqslant 1)$ for the function f if f is of class C^{p-1} on $\mathcal{N}$ but not of class C^p with the restrictions $f|_{\mathcal{N}^+\cup\mathcal{S}}$ and $f|_{\mathcal{N}^-\cup\mathcal{S}}$ at least p-times continuously differentiable in the sense of Definition 4.1.

Let $\mathcal{S}$ be a singular surface of zero order for f. By Definition 5.2 the functions $f\|_{\mathcal{N}^+\cup\mathcal{S}}$ and $f\|_{\mathcal{N}^-\cup\mathcal{S}}$ have continuous extensions, f_+ and f_- respectively, to open sets containing $\mathcal{S}$. We consider the difference in the values

* The term "surface" is used rather than "hypersurface", although we are dealing with four-dimensional space-time.

of f_- and f_+ on $\mathscr{S}$. Remembering that $\mathscr{S}$ is composed of the set-theoretic sum of a one-parameter family of surfaces (recall Definition 1.1), we form the difference

$$f_-(t, \varphi^1, \varphi^2, \varphi^3) - f_+(t, \varphi^1, \varphi^2, \varphi^3)$$

for an arbitrary instant t.

In accordance with Lemma 4.1 and the properties (5.12) and (5.13), this difference does not depend on the choice of extensions f_- and f_+ and is therefore determined uniquely by f. We denote this difference by $[\![f]\!]$. It is clear that $[\![f]\!]$ is a function defined only on $\mathscr{S}$. Hence we may write

$$[\![f]\!](t, l^1, l^2) \equiv f_-(t, \boldsymbol{\varphi}(t, l^1, l^2)) - f_+(t, \boldsymbol{\varphi}(t, l^1, l^2)). \tag{5.14}$$

The function defined by (5.14) is called the **jump (discontinuity) of f on a singular surface $\mathscr{S}$**. For a given t the left-hand side of (5.14) becomes a function of the surface parameters l^1, l^2 only, and the function so obtained is called the jump of f on $\mathscr{S}(t)$. The surface $\mathscr{S}(t)$ is often called a singular surface although it is actually a discontinuity surface of f treated as a function of the spatial variables x^1, x^2, x^3 only, that is of the function $f(t, \cdot)$.

Example 5.1

In the region $\mathscr{N} := \{(t, \mathbf{x}) \in \mathscr{E}^4 : t > 0, x_1 > 0, x_2 > 0\}$ we define the function

$$f(t, \mathbf{x}) = \begin{cases} -t^2 \mathbf{x} \cdot \mathbf{x} & \text{for } t^2 > x_1^2 + x_2^2, \\ t\mathbf{x} \cdot \mathbf{x} & \text{for } t^2 \leqslant x_1^2 + x_2^2, \end{cases}$$

where x_1, x_2, x_3 are the Cartesian coordinates of the point $\mathbf{x}$. Clearly, $f(t, \mathbf{x})$ is discontinuous on the moving cylindrical surface $\mathscr{S}(t): t^2 = x_1^2 + x_2^2$. In each of the regions

$$\mathscr{N}^+ \equiv \{(t, \mathbf{x}) \in \mathscr{N} : t^2 < x_1^2 + x_2^2\},$$
$$\mathscr{N}^- \equiv \{(t, \mathbf{x}) \in \mathscr{N} : t^2 > x_1^2 + x_2^2\}$$

the function is continuous and continuously differentiable. The jump of the function is given by the difference

$$[\![f]\!] = -t^2 \mathbf{x} \cdot \mathbf{x} - t\mathbf{x} \cdot \mathbf{x}$$

where $t^2 = x_1^2 + x_2^2$.

Introducing the convected parametrization of a moving cylinder (see Example 1.1) by means of

$$x_1 = t \cos l^1, \quad x_2 = t \sin l^1, \quad x_3 = l^2,$$

we obtain

$$[\![f]\!](t, l^1, l^2) = -(t^2 + t)(t^2 + (l^2)^2). \quad \square$$

6 COMPATIBILITY CONDITIONS ON A DISCONTINUITY SURFACE

We continue to discuss the function f for which $\mathscr{S}$ is a zero-order singular surface, but we make the additional assumption that $f||_{\mathscr{N}^+\cup\mathscr{S}}$ and $f||_{\mathscr{N}^-\cup\mathscr{S}}$ are continuously differentiable. Then we can apply (2.4) and (2.9) to the functions f_+ and f_- defined on open sets containing $\mathscr{S}$. First, we observe that the interior derivatives of $[\![f]\!]$ can be expressed

$$[\![f]\!]_{,\alpha} = (f_- - f_+)\tilde{}_{,\alpha} = \tilde{f}_{-,\alpha} - \tilde{f}_{+,\alpha}, \tag{6.1}$$

where, in accordance with (2.1), we write

$$\tilde{f}_-(t, l^1, l^2) \equiv f_-\big(t, \boldsymbol{\varphi}(t, l^1, l^2)\big), \quad \tilde{f}_+(t, l^1, l^2) \equiv f_+\big(t, \boldsymbol{\varphi}(t, l^1, l^2)\big).$$

For the displacement derivative we have

$$\frac{\delta}{\delta t}[\![f]\!] = \frac{\delta}{\delta t}(f_- - f_+) = \frac{\delta \tilde{f}_-}{\delta t} - \frac{\delta \tilde{f}_+}{\delta t}. \tag{6.2}$$

Inserting f_- and f_+ successively into (2.4) and subtracting the results, we obtain

$$\mathbf{grad} f_- - \mathbf{grad} f_+ = (\tilde{f}_- - \tilde{f}_+)_{,\alpha}\boldsymbol{\varphi}^\alpha + \frac{\partial}{\partial n}(f_- - f_+)\mathbf{n} \tag{6.3}$$

on $\mathscr{S}$.

Since $\partial f_-/\partial n$ is an extension of the normal derivative of $f||_{\mathscr{N}^-\cup\mathscr{S}}$ and $\mathbf{grad} f_-$ is an extension of the gradient of $f||_{\mathscr{N}^-\cup\mathscr{S}}$ the following relations hold on $\mathscr{S}$:

$$\begin{gathered}[(\mathbf{grad} f)_- - (\mathbf{grad} f)_+]|_{\mathscr{S}} = [\mathbf{grad} f_- - \mathbf{grad} f_+]|_{\mathscr{S}} \equiv [\![\mathbf{grad} f]\!], \\ \left[\left(\frac{\partial f}{\partial n}\right)_- - \left(\frac{\partial f}{\partial n}\right)_+\right]\bigg|_{\mathscr{S}} = \left[\frac{\partial f_-}{\partial n} - \frac{\partial f_+}{\partial n}\right]\bigg|_{\mathscr{S}} \equiv \left[\!\!\left[\frac{\partial f}{\partial n}\right]\!\!\right].\end{gathered} \tag{6.4}$$

On use of (6.3) and (6.4) together with (6.1) we obtain the first compatibility condition (for the gradient of f) on a singular surface, namely

$$[\![\mathbf{grad} f]\!] = [\![f]\!]_{,\alpha}\boldsymbol{\varphi}^\alpha + \left[\!\!\left[\frac{\partial f}{\partial n}\right]\!\!\right]\mathbf{n}. \tag{6.5}$$

Similarly, inserting f_- and f_+ into (2.9) in turn and subtracting the results, we obtain

$$\frac{\partial f_-}{\partial t} - \frac{\partial f_+}{\partial t} = \frac{\delta \tilde{f}_-}{\delta t} - \frac{\delta \tilde{f}_+}{\delta t} - u_n\left(\frac{\partial f_-}{\partial n} - \frac{\partial f_+}{\partial n}\right).$$

For the time derivative we have

$$\left[\left(\frac{\partial f}{\partial t}\right)_- - \left(\frac{\partial f}{\partial t}\right)_+\right]\bigg|_{\mathscr{S}} = \left(\frac{\partial f_-}{\partial t} - \frac{\partial f_+}{\partial t}\right)\bigg|_{\mathscr{S}} \equiv \left[\!\!\left[\frac{\partial f}{\partial t}\right]\!\!\right] \tag{6.6}$$

analogously to (6.4). It follows that the second compatibility condition (for the time derivative of f) on a singular surface is

$$\left[\!\left[\frac{\partial f}{\partial t}\right]\!\right] = \frac{\delta}{\delta t}[\![f]\!] - u_n\left[\!\left[\frac{\partial f}{\partial n}\right]\!\right]. \tag{6.7}$$

Turning next to the compatibility conditions for the second derivatives we make the further assumption that $f||_{\mathscr{N}^+\cup\mathscr{S}}$ and $f||_{\mathscr{N}^-\cup\mathscr{S}}$ are twice continuously differentiable (in the sense of Definition 4.1). For the second derivatives of the extensions f_+ and f_-, which are continuous in their domains, we may apply (3.9), (3.14) and (3.15). After use of (6.1) and (6.2) we obtain the following set of compatibility conditions for the second partial derivatives of f on a surface of discontinuity $\mathscr{S}$:

$$\begin{aligned}
[\![\mathrm{grad}\,\mathrm{grad} f]\!] &= \left\{[\![f]\!]_{;\nu\alpha} - \left[\!\left[\frac{\partial f}{\partial n}\right]\!\right] b_{\alpha\nu}\right\}\boldsymbol{\varphi}^\nu\otimes\boldsymbol{\varphi}^\alpha \\
&\quad + \left\{\left[\!\left[\frac{\partial f}{\partial n}\right]\!\right]_{,\alpha} + [\![f]\!]_{,\nu} b^\nu_\alpha\right\}(\mathbf{n}\otimes\boldsymbol{\varphi}^\alpha + \boldsymbol{\varphi}^\alpha\otimes\mathbf{n}) \\
&\quad + \left[\!\left[\frac{\partial^2 f}{\partial n^2}\right]\!\right]\mathbf{n}\otimes\mathbf{n}, \\
\left[\!\left[\mathrm{grad}\frac{\partial f}{\partial t}\right]\!\right] &= \left\{\frac{\delta}{\delta t}[\![f]\!] - u_n\left[\!\left[\frac{\partial f}{\partial n}\right]\!\right]\right\}_{,\alpha}\boldsymbol{\varphi}^\alpha \\
&\quad + \left\{\frac{\delta}{\delta t}\left[\!\left[\frac{\partial f}{\partial n}\right]\!\right] + u_{n,\nu}[\![f]\!]_{,\alpha}g^{\nu\alpha} - u_n\left[\!\left[\frac{\partial^2 f}{\partial n^2}\right]\!\right]\right\}\mathbf{n}, \\
\left[\!\left[\frac{\partial^2 f}{\partial t^2}\right]\!\right] &= \frac{\delta}{\delta t}\left\{\frac{\delta}{\delta t}[\![f]\!] - u_n\left[\!\left[\frac{\partial f}{\partial n}\right]\!\right]\right\} \\
&\quad - u_n\left\{\frac{\delta}{\delta t}\left[\!\left[\frac{\partial f}{\partial n}\right]\!\right] - u_n\left[\!\left[\frac{\partial^2 f}{\partial n^2}\right]\!\right] + u_{n,\nu}[\![f]\!]_{,\alpha}g^{\nu\alpha}\right\}.
\end{aligned} \tag{6.8}$$

A particular case of $(6.8)_1$, corresponding to (3.10), is

$$[\![\mathrm{grad}\cdot\mathrm{grad} f]\!] = [\![f]\!]_{;\nu\alpha}g^{\nu\alpha} - 2K_m\left[\!\left[\frac{\partial f}{\partial n}\right]\!\right] + \left[\!\left[\frac{\partial^2 f}{\partial n^2}\right]\!\right]. \tag{6.9}$$

In the literature (see Thomas, 1957) the first relation in (6.8) is called the **geometrical compatibility condition of order two.** The other two are called **kinematic compatibility conditions.**

6.1 Compatibility Conditions on a Surface of Higher Order

Once we have a complete set of compatibility conditions for a function f on a singular surface of order zero it is easy to derive conditions for a singular surface of order one. To this end it is enough to assume that f is a continuous function on the whole set $\mathcal{N}$. Then

$$[\![\operatorname{grad} f]\!] = \left[\!\left[\frac{\partial f}{\partial n}\right]\!\right]\mathbf{n}, \qquad \left[\!\left[\frac{\partial f}{\partial t}\right]\!\right] = -u_n\left[\!\left[\frac{\partial f}{\partial n}\right]\!\right], \tag{6.10}$$

for which we obtain

$$u_n[\![\operatorname{grad} f]\!] = -\left[\!\left[\frac{\partial f}{\partial t}\right]\!\right]\mathbf{n}. \tag{6.11}$$

This is known as Maxwell's theorem. Equations (6.8) simplify to

$$\begin{aligned}
[\![\operatorname{grad}\operatorname{grad} f]\!] &= -\left[\!\left[\frac{\partial f}{\partial n}\right]\!\right] b_{\nu\alpha}\boldsymbol{\varphi}^{\nu}\otimes\boldsymbol{\varphi}^{\alpha} + \left[\!\left[\frac{\partial f}{\partial n}\right]\!\right]_{,\alpha}(\mathbf{n}\otimes\boldsymbol{\varphi}^{\alpha}+\boldsymbol{\varphi}^{\alpha}\otimes\mathbf{n}) \\
&\quad + \left[\!\left[\frac{\partial^2 f}{\partial n^2}\right]\!\right]\mathbf{n}\otimes\mathbf{n}, \\
\left[\!\left[\operatorname{grad}\frac{\partial f}{\partial t}\right]\!\right] &= -\left\{u_n\left[\!\left[\frac{\partial f}{\partial n}\right]\!\right]\right\}_{,\alpha}\boldsymbol{\varphi}^{\alpha} + \left\{\frac{\delta}{\delta t}\left[\!\left[\frac{\partial f}{\partial n}\right]\!\right] - u_n\left[\!\left[\frac{\partial^2 f}{\partial n^2}\right]\!\right]\right\}\mathbf{n}, \\
\left[\!\left[\frac{\partial^2 f}{\partial t^2}\right]\!\right] &= -2u_n\frac{\delta}{\delta t}\left[\!\left[\frac{\partial f}{\partial n}\right]\!\right] - \left[\!\left[\frac{\partial f}{\partial n}\right]\!\right]\frac{\delta u_n}{\delta t} + u_n^2\left[\!\left[\frac{\partial^2 f}{\partial n^2}\right]\!\right].
\end{aligned} \tag{6.12}$$

Next, assuming that $\mathcal{S}$ is a singular surface of order two for f, we obtain

$$\left[\!\left[\frac{\partial f}{\partial t}\right]\!\right] = \left[\!\left[\frac{\partial f}{\partial n}\right]\!\right] = 0 \tag{6.13}$$

and hence

$$\begin{aligned}
[\![\operatorname{grad}\operatorname{grad} f]\!] &= \left[\!\left[\frac{\partial^2 f}{\partial n^2}\right]\!\right]\mathbf{n}\otimes\mathbf{n}, \\
\left[\!\left[\operatorname{grad}\frac{\partial f}{\partial t}\right]\!\right] &= -u_n\left[\!\left[\frac{\partial^2 f}{\partial n^2}\right]\!\right]\mathbf{n}, \\
\left[\!\left[\frac{\partial^2 f}{\partial t^2}\right]\!\right] &= u_n^2\left[\!\left[\frac{\partial^2 f}{\partial n^2}\right]\!\right].
\end{aligned} \tag{6.14}$$

From (6.14) we deduce that

$$u_n^2[\![\operatorname{grad}\operatorname{grad} f]\!] = -u_n\left[\!\left[\operatorname{grad}\frac{\partial f}{\partial t}\right]\!\right]\otimes\mathbf{n} = \left[\!\left[\frac{\partial^2 f}{\partial t^2}\right]\!\right]\mathbf{n}\otimes\mathbf{n}. \tag{6.15}$$

Attention should be paid to the particular case of a singular surface of order one with a continuous time derivative, that is when $[\![\partial f/\partial t]\!] = 0$, but with $[\![\mathrm{grad} f]\!] \neq \mathbf{0}$; from (6.11) it is clear that this is possible only if $u_n = 0$. This is an example of a stationary singular surface. From (6.14) it follows that on a stationary singular surface of order two we have

$$\left[\!\!\left[\mathrm{grad}\frac{\partial f}{\partial t}\right]\!\!\right] = \mathbf{0}, \quad \left[\!\!\left[\frac{\partial^2 f}{\partial t^2}\right]\!\!\right] = 0 \tag{6.16}$$

in addition to $(6.14)_1$.

Finally, when $\mathscr{S}$ is a singular surface of order p ($\geqslant 2$), we apply (6.10) to determine the partial derivatives of f of order $p-1$. Consider

$$\underbrace{\mathrm{grad}\,\mathrm{grad}\,\ldots\,\mathrm{grad}}_{s\text{ times}}\frac{\partial^{p-s}f}{\partial t^{p-s}},$$

where $0 \leqslant s \leqslant p$. Using $(6.10)_1$ s times and $(6.10)_2$ $p-s$ times we arrive at

$$\left[\!\!\left[\underbrace{\mathrm{grad}\,\mathrm{grad}\,\ldots\,\mathrm{grad}}_{s\text{ times}}\frac{\partial^{p-s}f}{\partial t^{p-s}}\right]\!\!\right] = (-u_n)^{p-s}\left[\!\!\left[\frac{\partial^p f}{\partial n^p}\right]\!\!\right]\underbrace{\mathbf{n}\otimes\mathbf{n}\otimes\,\ldots\,\otimes\mathbf{n}}_{s\text{ times}}, \tag{6.17}$$

where

$$\frac{\partial^p f}{\partial n^p} = \left\{\frac{\partial}{\partial n}\underbrace{(\mathrm{grad}\,\mathrm{grad}\,\ldots\,\mathrm{grad} f)}_{p-1\text{ times}}\right\}\cdot\underbrace{(\mathbf{n}\otimes\mathbf{n}\otimes\,\ldots\,\otimes\mathbf{n})}_{p-1\text{ times}}, \tag{6.18}$$

the dot denoting full contraction.

If, additionally, we require that f should be $p+1$ times continuously differentiable on each of the sets $\mathscr{N}^+\cup\mathscr{S}$ and $\mathscr{N}^-\cup\mathscr{S}$ then we may use (6.12) to determine the $(p-1)$-th gradients of the second derivatives of f. Replacing f by $\underbrace{\mathrm{grad}\,\mathrm{grad}\,\ldots\,\mathrm{grad}}_{p-1\text{ times}} f$, and using (6.17), we obtain

$$\begin{aligned}
[\![\mathrm{grad}^{p+1} f]\!] = &-\left[\!\!\left[\frac{\partial^p f}{\partial n^p}\right]\!\!\right]\underbrace{\mathbf{n}\otimes\mathbf{n}\otimes\,\ldots\,\otimes\mathbf{n}}_{p-1\text{ times}}\otimes\boldsymbol{\varphi}^\nu\otimes\boldsymbol{\varphi}^\alpha b_{\nu\alpha}\\
&+\left[\!\!\left[\frac{\partial^2}{\partial n^2}\mathrm{grad}^{p-1} f\right]\!\!\right]\otimes\mathbf{n}\otimes\mathbf{n}\\
&+\left\{\left[\!\!\left[\frac{\partial^p f}{\partial n^p}\right]\!\!\right]\underbrace{\mathbf{n}\otimes\mathbf{n}\otimes\,\ldots\,\otimes\mathbf{n}}_{p-1\text{ times}}\right\}_{,\alpha}\otimes(\mathbf{n}\otimes\boldsymbol{\varphi}^\alpha+\boldsymbol{\varphi}^\alpha\otimes\mathbf{n}),
\end{aligned} \tag{6.19$_1$}$$

$$\begin{aligned}
\left[\!\!\left[\mathrm{grad}^p\frac{\partial f}{\partial t}\right]\!\!\right] = &-\left\{u_n\left[\!\!\left[\frac{\partial^p f}{\partial n^p}\right]\!\!\right]\underbrace{\mathbf{n}\otimes\mathbf{n}\otimes\,\ldots\,\otimes\mathbf{n}}_{p-1\text{ times}}\right\}_{,\alpha}\otimes\boldsymbol{\varphi}^\alpha\\
&+\frac{\delta f}{\delta t}\left\{\left[\!\!\left[\frac{\partial^p f}{\partial n^p}\right]\!\!\right]\underbrace{\mathbf{n}\otimes\mathbf{n}\otimes\,\ldots\,\otimes\mathbf{n}}_{p-1\text{ times}}\right\}\otimes\mathbf{n}\\
&-u_n\left[\!\!\left[\frac{\partial^2}{\partial n^2}\mathrm{grad}^{p-1} f\right]\!\!\right]\otimes\mathbf{n},
\end{aligned}$$

$$\left[\!\!\left[\frac{\partial^2}{\partial t^2}\operatorname{grad}^{p-1}f\right]\!\!\right] = -2u_n\frac{\delta}{\delta t}\left\{\left[\!\!\left[\frac{\partial^p f}{\partial n^p}\right]\!\!\right]\underbrace{\mathbf{n}\otimes\mathbf{n}\otimes\mathbf{n}\ \ldots\ \otimes\mathbf{n}}_{p-1 \text{ times}}\right\}$$
$$-\frac{\delta u_n}{\delta t}\left[\!\!\left[\frac{\partial^p f}{\partial n^p}\right]\!\!\right]\underbrace{\mathbf{n}\otimes\mathbf{n}\otimes\ \ldots\ \otimes\mathbf{n}}_{p-1 \text{ times}}$$
$$+u_n^2\left[\!\!\left[\frac{\partial^2}{\partial n^2}\operatorname{grad}^{p-1}f\right]\!\!\right]. \tag{6.19$_2$}$$

For $k \geqslant 1$ we have introduced the notation

$$\operatorname{grad}^k f = \operatorname{grad}\operatorname{grad}^{k-1}f. \tag{6.20}$$

Note that in accordance with (6.18) we have

$$\left[\!\!\left[\frac{\partial^2}{\partial n^2}\operatorname{grad}^{p-1}f\right]\!\!\right] = [\![\operatorname{grad}^{p+1}f]\!]\cdot(\mathbf{n}\otimes\mathbf{n}). \tag{6.21}$$

From (6.19) and (6.21) we deduce that

$$[\![\operatorname{grad}\cdot\operatorname{grad}^p f]\!] = -2K_m\left[\!\!\left[\frac{\partial^p f}{\partial n^p}\right]\!\!\right]\underbrace{\mathbf{n}\otimes\mathbf{n}\otimes\ \ldots\ \otimes\mathbf{n}}_{p-1 \text{ times}}+[\![\operatorname{grad}^{p+1}f]\!]\cdot(\mathbf{n}\otimes\mathbf{n}). \tag{6.22}$$

6.2 Hadamard's Lemma

In all recent publications dealing with the analysis of the motion of singular surfaces the starting point is the basic result of the classical treatise of Hadamard (1903) known as Hadamard's lemma. Using the notion of a smooth extension of a function defined on a non-open set (recall Definition 4.1), we can formulate Hadamard's lemma in the following way:

Lemma 6.1

If $\mathscr{S}$ is a discontinuity hypersurface of a function f, where $f||_{\mathscr{N}^+\cup\mathscr{S}}$ and $f||_{\mathscr{N}^-\cup\mathscr{S}}$ are once continuously differentiable with $\mathscr{N}^+$ and $\mathscr{N}^-$ satisfying (5.7)–(5.9), then for an arbitrary smooth curve $\mathscr{L}: \mathbf{y} = \mathbf{\Lambda}(s)$ on $\mathscr{S}$ we have

$$\frac{\mathrm{d}f_+}{\mathrm{d}s} = \operatorname{grad}_y f_+\cdot\frac{\mathrm{d}\mathbf{\Lambda}}{\mathrm{d}s}, \qquad \frac{\mathrm{d}f_-}{\mathrm{d}s} = \operatorname{grad}_y f_-\cdot\frac{\mathrm{d}\mathbf{\Lambda}}{\mathrm{d}s},$$

where f_- is an arbitrary smooth extension of the function $f||_{\mathscr{N}^-\cup\mathscr{S}}$ and f_+ is a smooth extension of $f||_{\mathscr{N}^+\cup\mathscr{S}}$. These two relations yield

$$\frac{\mathrm{d}}{\mathrm{d}s}[\![f]\!] = [\![\operatorname{grad}_y f]\!]\cdot\frac{\mathrm{d}\mathbf{\Lambda}}{\mathrm{d}s}, \tag{6.23}$$

where $\operatorname{grad}_y f$ denotes the gradient with respect to the variables of the function $f(y)$.

To prove the validity of the first two relations it is enough to apply the chain rule directly to the functions f_+ and f_-. Then (6.23) follows from the definition of the jump of a function and its gradient (see (5.14) and $(6.4)_1$). Note that in the statement of the lemma no assumption is made about the dimension of the hypersurface $\mathscr{S}$.

6.3 Basic Results in Index Notation

In index notation (6.5) has the form

$$\left[\!\!\left[\frac{\partial f}{\partial x^i}\right]\!\!\right] = G_{ik}\left\{[\![f]\!]_{,\alpha}\varphi^{k\alpha} + \left[\!\!\left[\frac{\partial f}{\partial n}\right]\!\!\right] n^k\right\}, \tag{6.24}$$

(6.8) becomes

$$\begin{aligned}
\left[\!\!\left[\frac{\partial f_{;k}}{\partial t}\right]\!\!\right] &= G_{ki}\left\{\left\{\frac{\delta}{\delta t}[\![f]\!] - u_n\left[\!\!\left[\frac{\partial f}{\partial n}\right]\!\!\right]\right\}_{,\alpha}\varphi^{i\alpha}\right. \\
&\quad \left. + \left\{\frac{\delta}{\delta t}\left[\!\!\left[\frac{\partial f}{\partial n}\right]\!\!\right] + u_{n,\gamma}[\![f]\!]_{,\alpha}g^{\gamma\alpha} - u_n\left[\!\!\left[\frac{\partial^2 f}{\partial n^2}\right]\!\!\right]\right\} n^i\right\}, \\
[\![f_{;kl}]\!] &= G_{ki}G_{lj}\left\{\left\{[\![f]\!]_{;(\nu\alpha)} - \left[\!\!\left[\frac{\partial f}{\partial n}\right]\!\!\right] b_{(\nu\alpha)}\right\}\varphi^{i(\alpha}\varphi^{j\nu)}\right. \\
&\quad \left. + 2\left\{\left[\!\!\left[\frac{\partial f}{\partial n}\right]\!\!\right]_{,\alpha} + [\![f]\!]_{,\gamma} b^{\gamma}_{\alpha}\right\} n^{(i}\varphi^{j)\alpha} + \left[\!\!\left[\frac{\partial^2 f}{\partial n^2}\right]\!\!\right] n^i n^j\right\},
\end{aligned} \tag{6.25}$$

and (6.9) yields

$$[\![f_{;ij}G^{ij}]\!] = [\![f]\!]_{;\nu\alpha}g^{\nu\alpha} - 2K_m\left[\!\!\left[\frac{\partial f}{\partial n}\right]\!\!\right] + \left[\!\!\left[\frac{\partial^2 f}{\partial n^2}\right]\!\!\right]. \tag{6.26}$$

The formulae (6.10) and (6.11) become

$$[\![f_{;i}]\!] = G_{ik}\left[\!\!\left[\frac{\partial f}{\partial n}\right]\!\!\right] n^k, \qquad -u_n[\![f_{;i}]\!] = G_{ik}\left[\!\!\left[\frac{\partial f}{\partial t}\right]\!\!\right] n^k \tag{6.27}$$

and (6.12) yields

$$\begin{aligned}
[\![f_{;kl}]\!] &= -G_{ik}G_{jl}\left\{\left[\!\!\left[\frac{\partial f}{\partial n}\right]\!\!\right] b_{(\nu\alpha)}\varphi^{i(\nu}\varphi^{j\alpha)} - 2\left[\!\!\left[\frac{\partial f}{\partial n}\right]\!\!\right]_{,\alpha} n^{(i}\varphi^{j)\alpha}\right. \\
&\quad \left. - \left[\!\!\left[\frac{\partial^2 f}{\partial n^2}\right]\!\!\right] n^i n^j\right\}, \\
\left[\!\!\left[\frac{\partial f_{;k}}{\partial t}\right]\!\!\right] &= -G_{ik}\left\{\left\{u_n\left[\!\!\left[\frac{\partial f}{\partial n}\right]\!\!\right]\right\}_{,\alpha}\varphi^{i\alpha} - \left\{\frac{\delta}{\delta t}\left[\!\!\left[\frac{\partial f}{\partial n}\right]\!\!\right] - u_n\left[\!\!\left[\frac{\partial^2 f}{\partial n^2}\right]\!\!\right]\right\} n^i\right\}.
\end{aligned} \tag{6.28}$$

Equations (6.14) and (6.15) give

$$[\![f_{;kl}]\!] = G_{ik}G_{jl}\left[\!\!\left[\frac{\partial^2 f}{\partial n^2}\right]\!\!\right]n^i n^j \tag{6.29}$$

and

$$u_n^2[\![f_{;kl}]\!] = -u_n G_{lj}\left[\!\!\left[\frac{\partial f_{;k}}{\partial t}\right]\!\!\right]n^j = G_{kl}G_{lj}\left[\!\!\left[\frac{\partial^2 f}{\partial t^2}\right]\!\!\right]n^i n^j. \tag{6.30}$$

Equations (6.17) and (6.18) become

$$\begin{aligned}&\left[\!\!\left[\frac{\partial^{p-s}f_{;ij\ldots l}}{\partial t^{p-s}}\right]\!\!\right] = (-u_n)^{p-s}\left[\!\!\left[\frac{\partial^p f}{\partial n^p}\right]\!\!\right]n_i n_j \ldots n_l,\\ &\frac{\partial^p f}{\partial n^p} = \frac{\partial}{\partial n}(f_{;ij\ldots l})n^i n^j \ldots n^l,\end{aligned} \tag{6.31}$$

while equations (6.19) yield

$$\begin{aligned}[\![f_{;ij\ldots dlk}]\!] &= 2\left\{\left[\!\!\left[\frac{\partial^p f}{\partial n^p}\right]\!\!\right]n_i n_j \ldots n_d\right\}_{;\alpha} n_{(l}\varphi_{k),\beta}g^{\alpha\beta}\\ &\quad -\left[\!\!\left[\frac{\partial^p f}{\partial n^p}\right]\!\!\right]n_i n_j \ldots n_d\varphi_{l,\nu}\varphi_{k,\alpha}b^{\nu\alpha} + [\![f_{;ij\ldots dsq}]\!]n^s n^q n_l n_k,\\ \left[\!\!\left[\frac{\partial f_{;ij\ldots dl}}{\partial t}\right]\!\!\right] &= -\left\{u_n\left[\!\!\left[\frac{\partial^p f}{\partial n^p}\right]\!\!\right]n_i n_j \ldots n_d\right\}_{;\alpha}\varphi_{l,\beta}g^{\alpha\beta}\\ &\quad +\frac{\delta}{\delta t}\left\{\left[\!\!\left[\frac{\partial^p f}{\partial n^p}\right]\!\!\right]n_i n_j \ldots n_d\right\}n_l - u_n[\![f_{;ij\ldots dsq}]\!]n^s n^q n_l,\\ \left[\!\!\left[\frac{\partial^2 f_{;ij\ldots d}}{\partial t^2}\right]\!\!\right] &= -2u_n\frac{\delta}{\delta t}\left\{\left[\!\!\left[\frac{\partial^p f}{\partial n^p}\right]\!\!\right]n_i n_j \ldots n_d\right\}\\ &\quad +\frac{\delta u_n}{\delta t}\left[\!\!\left[\frac{\partial^p f}{\partial n^p}\right]\!\!\right]n_i n_j \ldots n_d + u_n^2[\![f_{;ij\ldots dsq}]\!]n^s n^q.\end{aligned} \tag{6.32}$$

From (6.21) we obtain

$$\left[\!\!\left[\frac{\partial^2 f_{;ij\ldots d}}{\partial n^2}\right]\!\!\right] = [\![f_{;ij\ldots dsp}]\!]n^s n^p \tag{6.33}$$

and finally, from (6.22),

$$[\![f_{;ij\ldots dlk}]\!]G^{lk} = -2K_m\left[\!\!\left[\frac{\partial^p f}{\partial n^p}\right]\!\!\right]n_i n_j \ldots n_d + [\![f_{;ij\ldots dsp}]\!]n^s n^p. \tag{6.34}$$

If the coordinate system is non-Cartesian then in (6.32) $\delta/\delta t$ should be replaced by the invariant displacement derivative $\mathrm{d}/\delta t$ from (3.34). The same replacement should be made in (6.25) and (6.28) if instead of a scalar function f we are concerned with the components of a vector or tensor function and the non-Cartesian coordinate system.

6.4 Bibliographical Notes

The kinematic compatibility conditions (6.7) and (6.8) on a surface of discontinuity were first derived by Thomas (1957), and repeated in (1961), and were also given by Truesdell and Toupin (1960). In these works, just as in Wang and Truesdell (1973) and McCarthy (1975), the derivation of the compatibility conditions (6.8), (6.12), (6.14) and (6.19) was based on Hadamard's lemma. The method used here is different and is based on that in Chadwick and Powdrill (1965), starting from the formulae (2.4), (2.9), (3.9), (3.14) and (3.15) for the first and second derivatives of a smooth function f on a surface. This is why, in Section 4, we gave a full treatment of the smooth extension of a function defined on a non-open set. Lemma 4.1 on the uniqueness of the derivative, together with its proof, is published here for the first time.

The compatibility conditions (6.10), (6.14) and (6.17) are usually attributed to Hadamard (1903, pp. 85–87, 101–104). In Chadwick and Powdrill (1965) we find the derivation of the compatibility conditions in index notation in a Cartesian coordinate system. It seems that Maxwell (1873) was the first to derive compatibility conditions by differentiating a jump relation of the form (5.14) along a path lying on a surface. For more general treatments of compatibility conditions in space-time compare Cohen and Thomas (1983), Cohen and Wang (1982) and Anile (1984).

Chapter 3

General balance laws

Let Ψ be a measure of some thermodynamic quantity ascribed to a body $\mathscr{B}$. In classical textbooks dealing with the mechanics of continuous media it is assumed that Ψ, treated as set function, that is

$$\Psi = \Psi(\mathscr{P}), \quad \mathscr{P} \subset \mathscr{B},$$

is absolutely continuous with respect to the Lebesgue volume measure. Under the additional assumption of additivity on disjoint sets from the Boolean algebra of sub-bodies we obtain a representation for the function Ψ in the form of a volume integral of a density function ψ, namely

$$\Psi(\mathscr{B}) = \int_{\mathscr{B}} \psi \mathrm{d}v.$$

Modern engineering and technology problems, when formulated in the language of the theory of continua, involve effects which make the above assumptions on Ψ inappropriate. For example, surface effects play an essential role in the borderland between the chemistry, physics and mechanics of solids and fluids.

Surface effects are connected with the concentration of energy on surfaces. Although the notion of energy confined to a surface separating two media, or to a superficial layer, is a theoretical concept, it has proved a convenient idealization in solving numerous problems of scientific value.

In the classical approach surface quantities were regarded as two-dimensional analogues of a three-dimensional material medium. Our aim is to draw attention to the fact, which so far has rarely absorbed theoreticians studying continuous media, that a moving surface may carry not only disturbances, but also physical properties different from those of the surrounding media. As an example we consider the direct interaction of two different phases of a material. We can model a situation of this kind by the movement of a surface separating two well-behaved material media, while attributing to the surface the physical properties of a phase change. Other examples may be provided by phenomena such as the motion of surface dislocations, or the propagation of cracks. In fluid mechanics the surface tension of drops provides an example.

In general, in the description of a medium the existence of surface phenomena is connected with the occurrence of certain singular surfaces, which in chemical physics are called interfaces.

The above remarks induce us to weaken the assumption of absolute continuity of a function with respect to volume measure. In our subsequent considerations we therefore adopt the following integral representation for Ψ:

$$\Psi(\mathscr{B}) = \int_{\mathscr{B}} \psi \, dv + \int_{\mathscr{S}} \psi_{\mathscr{S}} \, da.$$

The existence of the second integral is evidence of the occurrence of surface effects associated with a concentration of the quantity Ψ on a (singular) surface $\mathscr{S}$.

In the analysis of phenomena involving singular surfaces two cases should be distinguished: (a) $\mathscr{S}$ is a material surface, (b) $\mathscr{S}$ is not a material surface, but a surface passing through the medium. In case (a) the same material particles remain on the surface during its motion, and the analysis concerns a film or layer, while in case (b) the analysis is applicable mainly to wave propagation problems and phase transition phenomena. Moreover, in a number of free boundary problems surfaces may be applied in modelling as well as in analysis.

7 DIFFERENTIATION OF VOLUME AND SURFACE INTEGRALS

Let a function $\boldsymbol{\chi}$ defined on $I \times \mathscr{P}$ describe a motion of the region $\mathscr{P}$ in $\mathscr{E}^3$. The set $\mathscr{P}$ will be treated as a reference region (or placement) whereas I, being an open interval of the real axis, will be treated as the duration interval of the motion. In effect, we have

$$\boldsymbol{\chi}\colon I \times \mathscr{P} \to \mathscr{E}^3. \tag{7.1}$$

As is customary we denote the points of $\mathscr{P}$ by $\mathbf{X}$ and the coordinates of $\mathbf{X}$ by X^K, $K = 1, 2, 3$. The position in $\mathscr{E}^3$ of the point $\mathbf{X}$ at an instant of time t will be denoted by $\mathbf{x}$ and the coordinates of $\mathbf{x}$ by x^k, $k = 1, 2, 3$.

We assume that $\mathscr{P}$ in $\mathscr{E}^3$ is bounded by a sufficiently smooth surface, ensuring the applicability of the divergence theorem.

We assume that $\boldsymbol{\chi}$ is a homeomorphism of class C^1 (at least) on $I \times \mathscr{P}$ and that it has a smooth continuation (see Definition 4.2) to the closure $I \times \overline{\mathscr{P}}$.

In order to ensure uniqueness of the first derivatives of $\boldsymbol{\chi}$ on $\mathscr{P}$ it should be assumed that $\mathscr{P} \subset \overline{\text{Int}\,\mathscr{P}}$ (Lemma 4.2).

We introduce the symbol J for the Jacobian of the transformation of (7.1), so that

$$J \equiv \det\left[\frac{\partial \chi^k}{\partial X^K}\right], \quad k, K = 1, 2, 3, \tag{7.2}$$

where $\boldsymbol{\chi} \equiv [\chi^k]$. In what follows we denote by $\boldsymbol{\chi}_t(\mathscr{P})$ the one-parameter family of set functions in $\mathscr{E}^3$ which is the image of the transformation $\boldsymbol{\chi}(t, \cdot)$: $\mathscr{P} \to \mathscr{E}^3$ for $t \in I$.

We may now state the following lemma.

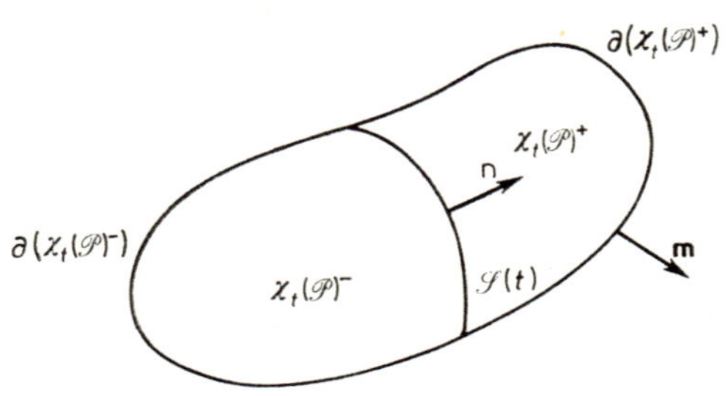

Fig. 7.1

Lemma 7.1

If the (scalar- or tensor-valued) function f is continuously differentiable on the set $\bigcup_{t \in I} \{t\} \times \boldsymbol{\chi}_t(\mathscr{P})$ and continuous on $\bigcup_{t \in I} \{t\} \times \overline{\boldsymbol{\chi}_t(\mathscr{P})}$ then

$$\frac{\mathrm{d}}{\mathrm{d}t} \int_{\boldsymbol{\chi}_t(\mathscr{P})} f\, \mathrm{d}v = \int_{\boldsymbol{\chi}_t(\mathscr{P})} f'\mathrm{d}v + \int_{\partial \boldsymbol{\chi}_t(\mathscr{P})} f\mathbf{v} \cdot \mathbf{m}\, \mathrm{d}a, \tag{7.3}$$

where $\mathbf{m}$ is the field of the outward unit normal vector to the boundary $\partial \boldsymbol{\chi}_t(\mathscr{P})$, f' denotes the time derivative

$$f' := \frac{\partial}{\partial t} f(t, \mathbf{x}), \tag{7.4}$$

and $\mathbf{v}$ is the velocity of displacement of the boundary points.

Proof

If the volume $\boldsymbol{\chi}_t(\mathscr{P})$ is replaced by the time-independent region $\mathscr{P}$ in the integration then, by use of (7.2), we obtain

$$\frac{\mathrm{d}}{\mathrm{d}t} \int_{\boldsymbol{\chi}_t(\mathscr{P})} f\, \mathrm{d}v = \int_{\mathscr{P}} \frac{\partial}{\partial t} (Jf)\, \mathrm{d}v_{\mathscr{P}}.$$

The integrand on the right-hand side is $\dot{J}f + J\dot{f}$, where the dot denotes the time derivative at fixed **X**. Let the function F be defined by

$$F(t, \mathbf{X}) := f(t, \boldsymbol{\chi}_t(\mathbf{X})) = f(t, \mathbf{x}).$$

Then

$$\dot{f} = \frac{\partial F}{\partial t}(t, \mathbf{X}) = \frac{\partial f}{\partial t} + \dot{\boldsymbol{\chi}} \cdot \operatorname{grad} f = f' + \dot{\boldsymbol{\chi}} \cdot \operatorname{grad} f. \tag{7.5}$$

The derivative of the Jacobian is given by

$$\dot{J} = J \operatorname{div} \mathbf{v}, \tag{7.6}$$

where $\operatorname{div} \mathbf{v} = v^k{}_{:k}$ and

$$\mathbf{v}(t, \mathbf{x}) = \mathbf{v}(t, \boldsymbol{\chi}_t(\mathbf{X})) \equiv \dot{\boldsymbol{\chi}}(t, \mathbf{X}) \quad \text{with } \mathbf{X} = \boldsymbol{\chi}_t^{-1}(\mathbf{x}), \tag{7.7}$$

$\boldsymbol{\chi}_t^{-1}$ denoting the function inverse to $\boldsymbol{\chi}(t, \cdot)$ for each $t \in I$.

It follows that

$$\int_{\mathscr{P}} (Jf)^{\cdot} \mathrm{d}v_{\mathscr{P}} = \int_{\mathscr{P}} J(f' + \dot{\boldsymbol{\chi}} \cdot \operatorname{grad} f + f \operatorname{div} \mathbf{v}) \mathrm{d}v_{\mathscr{P}}$$

$$= \int_{\boldsymbol{\chi}_t(\mathscr{P})} (f' + \mathbf{v} \cdot \operatorname{grad} f + f \operatorname{div} \mathbf{v}) \mathrm{d}v.$$

Application of the divergence theorem to the integral

$$\int_{\boldsymbol{\chi}_t(\mathscr{P})} (\mathbf{v} \cdot \operatorname{grad} f + f \operatorname{div} \mathbf{v}) \mathrm{d}v = \int_{\boldsymbol{\chi}_t(\mathscr{P})} \operatorname{div}(f\mathbf{v}) \mathrm{d}v = \int_{\partial\boldsymbol{\chi}_t(\mathscr{P})} f\mathbf{v} \cdot \mathbf{m} \mathrm{d}a$$

yields the required result (7.3). □

We now turn to a second lemma, important for subsequent work since it concerns the differentiation of surface integrals.

Let $\mathscr{S}(t)$ be a surface moving in a three-dimensional continuum which itself is subject to the motion $\boldsymbol{\chi}_t$. In order to describe this superposition of motions we assume the existence of a certain stationary regular reference surface $\mathscr{S}_0$ with parametrization L^Δ, $\Delta = 1, 2$. Then, for every instant t, the surface parameters l^α, $\alpha = 1, 2$, of the surface $\mathscr{S}(t)$ describe the position of a point (L^1, L^2) from the reference surface,* that is

$$l^\alpha = l^\beta(t, L^\Delta). \tag{7.8}$$

It is assumed that the transformation (7.8) does not change orientation, is continuously differentiable and is invertible on the set $I \times \mathscr{S}_0 \subset \boldsymbol{R} \times \mathscr{E}^2$,

---

* To understand this better we consider the following picture (see Scriven, 1960, and Moeckel, 1974): imagine that $\mathscr{S}(t)$ consists of fictitious particles characterized by pairs of numbers L^1 and L^2 whose position at time t is given by the parametric values (7.8).

so that $L^{\Delta} = L^{\Delta}(t, l^{\alpha})$. We define the velocity of motion of the point $L^{\Delta} = \text{const}$ in the surface $\mathscr{S}(t)$ by the relation

$$V^{\alpha} \equiv \dot{l}^{\alpha} = \frac{\partial l^{\alpha}}{\partial t}(t, L^{\Delta}). \tag{7.9}$$

The quantities V^{α} are the components of the velocity field of fictitious particles in the surface with respect to the basis vectors $\boldsymbol{\varphi}_{,\alpha}$, not the components of the absolute velocity in $\mathscr{E}^3$.

Let the motion of the surface $\mathscr{S}(t)$ in the continuum be given by (recall Definition 1.1)

$$\mathbf{x} = \boldsymbol{\varphi}(t, l^{\alpha}).$$

We denote the Jacobian of the transformation (7.8) by

$$J_{\mathscr{S}} \equiv \det[l^{\alpha}_{,\Delta}].$$

Using the chain rule we obtain

$$\boldsymbol{\Phi}_{,\Delta} = \boldsymbol{\varphi}_{,\alpha} l^{\alpha}_{,\Delta}, \qquad G_{\Delta\Gamma} \equiv \boldsymbol{\Phi}_{,\Delta} \cdot \boldsymbol{\Phi}_{,\Gamma} = l^{\alpha}_{,\Delta} l^{\beta}_{,\Gamma} g_{\alpha\beta},$$

where

$$\boldsymbol{\Phi}(t, L) \equiv \boldsymbol{\varphi}\big(t, l^{\alpha}(t, L^{\Delta})\big).$$

It follows that $G = \det[G_{\Delta\Gamma}]$ and $g = \det[g_{\alpha\beta}]$ are connected by

$$G = J_{\mathscr{S}}^2 g. \tag{7.10}$$

Lemma 7.2

If a function g with scalar or tensor values is continuously differentiable on $\bigcup\limits_{t \in I} \{t\} \times \mathscr{S}(t)$ then

$$\frac{\mathrm{d}}{\mathrm{d}t} \int\limits_{\mathscr{S}(t)} \mathrm{g}\, \mathrm{d}a = \int\limits_{\mathscr{S}(t)} \left\{ \frac{\partial \mathrm{g}}{\partial t} \mathrm{g}_{,\alpha} V^{\alpha} + \mathrm{g}\,(V^{\alpha}_{;\alpha} + c^{\alpha}_{;\alpha} - 2u_n K_m) \right\} \mathrm{d}a, \tag{7.11}$$

where $\partial \mathrm{g}/\partial t = \partial \mathrm{g}(t, l^{\alpha})/\partial t$, and c^{α} are the tangential components of the velocity of motion of the surface (recall (1.18)).

Proof

On changing the region of integration from $\mathscr{S}(t)$ to $\mathscr{S}_0$ in the left-hand integral in (7.11), we obtain

$$\int\limits_{\mathscr{S}(t)} \mathrm{g}\, \mathrm{d}a = \int\limits_{\mathscr{S}_0} \mathrm{g} \tilde{J}\, \mathrm{d}a_{\mathscr{S}_0}, \tag{7.12}$$

where $\tilde{J}$ is the Jacobian of the transformation of $\mathscr{S}_0$ into $\mathscr{S}(t)$ and $\mathrm{d}a_{\mathscr{S}_0}$ is an element of surface area in $\mathscr{S}_0$.

The surface element $\mathrm{d}a$ can be expressed as $\mathrm{d}a = g^{1/2}\mathrm{d}l^1\mathrm{d}l^2 = G^{1/2}\mathrm{d}L^1\mathrm{d}L^2$. In view of (7.10) it follows that $\mathrm{d}a = J_{\mathscr{S}} g^{1/2}\mathrm{d}L^1\mathrm{d}L^2 = J_{\mathscr{S}} g^{1/2}\mathrm{d}a_{\mathscr{S}_0}$. Hence $\tilde{J} = J_{\mathscr{S}} g^{1/2}$ for $J_{\mathscr{S}} > 0$.

In view of (7.12) the integral on the left-hand side of (7.11) may be written as

$$\frac{\mathrm{d}}{\mathrm{d}t}\int_{\mathscr{S}(t)} \mathrm{g}\,\mathrm{d}a = \int_{\mathscr{S}_0} (\mathrm{g}\tilde{J})^{\cdot}\mathrm{d}L^1\mathrm{d}L^2, \tag{7.13}$$

where the dot denotes differentiation with respect to t, (L^Δ) being constant. In order to calculate $(\tilde{J})^{\cdot}$ we note first that

$$\dot{J}_{\mathscr{S}} = \frac{\partial J_{\mathscr{S}}}{\partial l^\alpha_{,\Delta}}(l^\alpha_{,\Delta})^{\cdot} = J_{\mathscr{S}} L^\Delta_{,\alpha} V^\alpha_{,\Delta} = J_{\mathscr{S}} V^\alpha_{,\alpha} \tag{7.14}$$

and

$$(g^{1/2})^{\cdot} = \tfrac{1}{2} g^{1/2} g^{\alpha\beta}\dot{g}_{\alpha\beta}.$$

The derivative $\dot{g}_{\alpha\beta}$ is calculated from the formula

$$\dot{g}_{\alpha\beta} = \frac{\partial g_{\alpha\beta}}{\partial t} + V^\gamma g_{\alpha\beta,\gamma}$$

since $g_{\alpha\beta}$ is a function of t and l^γ and $\partial l^\gamma/\partial t = V^\gamma$. It remains to determine $\partial g_{\alpha\beta}/\partial t$. From the results of Example 2.1 we have

$$(g^{1/2})^{\cdot} = g^{1/2}(c^\alpha_{;\alpha} - u_n b^\alpha_\alpha + \Gamma^\alpha_{\alpha\gamma} V^\gamma). \tag{7.15}$$

By insertion of (7.14) and (7.15) into

$$(\tilde{J})^{\cdot} = \dot{J}_{\mathscr{S}} g^{1/2} + J_{\mathscr{S}}(g^{1/2})^{\cdot}$$

we arrive at

$$(\tilde{J})^{\cdot} = J_{\mathscr{S}} g^{1/2}(V^\alpha_{;\alpha} + c^\alpha_{;\alpha} - u_n b^\alpha_\alpha) = \tilde{J}(V^\alpha_{;\alpha} + c^\alpha_{;\alpha} - 2u_n K_m). \tag{7.16}$$

Putting this result into (7.13) we obtain

$$\frac{\mathrm{d}}{\mathrm{d}t}\int_{\mathscr{S}(t)} \mathrm{g}\,\mathrm{d}a = \int_{\mathscr{S}(t)} (\dot{\mathrm{g}} + \mathrm{g}V^\alpha_{;\alpha} + \mathrm{g}c^\alpha_{;\alpha} - 2u_n K_m \mathrm{g})\,\mathrm{d}a. \tag{7.17}$$

If we observe that

$$\dot{\mathrm{g}} = \frac{\partial \mathrm{g}}{\partial t} + \mathrm{g}_{,\alpha} V^\alpha,$$

the final integral relation takes the required form (7.11). □

In the case of a convected parametrization the components c^α vanish and (7.17) and (7.11) assume the simpler forms

$$\frac{\mathrm{d}}{\mathrm{d}t}\int\limits_{\mathscr{S}(t)} \mathrm{g}\,\mathrm{d}a = \int\limits_{\mathscr{S}(t)} (\dot{\mathrm{g}}+\mathrm{g}V^{\alpha}_{;\alpha}-2u_n K_m)\,\mathrm{g}\,\mathrm{d}a,$$
$$\frac{\mathrm{d}}{\mathrm{d}t}\int\limits_{\mathscr{S}(t)} \mathrm{g}\,\mathrm{d}a = \int\limits_{\mathscr{S}(t)} \left\{\frac{\partial \mathrm{g}}{\partial t}+\mathrm{g}_{,\alpha}V^{\alpha}+\mathrm{g}\,(V^{\alpha}_{;\alpha}-2u_n K_m)\right\}\mathrm{d}a \tag{7.18}$$

respectively. Note that the displacement time derivative $\delta/\delta t$ may be used in (7.11) (compare $(8.19)_2$ and the proof on pp. 101–102).

7.1 Integration of a Function with Surface Singularities

Here we consider differentiation of an integral defined on a family of regions $\chi_t(\mathscr{P})$ containing a moving surface of discontinuity $\mathscr{S}(t)$. We assume that for every instant t the open set $\chi_t(\mathscr{P})$ is divided by the surface $\mathscr{S}(t)$ into two disjoint regions with common boundary $\mathscr{S}(t)$; thus

$$\chi_t(\mathscr{P}) = \chi_t(\mathscr{P})^{+}\cup\chi_t(\mathscr{P})^{-}\cup\mathscr{S}(t),$$
$$\chi_t(\mathscr{P})^{+}\cap\chi_t(\mathscr{P})^{-} = \emptyset, \tag{7.19}$$
$$\mathscr{S}(t) \subset (\chi_t(\mathscr{P})^{+})^{\mathrm{d}}, \quad \mathscr{S}(t) \subset (\chi_t(\mathscr{P})^{-})^{\mathrm{d}},$$

where the index d denotes the derivative of the set (as in Section 5). The above assumptions are identical to those made for the set $\mathscr{D}_t$ in Section 5. The set $\chi_t(\mathscr{P})^{+}$ is that part of $\chi_t(\mathscr{P})$ into which the normal vector to $\mathscr{S}(t)$ is directed. We make use of the notation

$$\mathscr{N} := \bigcup_{t\in I} \{t\}\times\chi_t(\mathscr{P}),$$
$$\mathscr{S} := \bigcup_{t\in I} \{t\}\times\mathscr{S}(t). \tag{7.20}$$

As in Section 5 we define the decomposition of the set into two non-intersecting subsets $\mathscr{N}^{+}$ and $\mathscr{N}^{-}$ and a surface $\mathscr{S}$,

$$\mathscr{N} = \mathscr{N}^{+}\cup\mathscr{N}^{-}\cup\mathscr{S}, \tag{7.21}$$

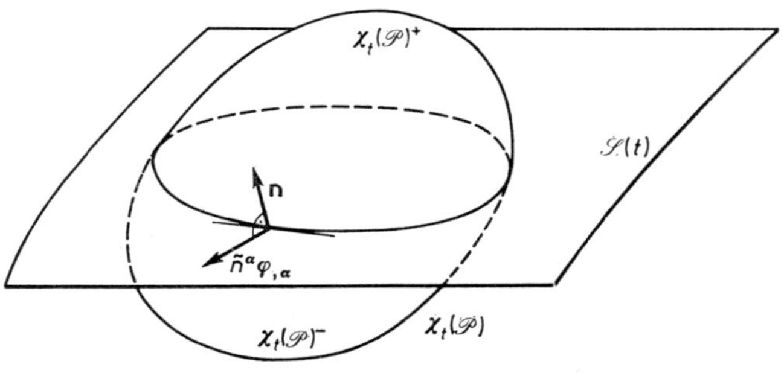

Fig. 7.2

where

$$\mathcal{N}^+ := \bigcup_{t\in I} \{t\}\times \boldsymbol{\chi}_t(\mathscr{P})^+, \qquad \mathcal{N}^- := \bigcup_{t\in I} \{t\}\times \boldsymbol{\chi}_t(\mathscr{P})^-.$$

Just as for the sets $\boldsymbol{\chi}_t(\mathscr{P})^+$ and $\boldsymbol{\chi}_t(\mathscr{P})^-$, $t \in I$, we can write

$$\mathscr{S} \subset (\mathcal{N}^+)^{\mathrm{d}}, \qquad \mathscr{S} \subset (\mathcal{N}^-)^{\mathrm{d}}. \tag{7.22}$$

The above property means that the non-open sets $\mathcal{N}^+ \cup \mathscr{S}$ and $\mathcal{N}^- \cup \mathscr{S}$ satisfy the relation (4.5) required in Lemma 1.1. We can now formulate

Lemma 7.3

If for a scalar- or vector-valued function f defined on $\mathcal{N}$ the surface $\mathscr{S}$ with properties (7.21) and (7.22) is a surface of discontinuity and, furthermore, f is bounded on $\mathscr{S}$ and the functions $f||_{\mathcal{N}^+\cup\mathscr{S}}$ and $f||_{\mathcal{N}^-\cup\mathscr{S}}$ are continuously differentiable in the sense of Definition 4.1, then

$$\frac{\mathrm{d}}{\mathrm{d}t}\int\limits_{\boldsymbol{\chi}_t(\mathscr{P})} f\,\mathrm{d}v = \int\limits_{\boldsymbol{\chi}_t(\mathscr{P})} f'\,\mathrm{d}v + \int\limits_{\partial\boldsymbol{\chi}_t(\mathscr{P})} f\dot{\boldsymbol{\chi}}\cdot\mathbf{m}\,\mathrm{d}a + \int\limits_{\mathscr{S}(t)} [\![f]\!]u_n\,\mathrm{d}a. \tag{7.23}$$

Proof

Since, according to Definition 4.1, the functions $f||_{\mathcal{N}^+\cup\mathscr{S}}$ and $f||_{\mathcal{N}^-\cup\mathscr{S}}$ are extendible they have smooth extensions f_+ and f_- respectively to open sets $\mathscr{U}_+$ and $\mathscr{U}_-$ such that

$$\mathscr{U}_+ \supset \mathcal{N}^+\cup\mathscr{S}, \qquad \mathscr{U}_- \supset \mathcal{N}^-\cup\mathscr{S}.$$

We split up the integral on the left-hand side of (7.23) as

$$\int\limits_{\boldsymbol{\chi}_t(\mathscr{P})} f\,\mathrm{d}v = \int\limits_{\boldsymbol{\chi}_t(\mathscr{P})^+} f\,\mathrm{d}v + \int\limits_{\boldsymbol{\chi}_t(\mathscr{P})^-} f\,\mathrm{d}v + \int\limits_{\mathscr{S}(t)} f\,\mathrm{d}v.$$

The last integral vanishes since the volume measure of the set $\mathscr{S}(t)$ is zero. From the definitions of f_+ and f_- it can be seen that

$$\begin{aligned} f|_{\{t\}\times\boldsymbol{\chi}_t(\mathscr{P})^+} &= f_+|_{\{t\}\times\boldsymbol{\chi}_t(\mathscr{P})^+},\\ f|_{\{t\}\times\boldsymbol{\chi}_t(\mathscr{P})^-} &= f_-|_{\{t\}\times\boldsymbol{\chi}_t(\mathscr{P})^-}. \end{aligned} \tag{7.24}$$

Thus we may write

$$\int\limits_{\boldsymbol{\chi}_t(\mathscr{P})} f\,\mathrm{d}v = \int\limits_{\boldsymbol{\chi}_t(\mathscr{P})^+} f_+\,\mathrm{d}v + \int\limits_{\boldsymbol{\chi}_t(\mathscr{P})^-} f_-\,\mathrm{d}v.$$

Each of the functions f_+ and f_- satisfies the assumptions of Lemma 7.1 so that

$$\frac{\mathrm{d}}{\mathrm{d}t}\int\limits_{\boldsymbol{\chi}_t(\mathscr{P})^+} f_+\,\mathrm{d}v = \int\limits_{\boldsymbol{\chi}_t(\mathscr{P})^+} f_+'\,\mathrm{d}v + \int\limits_{\partial(\boldsymbol{\chi}_t(\mathscr{P})^+)} f_+\dot{\boldsymbol{\chi}}\cdot\mathbf{m}\,\mathrm{d}a,$$

$$\frac{\mathrm{d}}{\mathrm{d}t}\int\limits_{\boldsymbol{\chi}_t(\mathscr{P})^-} f_-\,\mathrm{d}v = \int\limits_{\boldsymbol{\chi}_t(\mathscr{P})^-} f_-'\,\mathrm{d}v + \int\limits_{\partial(\boldsymbol{\chi}_t(\mathscr{P})^-)} f_-\dot{\boldsymbol{\chi}}\cdot\mathbf{m}\,\mathrm{d}a.$$

The boundary of the region $\chi_t(\mathscr{P})^+$ is composed of two surfaces, $\mathscr{S}^+ := \partial(\chi_t(\mathscr{P})^+) - \mathscr{S}(t)$ and $\mathscr{S}(t)$, moving with different velocities. The velocity of displacement of the surface $\mathscr{S}^+$ is equal to $\mathbf{v} = \dot{\chi}_t|_{\mathscr{S}^+}$ while that of the surface $\mathscr{S}(t)$ is $\mathbf{c}$ (see (1.13) and Figure 7.2). A similar observation can be made about the boundary of the region $\chi_t(\mathscr{P})^-$ and its decomposition into two surfaces, $\mathscr{S}^- := \partial(\chi_t(\mathscr{P})^-) - \mathscr{S}(t)$ and $\mathscr{S}(t)$, moving with different velocities.

We now decompose the surface integrals as

$$\int_{\partial(\chi_t(\mathscr{P})^+)} f_+ \dot{\chi}\cdot\mathbf{m}\,\mathrm{d}a = \int_{\mathscr{S}^+} f_+ \dot{\chi}\cdot\mathbf{m}\,\mathrm{d}a + \int_{\mathscr{S}(t)} f_+ \mathbf{c}\cdot\mathbf{m}\,\mathrm{d}a,$$
$$\int_{\partial(\chi_t(\mathscr{P})^-)} f_- \dot{\chi}\cdot\mathbf{m}\,\mathrm{d}a = \int_{\mathscr{S}^-} f_- \dot{\chi}\cdot\mathbf{m}\,\mathrm{d}a + \int_{\mathscr{S}(t)} f_- \mathbf{c}\cdot\mathbf{m}\,\mathrm{d}a. \tag{7.25}$$

Since the normal vector $\mathbf{m}$ external to the region $\partial(\chi_t(\mathscr{P})^+)$ on the surface $\mathscr{S}(t)$ is opposite in direction to the vector $\mathbf{n}$ normal to $\mathscr{S}(t)$ the first equation in (7.25) may be written

$$\int_{\partial(\chi_t(\mathscr{P})^+)} f_+ \dot{\chi}\cdot\mathbf{m}\,\mathrm{d}a = \int_{\mathscr{S}^+} f_+ \dot{\chi}\cdot\mathbf{m}\,\mathrm{d}a - \int_{\mathscr{S}(t)} f_+ \mathbf{c}\cdot\mathbf{n}\,\mathrm{d}a.$$

In the second equation in (7.25) $\mathbf{m}$ and $\mathbf{n}$ have the same direction on $\mathscr{S}(t)$. Equations (7.24) and the assumptions of Lemma 7.3 allow us to write

$$\frac{\mathrm{d}}{\mathrm{d}t}\int_{\chi_t(\mathscr{P})} f\,\mathrm{d}v = \int_{\chi_t(\mathscr{P})} f'\,\mathrm{d}v + \int_{\partial\chi_t(\mathscr{P})-\mathscr{S}(t)} f\dot{\chi}\cdot\mathbf{m}\,\mathrm{d}a + \int_{\mathscr{S}(t)} (f_- - f_+)\mathbf{c}\cdot\mathbf{n}\,\mathrm{d}a.$$

Using the definition (5.14) of the jump of a function f and the equality $u_n = \mathbf{c}\cdot\mathbf{n}$ we obtain (7.23) or, equivalently,

$$\frac{\mathrm{d}}{\mathrm{d}t}\int_{\chi_t(\mathscr{P})-\mathscr{S}(t)} f\,\mathrm{d}v = \int_{\chi_t(\mathscr{P})-\mathscr{S}(t)} f'\mathrm{d}v + \int_{\partial\chi_t(\mathscr{P})-\mathscr{S}(t)} f\dot{\chi}\cdot\mathbf{m}\,\mathrm{d}a + \int_{\mathscr{S}(t)} [\![f]\!]\mathbf{c}\cdot\mathbf{n}\,\mathrm{d}a. \;\square \tag{7.26}$$

The following version of the divergence theorem for a region intersected by a surface of discontinuity is useful in many problems.

Lemma 7.4

If the vector function $\mathbf{f}$ satisfies the assumptions of Lemma 7.3 then

$$\int_{\chi_t(\mathscr{P})} \operatorname{div}\mathbf{f}\,\mathrm{d}v = \int_{\partial\chi_t(\mathscr{P})-\mathscr{S}(t)} \mathbf{f}\cdot\mathbf{m}\,\mathrm{d}a + \int_{\mathscr{S}(t)} [\![\mathbf{f}]\!]\cdot\mathbf{n}\,\mathrm{d}a, \tag{7.27}$$

where, from the formal point of view, the volume integral is computed everywhere except for points of non-differentiability of f, that is over $\chi_t(\mathscr{P})-\mathscr{S}(t)$

Proof

Following part of the proof of Lemma 7.3 we have

$$\int_{\chi_t(\mathscr{P})} \operatorname{div}\mathbf{f}\,dv = \int_{\chi_t(\mathscr{P})-\mathscr{S}(t)} \operatorname{div}\mathbf{f}\,dv = \int_{\chi_t(\mathscr{P})^+} \operatorname{div}\mathbf{f}\,dv + \int_{\chi_t(\mathscr{P})^-} \operatorname{div}\mathbf{f}\,dv$$

$$= \int_{\partial(\chi_t(\mathscr{P})^+)} \mathbf{f}_+ \cdot \mathbf{m}\,da + \int_{\partial(\chi_t(\mathscr{P})^-)} \mathbf{f}_- \cdot \mathbf{m}\,da$$

$$= \int_{\partial\chi_t(\mathscr{P})-\mathscr{S}(t)} \mathbf{f}\cdot\mathbf{m}\,da + \int_{\mathscr{S}(t)} [\![\mathbf{f}]\!]\cdot\mathbf{n}\,da. \quad \square$$

We observe that in the proof of Lemma 7.3 there is an implicit derivation of (7.27) for the function $f\mathbf{v}$.

7.2 Bibliographical Notes

The formula (7.3) concerning the differentiation of a volume integral of a smooth function was given by Reynolds (1903), and the first proofs came from Spielrein (1916) and Hadamard (1927, p. 504). It is known as the transport theorem (see Truesdell and Toupin, 1960; Truesdell, 1977; Eringen, 1975).

The formula for the differentiation of surface integrals in the general form (7.11) has not been proved previously, but the special case $(7.18)_2$ can be found in Moeckel (1974) together with a proof valid for when the surface $\mathscr{S}(t)$ is represented in a convected parametrization. The idea of introducing the reference surface $\mathscr{S}_0$ is due to Scriven (1960) who was the first to calculate the time derivative of the surface integral (7.11) without, however, bringing it into the form $(7.18)_2$; this was done by Ghez (1966).

The rule (7.23) for differentiating volume integrals of a function with surface singularities, which is known as the transport theorem for a region containing a singular surface, was proved by Thomas (1949); see also Thomas (1961). In Truesdell and Toupin (1960) and Eringen (1975) the formulae corresponding to (7.23) and (7.27) differ in that they have a minus sign in front of the surface integral over $\mathscr{S}(t)$. In their definition of a jump of a function these authors adopted the reverse order for the terms on the right-hand side of (5.14).

8 GENERAL BALANCE EQUATIONS

Anticipating the occurrence of various sources, fluxes and effluxes in the description of a material medium we write the general balance law for a continuous body $\mathscr{B}$ occupying the region $\boldsymbol{\chi}_t(\mathscr{B})$ with sufficiently smooth boundary at an instant t in the form

$$\frac{\mathrm{d}}{\mathrm{d}t}\Psi_t = -W(\Psi_t)+P(\Psi_t)+R(\Psi_t). \tag{8.1}$$

The symbols Ψ_t, W, P and R denote time-dependent measures of certain thermodynamic quantities defined in $\boldsymbol{\chi}_t(\mathscr{B})$. Here, $W(\Psi_t)$ is an efflux of Ψ_t through the boundary $\partial\boldsymbol{\chi}_t(\mathscr{B})$, $P(\Psi_t)$ is the production of Ψ_t in $\boldsymbol{\chi}_t(\mathscr{B})$ and $R(\Psi_t)$ is the source (supply) of Ψ_t in $\boldsymbol{\chi}_t(\mathscr{B})$. In general, W and P are given as constitutive quantities defined for a given body $\mathscr{B}$, while R represents an external action on $\mathscr{B}$.

This decomposition is not universal and therefore not unique. For example, if Ψ_t is the entropy of the material body $\mathscr{B}$ at time t then $W(\Psi_t)$ is the flux of entropy, $P(\Psi_t)$ is the production of entropy, and $R(\Psi_t)$ is the supply of entropy to $\mathscr{B}$ by the environment; in many problems $R(\Psi_t)$ is assumed to be zero.

Consider the situation in which a region* $\boldsymbol{\chi}_t(\mathscr{B})$ occupied by the medium at $t \in I$ contains a moving surface $\mathscr{S}_{\mathscr{B}}(t)$. Let an arbitrary non-empty open region $\mathscr{P}$ within $\mathscr{B}$ with a sufficiently smooth boundary be called a **subbody.**

The function $\boldsymbol{\chi}_t\colon \mathscr{B} \to \mathscr{E}^3$ is assumed to be sufficiently smooth in the sense that its continuation to $\mathscr{B}\cup\partial\mathscr{B}$ preserves the regularity of the boundary required for the divergence theorem, together with the regularity of every surface within $\mathscr{B}$. The transformation $\boldsymbol{\chi}_t$ will determine a moving region $\boldsymbol{\chi}_t(\mathscr{P})$ for every instant t: $\boldsymbol{\chi}_t(\mathscr{P})$ is called the location of the subbody $\mathscr{P}$ in the motion $\boldsymbol{\chi}$ at the instant t. If $\mathscr{S}(t)$ is part of the surface $\mathscr{S}_{\mathscr{B}}(t)$ contained in the region $\boldsymbol{\chi}_t(\mathscr{P})$, then we shall assume that $\mathscr{S}(t)$ divides this region into two sub-regions, $\boldsymbol{\chi}_t(\mathscr{P})^-$ and $\boldsymbol{\chi}_t(\mathscr{P})^+$, for which $\mathscr{S}(t)$ is a common boundary. We also assume that the decomposition has the property (7.19) and the whole surface $\mathscr{S}$ (see $(7.20)_2$) has the properties (7.21) and (7.22).

We denote the boundary $\partial\big(\boldsymbol{\chi}_t(\mathscr{P})^+\big)\cap\partial\boldsymbol{\chi}_t(\mathscr{P})$ by $\mathscr{S}^+(t)$. Then

$$\mathscr{S}^+(t) = \partial\big(\boldsymbol{\chi}_t(\mathscr{P})^+\big)-\mathscr{S}(t). \tag{8.2}$$

Similarly, we may write

$$\mathscr{S}^-(t) = \partial\big(\boldsymbol{\chi}_t(\mathscr{P})^-\big)-\mathscr{S}(t)$$

and

$$\partial\big(\boldsymbol{\chi}_t(\mathscr{P})^+\big) = \mathscr{S}^+(t)\cup\mathscr{S}(t), \qquad \partial\big(\boldsymbol{\chi}_t(\mathscr{P})^-\big) = \mathscr{S}^-(t)\cup\mathscr{S}(t).$$

---

* In Section 7 we defined a moving region $\mathscr{P}$ and a function defined on it (see (7.1)). For our present purposes we shall extend the region $\mathscr{P}$ to a certain set $\mathscr{B}$, called a body (or medium), containing $\mathscr{P}$. Further, we assume that the set $\mathscr{B}$ has the properties ascribed to the region $\mathscr{P}$ in Section 7.1.

The notation $\partial\mathscr{S}(t)$ is introduced now for the curve on $\mathscr{S}_{\mathscr{B}}(t)$ bounding $\mathscr{S}(t)$. The following relations hold:

$$\begin{aligned} &\mathscr{S}^{-}(t)\cap\mathscr{S}^{+}(t) = \partial\mathscr{S}(t), \\ &\partial\boldsymbol{\chi}_t(\mathscr{P}) = \mathscr{S}_0^{-}(t)\cup\partial\mathscr{S}(t)\cup\mathscr{S}_0^{+}(t), \end{aligned} \tag{8.3}$$

where

$$\mathscr{S}_0^{+}(t) = \mathscr{S}^{+}(t)-\partial\mathscr{S}(t), \qquad \mathscr{S}_0^{-}(t) = \mathscr{S}^{-}(t)-\partial\mathscr{S}(t). \tag{8.4}$$

We denote by $\tilde{n}_\alpha$, $\alpha = 1, 2$, the components of the vector tangent to the surface $\mathscr{S}_{\mathscr{B}}(t)$ which forms the outward unit normal to the curve $\partial\mathscr{S}(t)$. As in Section 7.2 the outward unit normal vector to $\mathscr{S}^{+}(t)$ and $\mathscr{S}^{-}(t)$ is denoted by $\mathbf{m}$.

In what follows we make use of the assumptions:

Hypothesis 8.1

The balance law (8.1) holds for every subbody $\mathscr{P}$.

Hypothesis 8.2

The thermodynamic quantities Ψ, W, P and R have the following integral representations for an arbitrary subbody $\mathscr{P}$ and an instant t in the motion $\boldsymbol{\chi}$:

$$\Psi_t = \int\limits_{\chi_t(\mathscr{P})-\mathscr{S}(t)} \psi\,\mathrm{d}v + \int\limits_{\mathscr{S}(t)} \psi_{\mathscr{S}}\,\mathrm{d}a, \tag{8.5$_1$}$$

where ψ is a continuous and continuously differentiable function on $\bigcup\limits_{t\in I}\{t\}\times\boldsymbol{\chi}_t(\mathscr{P})^{+}\cup\mathscr{S}(t)$ and $\bigcup\limits_{t\in I}\{t\}\times\boldsymbol{\chi}_t(\mathscr{P})^{-}\cup\mathscr{S}(t)$ while $\psi_{\mathscr{S}}$ is continuously differentiable on $\bigcup\limits_{t\in I}\{t\}\times\mathscr{S}(t)$;

$$P(\Psi_t) = \int\limits_{\chi_t(\mathscr{P})-\mathscr{S}(t)} p\,\mathrm{d}v + \int\limits_{\mathscr{S}(t)} p_{\mathscr{S}}\,\mathrm{d}a, \tag{8.5$_2$}$$

where p is a continuous function on $\bigcup\limits_{t\in I}\{t\}\times\boldsymbol{\chi}_t(\mathscr{P})^{+}\cup\mathscr{S}(t)$ and $\bigcup\limits_{t\in I}\{t\}\times\boldsymbol{\chi}_t(\mathscr{P})^{-}\cup\mathscr{S}(t)$ while $p_{\mathscr{S}}$ is continuous on $\bigcup\limits_{t\in I}\{t\}\times\mathscr{S}(t)$;

$$W(\Psi_t) = \int\limits_{\partial\chi_t(\mathscr{P})-\partial\mathscr{S}(t)} w\,\mathrm{d}a + \int\limits_{\partial\mathscr{S}(t)} w_{\mathscr{S}}\,\mathrm{d}l, \tag{8.5$_3$}$$

where w is continuous on $\bigcup\limits_{t\in I}\{t\}\times\mathscr{S}_0^{+}(t)$ and $\bigcup\limits_{t\in I}\{t\}\times\mathscr{S}_0^{-}(t)$ while $w_{\mathscr{S}}$ is continuous on $\bigcup\limits_{t\in I}\{t\}\times\partial\mathscr{S}(t)$;

$$R(\Psi_t) = \int\limits_{\chi_t(\mathscr{P})-\mathscr{S}(t)} r\,\mathrm{d}v + \int\limits_{\mathscr{S}(t)} r_{\mathscr{S}}\,\mathrm{d}a, \tag{8.5$_4$}$$

where r is continuous on $\bigcup_{t\in I}\{t\}\times\boldsymbol{\chi}_t(\mathscr{P})^+\cup\mathscr{S}(t)$ and $\bigcup_{t\in I}\{t\}\times\boldsymbol{\chi}_t(\mathscr{P})^-\cup\mathscr{S}(t)$, while $r_{\mathscr{S}}$ is continuous on $\bigcup_{t\in I}\{t\}\times\mathscr{S}(t)$.

Hypothesis 8.3

The surface density w and the curvilinear density $w_{\mathscr{L}}$ have the representations

$$w = w^k m_k = \mathbf{w}\cdot\mathbf{m}, \qquad w_{\mathscr{L}} = w^{\alpha}_{\mathscr{L}}\tilde{n}_{\alpha}. \tag{8.6}$$

The first hypothesis implies that strictly non-local effects are rejected. In strictly non-local theories this hypothesis is not adopted and only the global balance laws for the whole body are postulated. The integral representations required in the second hypothesis assert absolute continuity of the thermodynamic quantities with respect to the volume, surface and linear measures, with the simultaneous requirement of additivity or σ-additivity of these quantities as set functions. The third hypothesis, of a rather technical nature, replaces the well-known Cauchy lemma on the existence of a stress tensor, which is not obtainable from the postulated form of the balance law.

Using the representations (8.5) we can write (8.1) as

$$\frac{\mathrm{d}}{\mathrm{d}t}\int\limits_{\boldsymbol{\chi}_t(\mathscr{P})-\mathscr{S}(t)}\psi\,\mathrm{d}v+\frac{\mathrm{d}}{\mathrm{d}t}\int\limits_{\mathscr{S}(t)}\psi_{\mathscr{S}}\,\mathrm{d}a = -\int\limits_{\partial\boldsymbol{\chi}_t(\mathscr{P})-\partial\mathscr{S}(t)} w\,\mathrm{d}a-\int\limits_{\partial\mathscr{S}(t)} w_{\mathscr{L}}\,\mathrm{d}l$$
$$+\int\limits_{\boldsymbol{\chi}^t(\mathscr{P})-\mathscr{S}(t)} p\,\mathrm{d}v+\int\limits_{\mathscr{S}(t)} p_{\mathscr{S}}\,\mathrm{d}a+\int\limits_{\boldsymbol{\chi}_t(\mathscr{P})-\mathscr{S}(t)} r\,\mathrm{d}v+\int\limits_{\mathscr{S}(t)} r_{\mathscr{S}}\,\mathrm{d}v. \tag{8.1$_a$}$$

We now state the basic result of this chapter.

Theorem 8.1

Let $\mathscr{P}$ be an arbitrary subbody whose location $\boldsymbol{\chi}_t(\mathscr{P})$ at an instant t is intersected by a surface of discontinuity $\mathscr{S}(t)$ of the function ψ. The general balance law holds if and only if

$$\int\limits_{\boldsymbol{\chi}_t(\mathscr{P})-\mathscr{S}(t)}\left(\frac{\partial\psi}{\partial t}-p-r\right)\mathrm{d}v+\int\limits_{\partial\boldsymbol{\chi}_t(\mathscr{P})-\partial\mathscr{S}(t)}(\psi\mathbf{v}+\mathbf{w})\cdot\mathbf{m}\,\mathrm{d}a$$
$$+\int\limits_{\partial\mathscr{S}(t)} w^{\alpha}_{\mathscr{L}}\tilde{n}_{\alpha}\,\mathrm{d}l-\int\limits_{\mathscr{S}(t)}(p_{\mathscr{S}}+r_{\mathscr{S}})\,\mathrm{d}a+\int\limits_{\mathscr{S}(t)}\left\{\frac{\partial\psi_{\mathscr{S}}}{\partial t}+(\psi_{\mathscr{S}}V^{\alpha})_{;\alpha}\right.$$
$$\left.+\psi_{\mathscr{S}}(c^{\alpha}_{;\alpha}-2u_nK_m)+[\![\psi]\!]u_n\right\}\mathrm{d}a = 0. \tag{8.7}$$

Proof

Because of $(8.5)_1$ we can write

$$\frac{\mathrm{d}}{\mathrm{d}t}\Psi_t = \frac{\mathrm{d}}{\mathrm{d}t}\int\limits_{\boldsymbol{\chi}_t(\mathscr{P})-\mathscr{S}(t)}\psi\,\mathrm{d}v+\frac{\mathrm{d}}{\mathrm{d}t}\int\limits_{\mathscr{S}(t)}\psi_{\mathscr{S}}\,\mathrm{d}a. \tag{8.8}$$

The results of Lemma 7.3 and its proof together with (7.26) allow us to write the first integral in the form

$$\frac{\mathrm{d}}{\mathrm{d}t}\int_{\chi_t(\mathscr{P})-\mathscr{S}(t)} \psi \,\mathrm{d}v = \int_{\chi_t(\mathscr{P})-\mathscr{S}(t)} \psi' \mathrm{d}v + \int_{\partial\chi_t(\mathscr{P})-\partial\mathscr{S}(t)} \psi \dot{\boldsymbol{\chi}} \cdot \mathbf{m}\,\mathrm{d}a + \int_{\mathscr{S}(t)} [\![\psi]\!] \mathbf{c}\cdot\mathbf{n}\,\mathrm{d}a.$$

On the right-hand side of this the set $\partial\chi_t(\mathscr{P})-\partial\mathscr{S}(t)$ may be replaced by $\mathscr{S}_0^+(t)\cup\mathscr{S}_0^-(t)$ since the sets differ by the curve $\partial\mathscr{S}(t)$ whose surface measure is zero. The second integral in (8.8) is reduced, by Lemma 7.2, to the form

$$\frac{\mathrm{d}}{\mathrm{d}t}\int_{\mathscr{S}(t)} \psi_{\mathscr{S}}\,\mathrm{d}a = \int_{\mathscr{S}(t)} \left\{\frac{\partial\psi_{\mathscr{S}}}{\partial t} + (\psi_{\mathscr{S}}V^{\alpha})_{;\alpha} + \psi_{\mathscr{S}}(c^{\alpha}_{;\alpha} - 2u_n K_m)\right\}\mathrm{d}a.$$

Summing the two expressions and inserting into (8.8) and then inserting (8.8) together with the other integral representations from (8.5) into (8.1), we obtain

$$\int_{\chi_t(\mathscr{P})-\mathscr{S}(t)} (\psi'-p-r)\,\mathrm{d}v + \int_{\partial\chi_t(\mathscr{P})-\mathscr{S}(t)} (\psi\dot{\boldsymbol{\chi}}\cdot\mathbf{m}+w)\,\mathrm{d}a$$
$$+\int_{\mathscr{S}(t)} \left\{\frac{\partial\psi_{\mathscr{S}}}{\partial t} + (\psi_{\mathscr{S}}V^{\alpha})_{;\alpha} + \psi_{\mathscr{S}}(c^{\alpha}_{;\alpha} - 2u_n K_m) + [\![\psi]\!]u_n\right\}\mathrm{d}a$$
$$-\int_{\mathscr{S}(t)} (p_{\mathscr{S}}+r_{\mathscr{S}})\,\mathrm{d}a + \int_{\partial\mathscr{S}(t)} w_{\mathscr{S}}\,\mathrm{d}l = 0. \tag{8.9}$$

Using the representations for the fluxes w and $w_{\mathscr{S}}$ and the fact that $\psi' = \partial\psi(t,\mathbf{x})/\partial t$, we obtain the required form (8.7). □

If we assume that the surface $\mathscr{S}(t)$ is given in a convected parametrization then (8.7) reduces to

$$\int_{\chi_t(\mathscr{P})-\mathscr{S}(t)} \left(\frac{\partial\psi}{\partial t}-p-r\right)\mathrm{d}v + \int_{\partial\chi_t(\mathscr{P})-\mathscr{S}(t)} (\psi\dot{\boldsymbol{\chi}}+\mathbf{w})\cdot\mathbf{m}\,\mathrm{d}a + \int_{\partial\mathscr{S}(t)} w^{\alpha}_{\mathscr{S}}\tilde{n}_{\alpha}\,\mathrm{d}l$$
$$-\int_{\mathscr{S}(t)} (p_{\mathscr{S}}+r_{\mathscr{S}})\,\mathrm{d}a + \int_{\mathscr{S}(t)} \left\{\frac{\partial\psi_{\mathscr{S}}}{\partial t} + (\psi_{\mathscr{S}}V^{\alpha})_{;\alpha} - 2u_n K_m\psi_{\mathscr{S}}\right.$$
$$\left. + [\![\psi]\!]u_n\right\}\mathrm{d}a = 0. \tag{8.10}$$

A particular case of the above result is given by the following corollary.

Corollary 8.1

If the region $\chi_t(\mathscr{P})$ does not contain a part of the singular surface $\mathscr{S}(t)$ then the balance law is reduced to

$$\int_{\chi_t(\mathscr{P})} \left\{ \frac{\partial \psi}{\partial t} + \operatorname{div}(\psi \dot{\chi} + \mathbf{w}) - p - r \right\} \mathrm{d}v = 0. \tag{8.11}$$

Proof

If the region $\chi_t(\mathscr{P})$ is not intersected by the surface $\mathscr{S}(t)$ then $\chi_t(\mathscr{P}) - \mathscr{S}(t) = \chi_t(\mathscr{P})$, $\partial\chi_t(\mathscr{P}) - \partial\mathscr{S}(t) = \partial\chi_t(\mathscr{P})$ and the integrals over $\partial\mathscr{S}(t)$ and $\mathscr{S}(t)$ vanish. Application of the divergence theorem to the integral

$$\int_{\partial\chi_t(\mathscr{P})} (\psi \mathbf{v} + \mathbf{w}) \cdot \mathbf{m} \mathrm{d}a,$$

where $\mathbf{v}(t, \mathbf{x}) = \dot{\chi}(t, \mathbf{X})$ with $\mathbf{X} = \chi_t^{-1}(\mathbf{x})$, reduces (8.10) to (8.11). □

Corollary 8.2

The local form of (8.1) for points not on a singular surface is

$$\frac{\partial \psi}{\partial t}(t, \mathbf{x}) + \operatorname{div}(\psi \mathbf{v} + \mathbf{w}) = p + r, \tag{8.12}$$

where use is made of the notation from (7.4) and (7.7).

Proof

Equation (8.11) must hold for every arbitrarily small subbody $\mathscr{P}$ containing no part of a singular surface. Then (8.12) follows from the assumption of continuity of the integrand. □

8.1 The Balance Equation on a Surface

We now derive an important compatibility equation on a singular surface resulting from the balance law (8.1). To this end we make the following hypothesis.

Hypothesis 8.4

For every open subset (subbody) $\mathscr{P}$ with a sufficiently smooth boundary and every instant t such that the open region $\chi_t(\mathscr{P})$ is intersected by a surface of discontinuity $\mathscr{S}(t)$, there is a descending family of open subsets (subbodies) $\{\mathscr{P}_n\}_0^\infty$ with the properties (7.19) for every n and such that

$$\begin{gathered} \mathscr{P} \supset \mathscr{P}_0 \supset \mathscr{P}_1 \supset \mathscr{P}_2 \supset \dots, \\ \bigcap_{n=0}^{\infty} \chi_t(\mathscr{P}_n) = \mathscr{S}(t), \\ \bigcap_{n=0}^{\infty} \partial\big(\chi_t(\mathscr{P}_n)^+\big) = \bigcap_{n=0}^{\infty} \partial\big(\chi_t(\mathscr{P}_n)^-\big) = \mathscr{S}(t) \cup \partial\mathscr{S}(t), \end{gathered} \tag{8.13}$$

the orientation of the surface $\bigcap\limits_{n=0}^{\infty} \partial(\boldsymbol{\chi}_t(\mathscr{P}_n)^-)$ being opposite to the orientation of $\mathscr{S}(t)$.

Theorem 8.2

If the functions $\partial\psi/\partial t$, p and r are bounded in $\boldsymbol{\chi}_t(\mathscr{P})$, the functions $\mathbf{v}$ and $\mathbf{w}$ are defined and continuous on each of the sets $\boldsymbol{\chi}_t(\mathscr{P})^+ \cup \mathscr{S}(t)$ and $\boldsymbol{\chi}_t(\mathscr{P})^- \cup \mathscr{S}(t)$, the components $w^\alpha_{\mathscr{S}}$ are differentiable on $\mathscr{S}(t)$ and the curve $\partial\mathscr{S}(t)$ admits the Gauss–Green theorem for curvilinear integrals then, on the surface of discontinuity $\mathscr{S}(t)$, the following equation holds:

$$\int\limits_{\mathscr{S}(t)} \left\{\frac{\partial \psi_{\mathscr{S}}}{\partial t} + (\psi_{\mathscr{S}} V^\alpha + w^\alpha_{\mathscr{S}})_{;\alpha} + \psi_{\mathscr{S}}(c^\alpha_{;\alpha} - 2u_n K_m)\right\} \mathrm{d}a$$
$$= \int\limits_{\mathscr{S}(t)} \{[\![\psi(\mathbf{v}-\mathbf{c})]\!] \cdot \mathbf{n} + [\![\mathbf{w}]\!] \cdot \mathbf{n}\} \,\mathrm{d}a + \int\limits_{\mathscr{S}(t)} (p_{\mathscr{S}} + r_{\mathscr{S}}) \,\mathrm{d}a. \tag{8.14}$$

Proof

Consider the sequence of sets

$$\boldsymbol{\chi}_t(\mathscr{C}_n) \equiv \boldsymbol{\chi}_t(\mathscr{P}_0) \cap \boldsymbol{\chi}_t(\mathscr{P}_1) \cap \boldsymbol{\chi}_t(\mathscr{P}_2) \cap \ldots \cap \boldsymbol{\chi}_t(\mathscr{P}_n).$$

Since $\mathscr{C}_n$ is an open set (a subbody) we may apply the results of Theorem 8.1 to the set $\boldsymbol{\chi}_t(\mathscr{C}_n)$.* This gives

$$\int\limits_{\boldsymbol{\chi}_t(\mathscr{C}_n)-\mathscr{S}(t)} \left(\frac{\partial\psi}{\partial t} - p - r\right) \mathrm{d}v + \int\limits_{\partial\boldsymbol{\chi}_t(\mathscr{C}_n)-\partial\mathscr{S}(t)} (\psi\mathbf{v} + \mathbf{w}) \cdot \mathbf{m} \,\mathrm{d}a + \int\limits_{\partial\mathscr{S}(t)} w^\alpha_{\mathscr{S}} \tilde{n}_\alpha \,\mathrm{d}l$$
$$+ \int\limits_{\mathscr{S}(t)} \left\{\frac{\partial\psi_{\mathscr{S}}}{\partial t} + (\psi_{\mathscr{S}} V^\alpha)_{;\alpha} + \psi_{\mathscr{S}}(c^\alpha_{;\alpha} - 2u_n K_m) + [\![\psi]\!] u_n\right\} \mathrm{d}a$$
$$- \int\limits_{\mathscr{S}(t)} (p_{\mathscr{S}} + r_{\mathscr{S}}) \,\mathrm{d}a = 0.$$

Because of (8.13) and the assumptions we can write

$$\lim_{n\to\infty} \int\limits_{\boldsymbol{\chi}_t(\mathscr{C}_n)-\mathscr{S}(t)} \left(\frac{\partial\psi}{\partial t} - p - r\right) \mathrm{d}v = 0,$$

$$\lim_{n\to\infty} \int\limits_{\partial\boldsymbol{\chi}_t(\mathscr{C}_n)-\partial\mathscr{S}(t)} (\psi\mathbf{v} + \mathbf{w}) \cdot \mathbf{n} \,\mathrm{d}a = - \int\limits_{\mathscr{S}(t)\cup\partial\mathscr{S}(t)} [\![\psi\mathbf{v} + \mathbf{w}]\!] \cdot \mathbf{n} \,\mathrm{d}a$$
$$= - \int\limits_{\mathscr{S}(t)} [\![\psi\mathbf{v} + \mathbf{w}]\!] \cdot \mathbf{n} \,\mathrm{d}a,$$

* It is assumed that $\boldsymbol{\chi}_t$ is a homeomorphism at least.

$$\int_{\partial\mathscr{S}(t)} w^{\alpha}_{\mathscr{S}}\tilde{n}_{\alpha}\,\mathrm{d}l = \int_{\mathscr{S}(t)} (w^{\alpha}_{\mathscr{S}})_{;\alpha}\,\mathrm{d}a.$$

Hence we arrive at the required relations (8.14). □

For a convected parametrization (8.14) becomes

$$\int_{\mathscr{S}(t)} \left\{\frac{\partial \psi_{\mathscr{S}}}{\partial t} + (\psi_{\mathscr{S}} V^{\alpha} + w^{\alpha}_{\mathscr{S}})_{;\alpha} - 2u_n K_m \psi_{\mathscr{S}}\right\}\mathrm{d}a$$

$$= \int_{\mathscr{S}(t)} [\![\psi(\mathbf{v}-\mathbf{c})]\!]\cdot\mathbf{n}\,\mathrm{d}a + \int_{\mathscr{S}(t)} [\![\mathbf{w}]\!]\cdot\mathbf{n}\,\mathrm{d}a + \int_{\mathscr{S}(t)} (p_{\mathscr{S}} + r_{\mathscr{S}})\,\mathrm{d}a. \tag{8.15}$$

Having obtained the integral form of the balance law on a surface we now pass to the local form. From the arbitrariness of the choice of the subset $\mathscr{P}$ we conclude that the part $\mathscr{S}(t)$ of the discontinuity surface $\mathscr{S}_{\mathscr{B}}(t)$ is also arbitrary. This observation establishes the following result:

Corollary 8.3 (Generalized Kotchine theorem)

For points on the surface of discontinuity $\mathscr{S}_{\mathscr{B}}(t)$ the general balance law $(8.1)_a$ is equivalent to the condition

$$\frac{\partial \psi_{\mathscr{S}}}{\partial t} + (\psi_{\mathscr{S}} V^{\alpha} + w^{\alpha}_{\mathscr{S}})_{;\alpha} + \psi_{\mathscr{S}}(c^{\alpha}_{;\alpha} - 2u_n K_m)$$

$$= [\![\psi(v_n - u_n)]\!] + [\![\mathbf{w}]\!]\cdot\mathbf{n} + p_{\mathscr{S}} + r_{\mathscr{S}}, \tag{8.16}$$

where v_n denotes the normal component of the velocity $\mathbf{v}$ on the surface, that is $v_n \equiv \mathbf{v}\cdot\mathbf{n}$.

If $\mathscr{S}_{\mathscr{B}}(t)$ is given in a convected parametrization equation (8.16) reduces to

$$\frac{\partial \psi_{\mathscr{S}}}{\partial t} + (\psi_{\mathscr{S}} V^{\alpha} + w^{\alpha}_{\mathscr{S}})_{;\alpha} - 2u_n K_m \psi_{\mathscr{S}}$$

$$= [\![\psi(v_n - u_n)]\!] + [\![\mathbf{w}]\!]\cdot\mathbf{n} + p_{\mathscr{S}} + r_{\mathscr{S}}. \tag{8.17}$$

Let

$$U := u_n - v_n = (\mathbf{c}-\mathbf{v})\cdot\mathbf{n}. \tag{8.18}$$

This is called the **local speed of propagation.** It is a measure of the normal speed of the surface $\mathscr{S}(t)$ relative to points in the motion $\boldsymbol{\chi}$ instantaneously on the surface. When U is used the compatibility condition (8.16) assumes one of the equivalent forms (see the definition of $\delta/\delta t$ and (2.7))

$$\frac{\partial \psi_{\mathscr{S}}}{\partial t} + (\psi_{\mathscr{S}} V^{\alpha} + w^{\alpha}_{\mathscr{S}})_{;\alpha} + \psi_{\mathscr{S}}(c^{\alpha}_{;\alpha} - 2u_n K_m)$$

$$= -[\![\psi U]\!] + [\![\mathbf{w}]\!]\cdot\mathbf{n} + p_{\mathscr{S}} + r_{\mathscr{S}}, \tag{8.19$_1$}$$

$$\frac{\delta\psi_{\mathscr{S}}}{\delta t}+\{\psi_{\mathscr{S}}(V^{\alpha}+c^{\alpha})+w^{\alpha}_{\mathscr{S}}\}_{;\alpha}-2u_n K_m \psi_{\mathscr{S}}$$

$$= -[\![\psi U]\!]+[\![\mathbf{w}]\!]\cdot\mathbf{n}+p_{\mathscr{S}}+r_{\mathscr{S}}. \qquad (8.19)_2$$

In order to obtain the well-known compatibility condition resulting from the Kotchine theorem it is necessary to set the quantities $\psi_{\mathscr{S}}$, $w_{\mathscr{S}}$, $p_{\mathscr{S}}$ and $r_{\mathscr{S}}$ equal to zero. Then (8.19) becomes

$$[\![\psi U]\!]-[\![\mathbf{w}]\!]\cdot\mathbf{n} = 0. \qquad (8.20)$$

8.2 Bibliographical Notes

The general balance law in the form (8.1), together with the suggested integral representations, may be found in a paper by Moeckel (1974). See also Ghez (1966). In the same paper there is also a derivation, in the case of a convected parametrization, of the integral forms of the balance law, (8.10) and (8.15), together with the local form (8.17).

Moeckel's paper uses a definition of a jump of a function which reverses the order of the terms on the right-hand side of (5.14). This changes the signs of the terms with the jump $[\![\cdot]\!]$ (see also Ghez, 1966).

The general equations (8.7), (8.14) and (8.16), which are valid for an arbitrary parametrization of $\mathscr{S}(t)$, have also been derived by Ghez (1966). Observe that in the proof of the balance laws we have not used the mass conservation law.

The local balance law for points situated on a singular surface, in its classical form (8.20), is due to Kotchine (1926). An article by Truesdell and Toupin (1960) contains a derivation of the general balance law without the surface sources (see also Wang and Truesdell, 1973; Eringen, 1975; and Jeffrey, 1964, 1965).

A slightly different treatment of surface effects in the general balance law is given by Wilmański (1977). This paper also contains a partial generalization of Kotchine's condition in which surface sources are added to the classical balance law (see also Wilmański, 1974a, b and 1975a, b).

The integral representations of the thermodynamic quantities Ψ, P, W, R, which are postulated in (8.5), are still open to question. So far there has been no complete axiomatic theory of thermodynamics which takes account of the existence of a moving surface of discontinuity in the region occupied by the body.

Papers on the thermodynamic theory of continua with surfaces published within the last eighteen years have provided a partial solution to the problem of accounting for surface effects. Articles by Fisher and Leitman (1968, 1970) concerned the axiomatic thermodynamics of non-deformable bodies, whereas

Williams (1972) took up selected problems connected with the postulated concentration of internal and kinetic energy on a surface. Other groups of surface phenomena were treated by Gurtin and Murdoch (1975, 1976) and Murdoch (1976, 1977).

A generalization of the Cauchy lemma, namely the condition for the existence of flux representations (8.6) in the case of weakly balanced interactions, is given by Gurtin and Martins (1976). See also Barański (1974), Gurtin (1972), Gurtin *et al.* (1968), and Hanyga (1985).

Chapter 4

The laws of dynamics and thermodynamics for deformable bodies

We now bring together the geometrical and mechanical considerations of the first three chapters in statements of physical laws or, more precisely, laws of dynamics and thermodynamics relating to deformable heat-conducting continuous media. After giving some basic definitions in Section 9 we formulate, in Section 10, the laws of mechanics and thermodynamics as equations of balance within a medium containing a moving surface and then derive the local counterparts of these laws. Full generality is maintained as far as possible, but attention is restricted to non-polar media, that is to media for which the Cauchy stress tensor is symmetric.

9 DEFORMABLE BODIES

In what follows we formulate the definitions of a body and a deformable body in precise mathematical terms. To this end we shall first require some basic mathematical notions.

9.1 Mathematical Preliminaries

Given a map $f\colon \mathscr{A} \to \mathscr{Y}$, the set $\mathscr{A}$ is denoted by Dom$f$ and called the **domain** of definition of the map, while the set $\mathscr{Y}$ is called its **co-domain.** The set $f(\mathscr{A})$, denoted $\mathrm{R}_g f$, is called the **range** of f or the **image** of $\mathscr{A}$ under f. The range of f is not necessarily the same as $\mathscr{Y}$. However, if $\mathrm{R}_g f = \mathscr{Y}$, then the map is called a **surjection** or is described as "onto". If the inverse $f^{-1}\colon \mathscr{Y} \to \mathscr{A}$ exists then f is called a **bijection.** In other words a bijection is one-to-one and onto.

If both $\mathscr{A}$ and $\mathscr{Y}$ are subsets of the same metric space then f is called an **isometric** bijection if it preserves distance between points.

If $\mathscr{A}$ and $\mathscr{Y}$ are each equipped with a volume measure, as, for example, are open subsets of three-dimensional Euclidean space, then f is called an **isochoric** map if it preserves the volume measure.

A bijection f: $\mathscr{A} \to \mathscr{Y}$ is called a **homeomorphism** if both f and f^{-1} are continuous with respect to appropriate topologies of $\mathscr{A}$ and $\mathscr{Y}$ respectively.

If $\mathscr{A}$ and $\mathscr{Y}$ are metric spaces with the metrics $\rho_{\mathscr{A}}$ and $\rho_{\mathscr{Y}}$, respectively, then the bijective map f: $\mathscr{A} \to \mathscr{Y}$ is called bi-lipschitzian if both f and f^{-1} satisfy the Lipschitz condition, that is there exist two positive constants $L(f)$ and $L(f^{-1})$ such that

$$\varrho_{\mathscr{A}}(a, b) \leqslant L(f^{-1})\varrho_{\mathscr{Y}}\big(f(a), f(b)\big) \leqslant L(f^{-1})L(f)\varrho_{\mathscr{A}}(a, b) \tag{9.1}$$

for any $a, b \in \mathscr{A}$. The constants appearing here are defined according to the formula

$$L(f) \equiv \sup\{\varrho_{\mathscr{Y}}\big(f(a), f(b)\big)\varrho_{\mathscr{A}}(a, b)^{-1}: a, b \in \mathscr{A},\ a \neq b\}. \tag{9.2}$$

It is obvious that a bi-lipschitzian map is a homeomorphism.

We have from Rademacher (1919) that a function f satisfying the Lipschitz condition (called a Lipschitz continuous function) is differentiable almost everywhere, that is the set of points of $\mathrm{Dom} f \subset \mathscr{E}^n$ at which f is not differentiable has measure zero.

A map f: $\mathscr{A} \to \mathscr{E}^3$ is said to belong to the class BLP if it is bi-lipschitzian and its Jacobian is positive wherever defined on $\mathscr{A}$.

If $\mathscr{B}$ is a non-empty set we say that the bijections f: $\mathscr{B} \to f(\mathscr{B}) \subset \mathscr{E}^3$ and g: $\mathscr{B} \to g(\mathscr{B}) \subset \mathscr{E}^3$ are homotopic within the class BLP if there exist a bijection h: $\mathscr{B} \to h(\mathscr{B}) \subset \mathscr{E}^3$ and a continuous map H: $[0, 1] \times h(\mathscr{B}) \to \mathscr{E}^3$ such that $H(0, \cdot) = f \circ h^{-1}(\cdot)$, $H(1, \cdot) = g \circ h^{-1}(\cdot)$ and $H(\tau, \cdot) \in \mathrm{BLP}$ for any $\tau \in [0, 1]$.

Two maps f: $\mathscr{A} \to \mathscr{Y}$ and g: $\mathscr{A} \to \mathscr{Y}$ with $\mathscr{A} \subset \mathscr{E}^3$ are homotopic if there exists a continuous map h: $[0, 1] \times \mathscr{A} \to \mathscr{Y}$ such that

$$h(0, \cdot) = f(\cdot) \quad \text{and} \quad h(1, \cdot) = g(\cdot).$$

Let $\mathscr{B}$ be a non-empty set and $\mathscr{M}$ a non-empty family of subsets of $\mathscr{B}$. The family $\mathscr{M}$ is called a countably-additive (σ-additive) ring* if $\mathscr{M}$ is closed under the formation of differences and countable unions, so that the following conditions are satisfied:

---

* Łojasiewicz (1973) used the term σ-additive body; cf. Taylor (1965).

(i) if $\mathscr{P}_i \in \mathscr{M}$, $i = 1, 2, \ldots$, then $\bigcup_{i=1}^{\infty} \mathscr{P}_i \in \mathscr{M}$,

(ii) if $\mathscr{P}$, $\mathscr{U} \in \mathscr{M}$ then $\mathscr{P} - \mathscr{U} \in \mathscr{M}$.

If $\mathscr{M}$ is a σ-additive ring of subsets of $\mathscr{B}$ then a non-negative, real-valued set function M defined on $\mathscr{M}$ is called a measure if it satisfies the conditions

$$M\left(\bigcup_{i=1}^{\infty} \mathscr{P}_i\right) = \sum_{i=1}^{\infty} M(\mathscr{P}_i), \qquad M(\mathscr{P}) < \infty$$

for some $\mathscr{P} \in \mathscr{M}$ and any family $\{\mathscr{P}_i\}_{i=1}^{\infty}$ of pairwise disjoint sets from $\mathscr{M}$, that is $\mathscr{P}_i \in \mathscr{M}$ for all i and $\mathscr{P}_i \cap \mathscr{P}_j = \emptyset$ for $i \neq j$, where $\emptyset$ denotes the empty set. It can easily be shown that on the empty set the function M vanishes: consider an arbitrary $\mathscr{D} \in \mathscr{M}$; then, since $\mathscr{D} \cap (\mathscr{P} - \mathscr{D}) = \emptyset$, we may write $M(\mathscr{D} \cup (\mathscr{P} - \mathscr{D})) = M(\mathscr{D}) + M(\mathscr{P} - \mathscr{D})$, and, on taking $\mathscr{D} = \emptyset$, we deduce that $M(\mathscr{P}) = M(\emptyset) + M(\mathscr{P})$, from which $M(\emptyset) = 0$ follows.

If both for a set $\mathscr{B}$ and for a σ-additive ring $\mathscr{M}$ of its subsets the measure M is defined then the set $\mathscr{B}$ is called a **measurable space.**

In defining a subbody we shall use the notion of a set with finite perimeter. We recall that a measurable set $\mathscr{A} \subset \mathscr{E}^n$ (with respect to the n-dimensional Lebesgue measure) is called a set with finite perimeter if its characteristic function $\chi_{\mathscr{A}}$ is of class $\mathrm{BV}(\mathscr{E}^n)$, which means that $\chi_{\mathscr{A}}$ has all the distributive derivatives of order one as regular and finite Borel measures. The total variation of the first derivative $\mathrm{D}\chi_{\mathscr{A}}$, which is a vector measure, is called the **perimeter** of the set $\mathscr{A}$, denoted Per $\mathscr{A}$. In symbols we have

$$\operatorname{Per}\mathscr{A} := \sup_{g_i \in C_0^1(\Omega)} \left\{ \sum_{i=1}^{n} \int \chi_{\mathscr{A}}(\mathbf{x}) \frac{\partial g_i}{\partial x^i}(\mathbf{x})\, d\mathbf{x} : \sum_{i=1}^{n} g_i^2 \leqslant 1 \right\},$$

where $\Omega \supset \mathscr{A}$ is an open subset of $\mathscr{E}^n$, while $C_0^1(\Omega)$ denotes the class of real-valued functions continuously differentiable on Ω and with compact supports.

It is worth noting that the property of finite perimeter is an invariant of bi-lipschitzian maps of the class BLP.

9.2 Main Definitions

In this book we are concerned with the deformation and motion of bodies consisting of continuously distributed material. Fields of physical quantities distributed through the body, however, may be discontinuous. Therefore the basic conecpts of deformable body need the formulation in which singularities may occur and can be described properly.

We now give a strict definition of the notion of a body (continuum).

Definition 9.1

A **body** $\mathscr{B}$ is a set of elements, called particles or material points, which is equipped with the structure of a measurable space. A non-vanishing measure M is called the **mass measure** or, simply, **mass**.

Definition 9.2

A body $\mathscr{B}$ is called a **continuous medium** (or **continuum**) if it is equipped with a structure defined by means of a non-empty family C of maps satisfying the following conditions:

(a) the elements of C are bijections of $\mathscr{B}$ onto open subsets of the Euclidean space $\mathscr{E}^3$;

(b) if $\boldsymbol{\varkappa}, \boldsymbol{\gamma} \in C$, then the composition $\boldsymbol{\varkappa} \circ \boldsymbol{\gamma}^{-1}$ is of class BLP;

(c) if $\boldsymbol{\varkappa} \in C$ and a map $\boldsymbol{\lambda}: \boldsymbol{\varkappa}(\mathscr{B}) \to \mathscr{E}^3$ is an isometry and is of class BLP, then $\boldsymbol{\lambda} \circ \boldsymbol{\varkappa} \in C$;

(d) each two elements of C are homotopic within BLP.

Elements of C are termed **placements** of the body $\mathscr{B}$. The image $\boldsymbol{\varkappa}(\mathscr{B})$ of $\mathscr{B}$ under $\boldsymbol{\varkappa}$ is called the region occupied by the body $\mathscr{B}$ in the placement $\boldsymbol{\varkappa}$. The composition $\boldsymbol{\varkappa} \circ \boldsymbol{\gamma}^{-1}$ of two arbitrary maps from C is called a **displacement** of the body $\mathscr{B}$ from the placement $\boldsymbol{\gamma}$ to the placement $\boldsymbol{\varkappa}$.

We shall say that the continuous medium $\mathscr{B}$ is of class C^p, $p \geqslant 1$, if the displacements of the body $\mathscr{B}$ are of class C^p. In problems dealt with in subsequent chapters displacements will not generally be of class C^p on their domains. The lack of continuous differentiability of the displacement is connected with the occurrence of singular surfaces.

We note that conditions (b) and (d) imply that if $\boldsymbol{\gamma}, \boldsymbol{\varkappa} \in C$ are homotopic then there exists a continuous map H: $[0, 1] \times \boldsymbol{\varkappa}(\mathscr{B}) \to \mathscr{E}^3$ such that $H(0, \cdot) = \mathrm{id}_{\boldsymbol{\varkappa}(\mathscr{B})}$, $H(1, \cdot) = \boldsymbol{\gamma} \circ \boldsymbol{\varkappa}^{-1}$ and $H(\tau, \cdot) \in \mathrm{BLP}$ for any $\tau \in [0, 1]$. This means that the body may pass continuously from one placement to another.

Definition 9.3

A subset $\mathscr{P} \subset \mathscr{B}$ will be called a **subbody** if the set $\boldsymbol{\varkappa}(\mathscr{P})$ has finite perimeter for some placement $\boldsymbol{\varkappa} \in C$.

The definition framed in this way ensures that the divergence theorem will hold for any subbody. Moreover, the union, the difference and the intersection of sets with finite perimeters each constitute a set with finite perimeter.

Note that if Definition 9.3 holds with one placement then it holds with any placement from C.

The displacement class determined by the placement family C is denoted by Pr_C, that is

$$\mathrm{Pr}_C := \{\boldsymbol{\lambda} = \boldsymbol{\varkappa} \circ \boldsymbol{\gamma}^{-1} \colon \boldsymbol{\gamma}, \boldsymbol{\varkappa} \in C\} \subset \mathrm{BLP}. \tag{9.3}$$

The conditions of Definition 9.2 imply that C contains at least one bijection $\varkappa\colon \mathscr{B} \to \varkappa(\mathscr{B}) \subset \mathscr{E}^3$ and Pr_C contains all isometric bijections between $\varkappa(\mathscr{B})$ and open subsets of $\mathscr{E}^3$.

If the class Pr_C is composed only of isometric bijections, then the body $\mathscr{B}$ is termed **undeformable or rigid.** If the class Pr_C is composed only of isochoric mappings then the body $\mathscr{B}$ is termed **incompressible**. We now give a precise definition of a deformable body.

Definition 9.4

A continuum with a family of placements C and a class of displacements Pr_C is called a **deformable body** or **deformable continuum** if the conditions $\varkappa \in C$ and $\boldsymbol{\lambda} \in \mathrm{Pr}_C$ with $\mathrm{Dom}\,\boldsymbol{\lambda} = \varkappa(\mathscr{B})$ imply $\boldsymbol{\lambda} \circ \varkappa \in C$ and, moreover, the class Pr_C contains at least one non-isometric map in addition to the isometric bijections.

The assumption of the existence of non-isometric displacements within the class Pr_C corresponds to the intuitive understanding of the term "deformability", by which we mean the ability of an object to change its shape.

The fact that the class Pr_C contains all isometric bijections allows us to introduce the concept of configuration.

Although Definition 9.4 is phrased in terms of the displacement class Pr_C it actually implies restrictions on the placement class C of $\mathscr{B}$.

Definition 9.5

Two placements $\varkappa$ and $\boldsymbol{\gamma}$ from the family C of the body $\mathscr{B}$ determine the same **configuration** of the body if the displacement $\boldsymbol{\lambda} = \boldsymbol{\gamma} \circ \varkappa^{-1}$ is an isometric bijection.

In the Euclidean space isometries are represented in terms of translations and rotations. Roughly speaking, two placements $\varkappa$ and $\boldsymbol{\gamma}$ determine the same configuration of the body $\mathscr{B}$ if the region $\varkappa(\mathscr{B})$ occupied by $\mathscr{B}$ in the placement $\varkappa$ may be superimposed on the region $\boldsymbol{\gamma}(\mathscr{B})$ by means of a rigid translation and a rotation.

We observe that if we associate with every placement $\varkappa \in C$ a distance function $d_\varkappa\colon \mathscr{B} \times \mathscr{B} \to \boldsymbol{R}^+$ defined by

$$d_\varkappa(X, Y) = ||\varkappa(X) - \varkappa(Y)||, \qquad X, Y \in \mathscr{B}, \tag{9.4}$$

where $||\cdot||$ denotes the norm in the Euclidean space, then Definition 9.5 may be formulated in the following way.

Proposition 9.1

Two placements $\varkappa$ and $\boldsymbol{\gamma}$ from the family C determine the same configuration if and only if $d_\varkappa = d_\gamma$. □

Moreover, we can use Definition 9.5 to introduce a relation into the family C. More precisely, we can regard configurations as equivalence classes of placements in the following manner: we write $\boldsymbol{\varkappa} \sim \boldsymbol{\gamma}$ if placements $\boldsymbol{\varkappa}$ and $\boldsymbol{\gamma}$ determine the same configuration of a body. It can easily be verified that $\sim$ is an equivalence relation. The relation is

(i) reflexive: $\boldsymbol{\varkappa} \sim \boldsymbol{\varkappa}$ for each $\boldsymbol{\varkappa} \in C$,

(ii) symmetric: if $\boldsymbol{\varkappa} \sim \boldsymbol{\gamma}$ then $\boldsymbol{\gamma} \sim \boldsymbol{\varkappa}$ for each pair $\boldsymbol{\varkappa}, \boldsymbol{\gamma} \in C$,

(iii) transitive: if $\boldsymbol{\varkappa} \sim \boldsymbol{\gamma}$ and $\boldsymbol{\gamma} \sim \boldsymbol{\mu}$ then $\boldsymbol{\varkappa} \sim \boldsymbol{\mu}$ for each triple $\boldsymbol{\varkappa}, \boldsymbol{\gamma}, \boldsymbol{\mu} \in C$.

Property (i) is obvious, while for the proof of (ii) we have to observe that since $\boldsymbol{\lambda} = \boldsymbol{\gamma} \circ \boldsymbol{\varkappa}^{-1}$ is an isometric bijection and belongs to Pr_C, the inverse map $\boldsymbol{\lambda}^{-1} = \boldsymbol{\varkappa} \circ \boldsymbol{\gamma}^{-1}$ also belongs to Pr_C. This fact also enables us to prove property (iii).

The relation $\sim$ partitions the placement family C into equivalence classes (co-sets), which may be called configurations. We denote the set of all configurations by $\mathscr{K}$, so that $\mathscr{K} \equiv C/\sim$. According to Proposition 9.1 we may write equivalently

$$\mathscr{K} = \{d_{\boldsymbol{\varkappa}} : \boldsymbol{\varkappa} \in C\}. \tag{9.5}$$

On the set of configurations we introduce the distance function d such that for any two $d_{\boldsymbol{\varkappa}}, d_{\boldsymbol{\gamma}} \in \mathscr{K}$,

$$d(d_{\boldsymbol{\varkappa}}, d_{\boldsymbol{\gamma}}) := \ln \max \left\{ \sup_{\substack{X,Y \in \mathscr{B} \\ X \neq Y}} \frac{d_{\boldsymbol{\varkappa}}(X, Y)}{d_{\boldsymbol{\gamma}}(X, Y)}, \sup_{\substack{X,Y \in \mathscr{B} \\ X \neq Y}} \frac{d_{\boldsymbol{\gamma}}(X, Y)}{d_{\boldsymbol{\varkappa}}(X, Y)} \right\}. \tag{9.6}$$

It is not difficult to check that $d\colon \mathscr{K} \to \boldsymbol{R}^+$ satisfies the axioms of a metric if we note that

$$d(d_{\boldsymbol{\varkappa}}, d_{\boldsymbol{\gamma}}) = \ln \max \{L(\boldsymbol{\varkappa} \circ \boldsymbol{\gamma}^{-1}), L(\boldsymbol{\gamma} \circ \boldsymbol{\varkappa}^{-1})\},$$

where $L(\boldsymbol{\varkappa} \circ \boldsymbol{\gamma}^{-1})$ is the Lipschitz constant of the displacement $\boldsymbol{\varkappa} \circ \boldsymbol{\gamma}^{-1}$ (see (9.2)).

On the set of placements, on the other hand, we introduce the metric defined by

$$d_C(\boldsymbol{\varkappa}, \boldsymbol{\gamma}) := \sup_{X \in \mathscr{B}} \|\boldsymbol{\varkappa}(X) - \boldsymbol{\gamma}(X)\| \quad \text{for } \boldsymbol{\varkappa}, \boldsymbol{\gamma} \in C. \tag{9.7}$$

With metric topologies on the sets C and $\mathscr{K}$ we are now in a position to define some new concepts.

9.3 Motion and Deformation

Definition 9.6

A Lipschitz continuous mapping of an arbitrary time interval I into the family C is called a **motion of the body** $\mathscr{B}$. A function $P\colon I \to \mathscr{K}$ defined on I with values in the set $\mathscr{K}$ of configurations is called a **deformation process** for $\mathscr{B}$ if there exists at least one motion $\chi_{(\cdot)}\colon I \to C$ such that $P(t) = d_{\chi_t}$ for any $t \in I$.

It is obvious that a motion on I determines uniquely a deformation process, but not vice-versa, since to any given deformation process there corresponds an infinite number of motions.

It is convenient to choose and fix a reference placement $\varkappa_0$ and describe all possible motions of a body $\mathscr{B}$ with respect to $\varkappa_0$. In this way we may identify particles X of $\mathscr{B}$ with their positions $\mathbf{X} \equiv \varkappa_0(X)$ in the placement $\varkappa_0$, and the whole body $\mathscr{B}$ with the region $\varkappa_0(\mathscr{B})$ which it occupies in the reference placement.

Let χ_t, $t \in I$, be a motion of a deformable body $\mathscr{B}$. We define certain associated tensor fields, if they exist, on $I \times \varkappa_0(\mathscr{B})$ in the following way:

the **displacement tensor**

$$\mathbf{F}(t, \mathbf{X}) = \nabla(\chi_t \circ \varkappa_0^{-1})(\mathbf{X}), \qquad \mathbf{X} \in \varkappa_0(\mathscr{B}), \tag{9.8}$$

the **right Cauchy–Green tensor**

$$\mathbf{C}(t, \mathbf{X}) = \mathbf{F}^T(t, \mathbf{X})\,\mathbf{F}(t, \mathbf{X}), \tag{9.9}$$

the **left Cauchy–Green tensor**

$$\mathbf{B}(t, \mathbf{X}) = \mathbf{F}(t, \mathbf{X})\mathbf{F}^{\mathrm{T}}(t, \mathbf{X}), \tag{9.10}$$

the **velocity field of particles**

$$\dot{\chi}(t, \mathbf{X}) = \frac{\partial}{\partial t}(\chi_t \circ \varkappa_0^{-1})(\mathbf{X}), \tag{9.11}$$

the **acceleration field of particles**

$$\ddot{\chi}(t, \mathbf{X}) = \frac{\partial}{\partial t}\dot{\chi}(t, \mathbf{X}) = \frac{\partial^2}{\partial t^2}(\chi_t \circ \varkappa_0^{-1})(\mathbf{X}). \tag{9.12}$$

In the expressions above the symbol ∇ denotes the gradient operator with respect to $\mathbf{X} \in \varkappa_0(\mathscr{B})$.

We observe that the fields $\mathbf{F}$, $\mathbf{B}$, $\mathbf{C}$ and $\dot{\chi}$ exist almost everywhere since $\chi_{(\cdot)} \circ \varkappa_0^{-1}$ is a Lipschitz continuous function on $I \times \varkappa_0(\mathscr{B})$. The field $\ddot{\chi}$, however, exists if the motion χ_t is twice differentiable as a function of time t.

We also introduce the following fields on $I \times I \times \varkappa_0(\mathscr{B})$:

the **relative displacement tensor**

$$\mathbf{F}_t(\tau, \mathbf{X}) = \mathbf{F}(\tau, \mathbf{X})\mathbf{F}^{-1}(t, \mathbf{X}), \tag{9.13}$$

the **relative right Cauchy–Green tensor**

$$\mathbf{C}_t(\tau, \mathbf{X}) = \mathbf{F}_t^{\mathrm{T}}(\tau, \mathbf{X})\mathbf{F}_t(\tau, \mathbf{X}), \tag{9.14}$$

the k-th **Rivlin–Ericksen tensor**

$$\mathbf{A}_k(t, \mathbf{X}) = \frac{\partial^k}{\partial \tau^k} \mathbf{C}_t(\tau, \mathbf{X})\big|_{\tau=t}, \tag{9.15}$$

the **velocity gradient**

$$\mathbf{L}(t, \mathbf{X}) = \dot{\mathbf{F}}_t(t, \mathbf{X}) = \frac{\partial}{\partial \tau} \mathbf{F}_t(\tau, \mathbf{X})\big|_{\tau=t}, \tag{9.16}$$

the **deformation rate (stretching) tensor**

$$\mathbf{D} \equiv \tfrac{1}{2}\mathbf{A}_1 = \tfrac{1}{2}(\mathbf{L}+\mathbf{L}^{\mathrm{T}}), \tag{9.17}$$

the **spin (rate of rotation) tensor**

$$\mathbf{W} \equiv \tfrac{1}{2}(\mathbf{L}-\mathbf{L}^{\mathrm{T}}). \tag{9.18}$$

The following lemma will be important in our subsequent considerations.*

Lemma 9.1

If $\boldsymbol{\lambda}$ is an isometric bijection between open subsets of the Euclidean space $\mathscr{E}^3$ then there exists a uniquely determined tensor $\mathbf{Q}$ with values in Orth such that for any p and $q \in \operatorname{Dom} \boldsymbol{\lambda}$

$$\boldsymbol{\lambda}(p) = \boldsymbol{\lambda}(q) + \mathbf{Q}(p-q). \ \square \tag{9.19}$$

Here and later on Orth denotes the set of orthogonal tensors over $\mathscr{E}^3$.

For a fixed reference placement $\boldsymbol{\varkappa}_0$ and an arbitrary motion $\boldsymbol{\chi}_t$, $t \in I$, we define the motion $\hat{\boldsymbol{\chi}}_t$ of the body $\mathscr{B}$ relative to $\boldsymbol{\varkappa}_0$ by

$$\hat{\boldsymbol{\chi}}_t \equiv \boldsymbol{\chi}_t \circ \boldsymbol{\varkappa}_0^{-1}. \tag{9.20}$$

The function $\hat{\boldsymbol{\chi}}_t$, $t \in I$, describes the motion of the body $\mathscr{B}$ identified with the region $\boldsymbol{\varkappa}_0(\mathscr{B})$ which it occupies in the reference placement $\boldsymbol{\varkappa}_0$. In most practical applications one has to deal with situations in which a body is treated as a fixed subset of the Euclidean space.

We now set the main properties of a motion of the body $\mathscr{B}$ relative to a reference placement in the form of a lemma.

Lemma 9.2

If two motions $\boldsymbol{\chi}_t$ and $\boldsymbol{\gamma}_t$, $t \in I$, of the body $\mathscr{B}$ differ by a displacement $\boldsymbol{\lambda}_t = \boldsymbol{\gamma}_t \circ \boldsymbol{\chi}_t^{-1}$ then the motions $\hat{\boldsymbol{\chi}}_t$ and $\hat{\boldsymbol{\gamma}}_t$ differ by the same displacement $\boldsymbol{\lambda}_t$.

Proof

From the definition of $\boldsymbol{\lambda}$ and the relation (9.20) applied to both $\boldsymbol{\chi}_t$ and $\boldsymbol{\gamma}_t$ we obtain

---

* The proof of this lemma may be found in several publications. For the more advanced reader that in Noll (1973) is recommended.

$$\hat{\gamma}_t \circ \hat{\chi}_t^{-1} = \gamma_t \circ \varkappa_0^{-1} \circ (\chi_t \circ \varkappa_0^{-1})^{-1} = \gamma_t \circ \varkappa_0^{-1} \circ \varkappa_0 \circ \chi_t^{-1}$$
$$= \gamma_t \circ \chi_t^{-1} = \lambda_t. \ \square$$

As a result of Lemma 9.2 we have the following corollary:

Corollary 9.1

The motions $\hat{\chi}_t$ and $\hat{\gamma}_t$, $t \in I$, of $\mathscr{B}$ relative to a reference placement $\varkappa_0$ determine the same deformation process P_t, $t \in I$, if and only if the motions χ_t and γ_t, $t \in I$, do likewise. □

For any motion $\hat{\chi}_t$, $t \in I$, we define the distance function $d_{\hat{\chi}_t}$ (see (9.5)) by

$$d_{\hat{\chi}_t}(\mathbf{X}, \mathbf{Y}) = \|\hat{\chi}_t(\mathbf{X}) - \hat{\chi}_t(\mathbf{Y})\|, \qquad \mathbf{X}, \mathbf{Y} \in \varkappa_0(\mathscr{B}). \tag{9.21}$$

Corollary 9.2

Two motions $\hat{\chi}_t$ and $\hat{\gamma}_t$, $t \in I$, determine the same deformation process P_t, $t \in I$, if and only if

$$d_{\hat{\chi}_t} = d_{\hat{\gamma}_t}, \qquad t \in I. \ \square$$

We present the final result of this section in the form of the following theorem:

Theorem 9.1

Let two motions $\hat{\chi}_t$ and $\hat{\gamma}_t$, $t \in I$, of $\mathscr{B}$ relative to a reference placement $\varkappa_0$ determine the same deformation process P_t, $t \in I$. Then there exist functions $\mathbf{Q}$ and $\mathbf{a}$ on I, taking values in Orth and $\mathscr{V}$ respectively, such that for any $t \in I$ and any $\mathbf{X}, \mathbf{Y} \in \varkappa_0(\mathscr{B})$ the following relations hold on $\varkappa_0(\mathscr{B})$ for the parameters listed:

the second gradients of the squared distance functions associated with the motions are equal,

$$\nabla_2 \nabla_1 d^2_{\hat{\gamma}_t}(\mathbf{X}, \mathbf{Y}) = \nabla_2 \nabla_1 d^2_{\hat{\chi}_t}(\mathbf{X}, \mathbf{Y}); \tag{9.22}$$

the displacement tensors are related by

$$\underset{\gamma}{\mathbf{F}}(t, \mathbf{X}) = \mathbf{Q}(t) \underset{\chi}{\mathbf{F}}(t, \mathbf{X}); \tag{9.23}$$

the right Cauchy–Green tensors are equal,

$$\underset{\gamma}{\mathbf{C}}(t, \mathbf{X}) = \underset{\chi}{\mathbf{C}}(t, \mathbf{X}); \tag{9.24}$$

the velocity fields are related by

$$\dot{\gamma}(t, \mathbf{X}) - \mathbf{Q}(t)\dot{\chi}(t, \mathbf{X}) = \dot{\mathbf{a}}(t) + \mathbf{A}(t)\big(\hat{\gamma}(t, \mathbf{X}) - \mathbf{a}(t)\big); \tag{9.25}$$

the acceleration fields are related by

$$\ddot{\gamma}(t, \mathbf{X}) - \mathbf{Q}(t)\ddot{\chi}(t, \mathbf{X}) = \ddot{\mathbf{a}}(t) + 2\mathbf{A}(t)\big(\dot{\gamma}(t, \mathbf{X}) - \dot{\mathbf{a}}(t)\big)$$
$$+ \big(\dot{\mathbf{A}}(t) - \mathbf{A}^2(t)\big)\big(\hat{\gamma}(t, \mathbf{X}) - \mathbf{a}(t)\big); \tag{9.26}$$

the relative displacement tensors are related by

$$\underset{\gamma}{\mathbf{F}_t}(\tau, \mathbf{X}) = \mathbf{Q}(\tau)\underset{\chi}{\mathbf{F}_t}(\tau, \mathbf{X})\mathbf{Q}^{\mathsf{T}}(t); \tag{9.27}$$

the relative Cauchy–Green tensors are related by

$$\underset{\gamma}{\mathbf{C}_t}(\tau, \mathbf{X}) = \mathbf{Q}(t)\underset{\chi}{\mathbf{C}_t}(\tau, \mathbf{X})\mathbf{Q}^{\mathsf{T}}(t); \tag{9.28}$$

the k-th Rivlin–Ericksen tensors are related by

$$\underset{\gamma}{\mathbf{A}_k}(t, \mathbf{X}) = \mathbf{Q}(t)\underset{\chi}{\mathbf{A}_k}(t, \mathbf{X})\mathbf{Q}^{\mathsf{T}}(t); \tag{9.29}$$

the velocity gradients are related by

$$\underset{\gamma}{\mathbf{L}}(t, \mathbf{X}) = \mathbf{Q}(t)\underset{\chi}{\mathbf{L}}(t, \mathbf{X})\mathbf{Q}^{\mathsf{T}}(t)+\mathbf{A}(t); \tag{9.30}$$

the deformation-rate tensors of the motions are related according to

$$\underset{\gamma}{\mathbf{D}}(t, \mathbf{X}) = \mathbf{Q}(t)\underset{\chi}{\mathbf{D}}(t, \mathbf{X})\mathbf{Q}^{\mathsf{T}}(t); \tag{9.31}$$

the spin tensors of the two motions are related by

$$\underset{\gamma}{\mathbf{W}}(t, \mathbf{X}) = \mathbf{Q}(t)\underset{\chi}{\mathbf{W}}(t, \mathbf{X})\mathbf{Q}^{\mathsf{T}}(t)+\mathbf{A}(t). \tag{9.32}$$

In the above equations $\mathbf{A}$ represents the function on I with values in the set Sk of skew-symmetric tensors, defined by

$$\mathbf{A}(t) = \dot{\mathbf{Q}}(t)\mathbf{Q}^{\mathsf{T}}(t), \tag{9.33}$$

and $\mathscr{V}$ denotes a vector space over $\mathscr{E}^3$.

Proof

For motions $\hat{\boldsymbol{\chi}}_t$ and $\hat{\boldsymbol{\gamma}}_t$ related by an isometry $\boldsymbol{\lambda}_t$, Lemma 9.1 implies that there exists a function $\mathbf{Q}$ on I with values in Orth such that for any $\mathbf{X} \in \boldsymbol{\varkappa}_0(\mathscr{B})$

$$\hat{\boldsymbol{\gamma}}_t(\mathbf{X}) - \boldsymbol{\lambda}_t(\mathbf{Z}) = \mathbf{Q}(t)\left(\hat{\boldsymbol{\chi}}_t(\mathbf{X}) - \mathbf{Z}\right), \tag{9.34}$$

where $\mathbf{Z} \in \boldsymbol{\varkappa}_0(\mathscr{B})$ may be chosen quite arbitrarily. Define a function $\mathbf{a}$ on I by

$$\mathbf{a}(t) = \boldsymbol{\lambda}_t(\mathbf{Z}). \tag{9.35}$$

Then (9.22) follows as a simple consequence of Corollary 9.2. In order to obtain (9.23)–(9.30) we have to apply definitions (9.8)–(9.18), (9.34) and (9.35) together with the formal rules of differentiation.

Example 9.1

We calculate the second mixed gradient of the function $d^2_{\hat{\chi}_t}$ required in (9.22). First, calculating the gradient of $d^2_{\hat{\chi}_t}$ with respect to $\mathbf{X}$, we obtain

$$\nabla_1 d^2_{\hat{\chi}_t}(\mathbf{X}, \mathbf{Y}) = \nabla_1\{(\hat{\chi}(t, \mathbf{X})-\hat{\chi}(t, \mathbf{Y})) \cdot (\hat{\chi}(t, \mathbf{X})-\hat{\chi}(t, \mathbf{Y}))\}$$
$$= (\nabla\hat{\chi}(t, \mathbf{X}))^{\mathrm{T}}(\hat{\chi}(t, \mathbf{X})-\hat{\chi}(t, \mathbf{Y}))$$
$$+ (\hat{\chi}(t, \mathbf{X})-\hat{\chi}(t, \mathbf{Y}))\nabla\hat{\chi}(t, \mathbf{X}),$$

while for the second gradient calculated with respect to **Y**, we obtain

$$\nabla_2\nabla_1 d^2_{\hat{\chi}_t}(\mathbf{X}, \mathbf{Y}) = -(\nabla\hat{\chi}(t, \mathbf{X}))^{\mathrm{T}}\nabla\hat{\chi}(t, \mathbf{Y})-(\nabla\hat{\chi}(t, \mathbf{Y}))^{\mathrm{T}}\nabla\hat{\chi}(t, \mathbf{X}). \tag{9.36}$$

In view of the definition (9.8), equation (9.36) can be written in the form

$$\nabla_2\nabla_1 d^2_{\hat{\chi}_t}(\mathbf{X}, \mathbf{Y}) = -\mathbf{F}^{\mathrm{T}}(t, \mathbf{X})\mathbf{F}(t, \mathbf{Y})-\mathbf{F}^{\mathrm{T}}(t, \mathbf{Y})\mathbf{F}(t, \mathbf{X}).$$

On putting $\mathbf{Y} = \mathbf{X}$ we obtain

$$\nabla_2\nabla_1 d^2_{\hat{\chi}_t}(\mathbf{X}, \mathbf{X}) = -2\mathbf{F}^{\mathrm{T}}(t, \mathbf{X})\mathbf{F}(t, \mathbf{X}) = -2\mathbf{C}(t, \mathbf{X}). \ \square \tag{9.37}$$

The relations between the kinematic quantities of two equivalent motions show that in addition to the metric there are two objects of interest: the right Cauchy–Green tensor and the mixed gradient of the squared metric. They are both independent of the representation of the class of equivalent motions. For this reason these three quantities may be regarded as "intrinsic objects" of a deformable body.

From the objects listed in Theorem 9.1 only the spin, velocity gradient, velocity and acceleration evidently depend on the rotational speed of one motion relative to another equivalent to it. It is often said in such cases that these quantities are not frame indifferent, or that they are not objective. The remaining objects (with the exception of the two-point displacement tensor) have transformation laws which are typical for second-order tensors.

Finally, we remark that the above relations, such as (9.31), relate two different tensors, not the components of a single tensor in two different coordinate systems.

9.4 Bibliographical and Historical Notes

The concept of a body proposed in Definition 9.1 is sufficiently general as to account for both continuous and discrete systems. Definition 9.2 restricts subsequent attention to continuous systems. The definition of a continuous medium is original; some of its elements may be found in Kosiński (1983) and Banfi and Fabrizio (1979, 1981). The latter authors were the first to notice that it is possible to construct a proper material universum of subbodies on the basis of the concepts of a set with a finite perimeter (as a subbody) and bi-lipschitzian maps as displacements.

Part (d) of Definition 9.2, not included in the definition of Noll (1973), has been adopted from Rychlewski (1970). It expresses the obvious possibility of continuous transition from one body placement to another. Without assumption (d) a body could appear in space with two different orientations.

The definitions of placement and displacement come from Noll (1972); see also Noll (1973). The previous terminology of configurations for elements of the family C was not used in these works. We note that in many publications no distinction is made between placement and displacement; both maps are often termed configuration (see Truesdell and Noll, 1965).

Definition 9.3 comes from Banfi and Fabrizio (1979), whereas Definition 9.4, which makes use of certain properties of the class Pr_C postulated by Noll (1973), appears here for the first time.

The term configuration as used in Definition 9.5 has been reserved here for a notion intrinsically related to a body and not to its manifestation in Euclidean space.

We have not enough space to discuss the notion of an observer. In this respect we refer here to a nice presentation of this notion in the book by Ogden (1984) or to Murdoch (1983) and Hanyga (1985).

In many publications a great deal of effort is devoted to formulating and examining the consequences of the principle of material frame-indifference (see Truesdell and Noll, 1965; and, for some recent results on the subject, Murdoch, 1983; Ogden, 1984). Because of the intrinsic definition of configuration this principle will be satisfied automatically.

Proposition 9.1 ensures the equivalence of the definitions of configuration given here and by Noll (1972, 1973).

Definition 9.6 is original and ensures the differentiability of motions in time and space almost everywhere.

In most standard works on continuum mechanics the tensor $\mathbf{F}$ defined by (9.8) is called the deformation gradient (see, for example, Truesdell and Noll, 1965; Wang and Truesdell, 1973; and Eringen, 1975). In one of his early papers Noll (1955) called $\mathbf{F}$ the displacement tensor. We shall use the terminology of Noll (1972, 1973, 1978).

The second mixed gradient in (9.22) evaluated at $(\mathbf{X}, \mathbf{X})$ was called the configuration of the infinitesimal element of a body by Noll (1973). According to (9.37) this gradient is determined by the right Cauchy–Green tensor, and is symmetric and negative-definite. (Noll (1973, p. 73) states that the gradient is positive-definite, contradicting our derivation in Example 9.1).

Some references to the concept of a set with finite perimeter can be found in the literature dealing with functions of bounded variation of class BV. Here we refer to Federer (1966), Vol'pert (1967) and Hanyga (1985).

10 BALANCE AND CONSERVATION LAWS

In Chapter 3 we derived the general balance laws, and we shall now use these to obtain the following specific laws:

(i) balance of mass,
(ii) balance of momentum,
(iii) balance of moment of momentum.
(iv) balance of energy,
(v) balance of entropy.

This list shows that we shall not include coupled electromagnetic fields but confine attention to thermo-mechanical interactions. Electro-magnetic effects may be taken into account in these laws only in the form of additional sources or fluxes. However, we do not postulate any additional balance laws for such fields.

Let $\mathscr{B}$ be a deformable body and $\boldsymbol{\chi}_t$, $t \in I$, be its motion relative to a fixed reference placement $\boldsymbol{\varkappa}_0$ (see (9.20)). In what follows we shall identify the body $\mathscr{B}$ with the region $\boldsymbol{\varkappa}_0(\mathscr{B})$ which it occupies in the reference placement. In view of this identification we can treat the motion $\boldsymbol{\chi}_t$ directly and use the global balance law postulated in Section 8.

By a subbody $\mathscr{P}$ of $\mathscr{B}$ we mean any bounded subset of $\boldsymbol{\varkappa}_0(\mathscr{B})$ with finite perimeter.

In the following derivations we shall assume that there exists a moving singular surface $\mathscr{S}_{\mathscr{B}}(t)$, $t \in I$, in the region $\boldsymbol{\chi}_t(\mathscr{B})$ occupied by the body (see Section 8).

10.1 Balance of Mass

Let the quantity Ψ appearing in (8.1) represent the mass measure M of Definition 9.1. According to Hypothesis 8.2 the mass $M_t(\mathscr{P})$ of the subbody $\mathscr{P}$ in the motion $\boldsymbol{\chi}_t$ may be expressed in the form

$$\Psi_t \equiv M_t(\mathscr{P}) = \int\limits_{\boldsymbol{\chi}_t(\mathscr{P})-\mathscr{S}(t)} \varrho \, dv + \int\limits_{\mathscr{S}(t)} \varrho_{\mathscr{S}} \, da, \tag{10.1}$$

where ϱ is the mass density for the set $\boldsymbol{\chi}_t(\mathscr{P})^- \cup \boldsymbol{\chi}_t(\mathscr{P})^+ = \boldsymbol{\chi}_t(\mathscr{P}) - \mathscr{S}(t)$, and $\varrho_{\mathscr{S}}$ is the density of the mass concentrated on the moving singular surface $\mathscr{S}(t)$. For the mass production $P(M)$ we can write

$$P(M_t(\mathscr{P})) = \int\limits_{\boldsymbol{\chi}_t(\mathscr{P})-\mathscr{S}(t)} p_\varrho \, dv + \int\limits_{\mathscr{S}(t)} p_{\varrho\mathscr{S}} \, da, \tag{10.2}$$

in accordance with $(8.5)_2$, where p_ϱ is the density of mass production for the volume $\boldsymbol{\chi}_t(\mathscr{P})^+ \cup \boldsymbol{\chi}_t(\mathscr{P})^-$, and $p_{\varrho\mathscr{S}}$ represents the density of mass production on the moving singular surface $\mathscr{S}(t)$.

The mass flux (efflux) $w_\varrho = \mathbf{w}_\varrho \cdot \mathbf{m}$ across the boundary $\mathscr{S}_0(t)^+ \cup \mathscr{S}_0(t)^-$ and the mass flux $w_{\varrho\mathscr{S}} = w^\alpha_{\varrho\mathscr{S}}\tilde{n}_\alpha$ across the curve $\partial\mathscr{S}(t)$ together give rise to the total efflux of mass $W(M_t(\mathscr{P}))$ across the boundary $\partial\boldsymbol{\chi}_t(\mathscr{P})$:

$$W(M_t(\mathscr{P})) = \int_{\partial\boldsymbol{\chi}_t(\mathscr{P})-\mathscr{S}(t)} \mathbf{w}_\varrho \cdot \mathbf{m}\, da + \int_{\partial\mathscr{S}(t)} w^\alpha_{\varrho\mathscr{S}}\tilde{n}_\alpha dl. \tag{10.3}$$

The final term in (8.1) comes from the sources of mass $R(M_t(\mathscr{P}))$ in the region $\boldsymbol{\chi}_t(\mathscr{P})$, that is

$$R(M_t(\mathscr{P})) = \int_{\boldsymbol{\chi}_t(\mathscr{P})-\mathscr{S}(t)} r_\varrho dv + \int_{\mathscr{S}(t)} r_{\varrho\mathscr{S}} da, \tag{10.4}$$

where r_ϱ is the density of body mass supply, and $r_{\varrho\mathscr{S}}$ is the density of surface mass supply.

On introducing the above notation into (8.7) we obtain a global balance law for mass. Instead of writing down the global law we shall consider its local forms (8.12) and (8.16) for the case considered. In the theory of continua they are called **field equations.**

Proposition 10.1

For points of the region $\boldsymbol{\chi}_t(\mathscr{P})-\mathscr{S}(t)$ the mass balance law is equivalent to

$$\frac{\partial\varrho}{\partial t} + \operatorname{div}(\varrho\mathbf{v}+\mathbf{w}_\varrho) = p_\varrho + r_\varrho, \tag{10.5}$$

where all densities are treated as functions of t and $\mathbf{x}$ (see Section 7). For points of the singular surface $\mathscr{S}(t)$, $t \in I$, the mass balance law is equivalent to

$$\begin{aligned}&\frac{\partial\varrho_{\mathscr{S}}}{\partial t} + (\varrho_{\mathscr{S}}V^\alpha + w^\alpha_{\varrho\mathscr{S}})_{;\alpha} + \varrho_{\mathscr{S}}(c^\alpha_{;\alpha} - 2u_n K_m)\\ &= [\![\varrho(v_n-u_n)]\!] + [\![\mathbf{w}_\varrho]\!]\cdot\mathbf{n} + p_{\varrho\mathscr{S}} + r_{\varrho\mathscr{S}},\end{aligned} \tag{10.6}$$

where all functions in (10.6) depend on t and l^α. □

The proof of (10.5) and (10.6) is a simple application of Corollaries 8.2 and 8.3. We should remember at this point that if we wish to consider sets with finite perimeter as subbodies then the application of the results from Section 8 (in particular, those derived from the divergence theorem) should be made with care. In particular, the local balance law (8.12) is valid only for the density points of the set $\boldsymbol{\chi}_t(\mathscr{B})-\mathscr{S}_{\mathscr{B}}(t)$. It is worth noting, however, that all internal points of the set are density points.

The corollary below follows immediately from Proposition 10.1.

Corollary 10.1

If the region $\boldsymbol{\chi}_t(\mathscr{P})$ contains no sources of mass and mass production then the mass balance equations assume the forms

$$\frac{\partial \varrho}{\partial t} + \operatorname{div}(\varrho \mathbf{v} + \mathbf{w}_\varrho) = 0 \quad \text{in } \boldsymbol{\chi}_t(\mathscr{P}) - \mathscr{S}(t),$$
$$\frac{\partial \varrho_{\mathscr{S}}}{\partial t} + (\varrho_{\mathscr{S}} V^\alpha + w^\alpha_{\varrho\mathscr{S}})_{;\alpha} + \varrho_{\mathscr{S}}(c^\alpha_{;\alpha} - 2u_n K_m)$$
$$= [\![\varrho(v_n - u_n)]\!] + [\![\mathbf{w}_\varrho]\!] \cdot \mathbf{n} \quad \text{on } \mathscr{S}(t). \; \square \tag{10.7}$$

The classical mass conservation laws may be obtained by assuming that $\mathbf{w}_\varrho = \mathbf{0}$ and $w^\alpha_{\varrho\mathscr{S}} = 0$, $\alpha = 1, 2$.

Corollary 10.2

Mass $M_t(\mathscr{P})$ is conserved at each point of the region $\boldsymbol{\chi}_t(\mathscr{P})$ if

$$\frac{\partial \varrho}{\partial t} + \operatorname{div}(\varrho \mathbf{v}) = 0 \quad \text{in } \boldsymbol{\chi}_t(\mathscr{P}) - \mathscr{S}(t),$$
$$\frac{\partial \varrho_{\mathscr{S}}}{\partial t} + (\varrho_{\mathscr{S}} V^\alpha)_{;\alpha} + \varrho_{\mathscr{S}}(c^\alpha_{;\alpha} - 2u_n K_m) = [\![\varrho(v_n - u_n)]\!] \quad \text{on } \mathscr{S}(t). \tag{10.8}$$

Remark

Mass $M_t(\mathscr{P})$ is conserved in the region $\boldsymbol{\chi}_t(\mathscr{P})$ if

$$\int_{\boldsymbol{\chi}_t(\mathscr{P}) - \mathscr{S}(t)} (\operatorname{div} \mathbf{w}_\varrho - r_\varrho)\, dv + \int_{\mathscr{S}(t)} (w^\alpha_{\varrho\mathscr{S};\alpha} - [\![\mathbf{w}_\varrho]\!] \cdot \mathbf{n})\, da = 0.$$

Moreover, the conditions for the validity of (10.8) can be expressed in the forms

$$\operatorname{div} \mathbf{w}_\varrho - p_\varrho - r_\varrho = 0 \quad \text{in } \boldsymbol{\chi}_t(\mathscr{P}) - \mathscr{S}(t),$$
$$w^\alpha_{\varrho\mathscr{S};\alpha} - [\![\mathbf{w}_\varrho]\!] \cdot \mathbf{n} - p_{\varrho\mathscr{S}} - r_{\varrho\mathscr{S}} = 0 \quad \text{on } \mathscr{S}(t). \; \square$$

If there is no mass concentration on the surface, that is if $\varrho_{\mathscr{S}} = 0$, and the mass is conserved at each point then equation $(10.8)_2$ becomes the well-known **Stokes–Christoffel condition**

$$[\![\varrho(v_n - u_n)]\!] = 0. \tag{10.9}$$

By using the local speed of propagation U introduced in (8.18), the condition (10.9) may be written as

$$[\![\varrho U]\!] = 0. \tag{10.10}$$

Note that if the normal speeds of the surface $\mathscr{S}(t)$, $t \in I$, and the medium are equal, that is $u_n = v_n$, then (10.9) is satisfied identically for any ϱ.

In terms of the variables t and $\mathbf{X} = \boldsymbol{\varkappa}_0(X)$ equation $(10.8)_1$ may be written

$$\dot{\varrho} + \varrho \operatorname{div} \mathbf{v} = 0 \tag{10.11}$$

since

$$\dot{\varrho} = \frac{\partial}{\partial t} \varrho(t, \mathbf{x}) + \mathbf{v} \cdot \operatorname{grad} \varrho(t, \mathbf{x}).$$

Equivalently, with the help of (7.6), equation (10.11) may be written

$$(\varrho J)^{\cdot} = 0, \tag{10.12}$$

where $J = \det \mathbf{F}(t, \mathbf{X})$. Under the initial condition $\varrho(0, \mathbf{X}) = \varrho_0$ equation (10.12) integrates to

$$\varrho J = \varrho_0 J_0, \tag{10.13}$$

where $J_0 = \det \mathbf{F}(0, \mathbf{X})$.

We now return briefly to the particular situation governed by $\varrho_{\mathscr{S}} = 0$, (10.5) and (10.6). If the mass $M_t(\mathscr{P})$ is to be conserved in the region $\chi_t(\mathscr{P})$ as a whole then the following are counterparts of (10.10) and (10.11):

$$\begin{aligned}
&(\varrho J)^{\cdot} = J(p_\varrho + r_\varrho - \operatorname{div} \mathbf{w}_\varrho) \quad \text{in } \chi_t(\mathscr{P}) - \mathscr{S}(t), \\
&[\![\varrho U]\!] = p_{\varrho\mathscr{S}} + r_{\varrho\mathscr{S}} + [\![\mathbf{w}_\varrho]\!] \cdot \mathbf{n} - w^{\alpha}_{\varrho\mathscr{S};\alpha} \quad \text{in } \mathscr{S}(t), \\
&\int\limits_{\chi_t(\mathscr{P}) - \mathscr{S}(t)} (p_\varrho + r_\varrho - \operatorname{div} \mathbf{w}_\varrho) \, \mathrm{d}v \\
&\quad + \int\limits_{\mathscr{S}(t)} (p_{\varrho\mathscr{S}} + r_{\varrho\mathscr{S}} + [\![\mathbf{w}_\varrho]\!] \cdot \mathbf{n} - w^{\alpha}_{\varrho\mathscr{S};\alpha}) \, \mathrm{d}a = 0.
\end{aligned} \tag{10.14}$$

It is worth pointing out that in the non-local theory, where Hypothesis 8.1 is rejected, the local formulation of the mass balance law has the form (10.14) with one change, namely that the whole body $\mathscr{B}$ replaces the subbody $\mathscr{P}$.

Integrating $(10.14)_1$ with respect to time with the assumption that $J_0 = 1$ we obtain $\varrho J = \varrho_0 + \overline{m}_\varrho$, where

$$\overline{m}_\varrho(\mathbf{x}, t) := \int_0^t J(p_\varrho + r_\varrho - \operatorname{div} \mathbf{w}_\varrho)(\mathbf{x}, s) \, \mathrm{d}s \quad \text{with } \mathbf{x} = \chi_t(\mathbf{X}).$$

In what follows we shall denote the above integrand by m_ϱ. In the non-local theory the quantity m_ϱ is called the **localization residual** for the global conservation of mass in $\mathscr{P}$ if $(10.14)_3$ holds at any t. Finally, it should be pointed out that in this situation we have

$$\frac{\mathrm{d}}{\mathrm{d}t} \int\limits_{\chi_t(\mathscr{P}) - \mathscr{S}(t)} \varrho \, \mathrm{d}v = 0$$

in view of (10.14). Note that from the formal viewpoint, in $(10.14)_3$ and many other places the surface integral should be taken over the set $\mathscr{S}(t) \cap \chi_t(\mathscr{P})$, but we do not write this explicitly.

10.2 Balance of Momentum

Before deriving the equations which result from the balance of momentum law we recall the discussion in Section 7 concerning the time derivative of the surface integral. In particular, equation (7.9), which defines the velocity V^α of the motion $l^\alpha = l^\alpha(t, L^\Delta)$ of a point $(L^1, L^2) = \text{const}$ on the surface $\mathscr{S}(t)$, is fundamental to the following work.

10.2.1 Velocity of surface point

The absolute velocity $\mathbf{V}$ in $\mathscr{E}^3$ of a point momentarily situated on the singular surface $\mathscr{S}(t)$ with surface mass density $\varrho_{\mathscr{S}}$ may be decomposed into normal and tangential components. We write

$$\mathbf{V} = V_n\mathbf{n}+\tilde{V}^\alpha\boldsymbol{\varphi}_{,\alpha}. \tag{10.15}$$

We note that if $V_n \neq u_n$ then the point will leave the surface; in other words there will be an exchange of mass between the surface and the surrounding medium. For example, if $V_n > u_n > 0$ then the point leaves the surface and enters the region $\boldsymbol{\chi}_t(\mathscr{B})^+$, that is the region ahead of the surface $\mathscr{S}(t)$ in the positive direction of the normal vector. Since loss of continuity in the region is excluded, a point from the region behind the surface replaces that point which moves ahead of the surface.

If, on the other hand, $V_n = u_n$ then the point remains on the surface.

Let $(\mathbf{v})^+$ and $(\mathbf{v})^-$ denote the limit values of the velocity of points from the regions $\boldsymbol{\chi}_t(\mathscr{P})^+$ and $\boldsymbol{\chi}_t(\mathscr{P})^-$ respectively on $\mathscr{S}(t)$; their tangential components in the l^α directions are $\boldsymbol{\varphi}^\alpha\cdot(\mathbf{v})^+$ and $\boldsymbol{\varphi}^\alpha\cdot(\mathbf{v})^-$ respectively. If there is no slip on $\mathscr{S}(t)$ then

$$\tilde{V}^\alpha = \boldsymbol{\varphi}^\alpha\cdot(\mathbf{v})^+ = \boldsymbol{\varphi}^\alpha\cdot(\mathbf{v})^-.$$

In general, however, this equality does not hold, and neither does the equality between $\tilde{V}^\alpha$ and V^α implicit in (7.9). The equality $\tilde{V}^\alpha = V^\alpha$ holds in the case of a convected parametrization.*

The case in which surface points remain on the surface so that $\mathscr{S}(t)$ is a material surface is of considerable interest (see, for example, Scriven, 1960; Aris, 1962; Slattery, 1964; Ghez, 1966; and Ościk, 1982). Then

$$V_n = u_n. \tag{10.16}$$

Definition 10.1

A surface $\mathscr{S}(t)$ with the property (10.16) will be called a ***p*-material surface.**

It should be pointed out that a p-material surface is not a material surface in the strict and generally accepted sense since the normal components of the velocities $\mathbf{c}$ and $\mathbf{v}$ do not have to be equal.

---

* Compare with the last equation is Section 4 of Moeckel (1974).

A sheet of paper falling under gravity could be modelled as the motion of a p-material surface in a three-dimensional continuum, that is in a gas.

The combustion process of an explosive spread on a surface serves as an example of a physical phenomenon in which the normal velocity component V_n of surface points is not equal to u_n. During the process the combusted particles leave the surface and enter the surrounding medium.

The adsorption process, in which an extremely thin layer of molecules (of a gas, liquid or solution) adheres to the surface of a solid body or liquid with which it is in contact, may be modelled by means of a p-material surface. This is possible because in adsorption, in contrast to absorption, there is no diffusive penetration of molecules of one phase (the adsorbate) into the bulk of another (the adsorbant). Once the particles of the adsorbate appear on the surface of the adsorbant they remain there or, at most, migrate along the surface. In mathematical terms this means that $V_n = u_n$. On the other hand, particles of the adsorbate near the interface are adsorbed by the surface, and for such particles $\mathbf{v}\cdot\mathbf{n} \neq u_n$.

Another example of a p-material surface is the Gibbs surface, a fundamental concept in the classical treatment of thermodynamics of interfaces. Gibbs (1928) called it "a hypothetical dividing surface". The Gibbs surface separating two phases is not a true material surface in the case of liquid/gas or liquid/liquid interfaces. In such cases there is no jump transition between the two phases; there exists, however, an interfacial layer. The Gibbs surface is then only a hypothetical geometrical cross-section of that layer, necessary to define the volumes of the neighbouring (bulk) phases and to make descriptions of adsorption and other surface phenomena. The surface is a carrier of most phase transition phenomena.

For a p-material surface the parametric representation (1.1) takes the form

$$\mathscr{S}(t)\colon\ \mathbf{x} = \boldsymbol{\varphi}\big(t, l^{\alpha}(t, L^{\Delta})\big), \tag{10.17}$$

while the velocity of a surface point is defined by

$$\left.\frac{\partial\boldsymbol{\varphi}}{\partial t}\right|_{L^{\Delta}=\mathrm{const}} \equiv \dot{\boldsymbol{\varphi}} = \mathbf{c}+V^{\alpha}\boldsymbol{\varphi}_{,\alpha}. \tag{10.18}$$

Bearing in mind the decomposition (1.17), we write

$$\dot{\boldsymbol{\varphi}} = u_n\mathbf{n}+(V^{\alpha}+c^{\alpha})\boldsymbol{\varphi}_{,\alpha}. \tag{10.19}$$

We now consider the situation where a surface $\mathscr{S}_{\mathscr{P}}$ has the representation

$$\mathscr{S}_{\mathscr{P}}\colon\ \mathbf{X} = \tilde{\mathbf{L}}(L^1, L^2)$$

in the reference placement, and is such that, at any time $t \in I$, the actual surface $\mathscr{S}(t)$ is the image* of $\mathscr{S}_{\mathscr{P}}$, that is

$$\mathscr{S}(t) = \hat{\chi}_t(\mathscr{S}_{\mathscr{P}}), \qquad t \in I.$$

Then

$$\hat{\chi}(t, \tilde{\mathbf{L}}(L^1, L^2)) = \boldsymbol{\varphi}(t, l^\alpha(t, L^1, L^2)) \tag{10.20}$$

for any $t \in I$ and consequently $u_n = v_n$, which means that the surface is an absolutely material surface (see Section 13.1), and

$$\dot{\chi}(t, \tilde{\mathbf{L}}) = \dot{\boldsymbol{\varphi}} = \mathbf{V}, \quad \ddot{\chi}(t, \tilde{\mathbf{L}}) = \ddot{\boldsymbol{\varphi}} = \dot{\mathbf{V}}. \tag{10.21}$$

In any case, however, the tangential components of $\mathbf{V}$ in (10.15) are

$$\tilde{V}^\alpha = V^\alpha + c^\alpha.$$

If the parametrization (l^1, l^2) is convected then $c^\alpha = 0$ and, for example, (10.19) is replaced by

$$\dot{\boldsymbol{\varphi}} = u_n \mathbf{n} + V^\alpha \boldsymbol{\varphi}_{,\alpha} = \mathbf{V}. \tag{10.22}$$

The boundary $\mathscr{S}_{\mathscr{P}}$ of a subbody $\mathscr{P}$ in the placement $\boldsymbol{\varkappa}_0$, that is $\mathscr{S}_{\mathscr{P}} = \boldsymbol{\varkappa}_0(\mathscr{P})$, is a particular example of a material surface with properties (10.20)–(10.22), which we shall encounter several times later in this section.

Example 10.1

We determine the acceleration of a point of a p-material surface. According to (10.18) and (10.19) we may write

$$\dot{\mathbf{V}} := \left.\frac{\partial \dot{\boldsymbol{\varphi}}}{\partial t}\right|_{L^A = \text{const}} = \frac{\partial \dot{\boldsymbol{\varphi}}}{\partial t} + V^\alpha \frac{\partial \dot{\boldsymbol{\varphi}}}{\partial l^\alpha}.$$

Hence

$$\dot{\mathbf{V}} = \frac{\partial u_n}{\partial t}\mathbf{n} + u_n \frac{\partial \mathbf{n}}{\partial t} + \frac{\partial}{\partial t}(V^\alpha + c^\alpha)\boldsymbol{\varphi}_{,\alpha} + (V^\alpha + c^\alpha)\frac{\partial \boldsymbol{\varphi}_{,\alpha}}{\partial t}$$
$$+ V^\beta \{u_{n,\beta}\mathbf{n} + u_n \mathbf{n}_{,\beta} + (V^\alpha + c^\alpha)_{,\beta}\boldsymbol{\varphi}_{,\alpha} + (V^\alpha + c^\alpha)\boldsymbol{\varphi}_{,\alpha\beta}\}.$$

Using the identity

$$\frac{\partial \mathbf{n}}{\partial t} = -(u_{n,\alpha} + b_{\gamma\alpha} c^\gamma)\boldsymbol{\varphi}^\alpha,$$

* From Definition 9.2 it follows that during the motion $\hat{\chi}_t$, $t \in I$, regions are mapped onto regions, surfaces onto surfaces and curves onto curves.

obtained with the help of (2.7) and (2.15), together with (1.7) we obtain

$$\dot{\mathbf{V}} = \left\{\frac{\partial u_n}{\partial t} + V^\alpha u_{n,\alpha} + (V^\alpha + c^\alpha)\,[u_{n,\alpha} + b_{\alpha\beta}(V^\beta + c^\beta)]\right\}\mathbf{n}$$

$$+\left\{(V^\alpha + c^\alpha)(c^\beta_{;\alpha} - 2u_n b^\beta_\alpha) + \frac{\partial}{\partial t}(V^\beta + c^\beta) + V^\alpha(V^\beta + c^\beta)_{;\alpha}\right.$$

$$\left. - u_n u_{n,\alpha} g^{\alpha\beta}\right\}\boldsymbol{\varphi}_{,\beta}. \;\square$$

Example 10.2

In accordance with (7.9) we set

$$V^{\alpha'} = \frac{\partial \hat{l}^{\alpha'}}{\partial \tau}(\tau, L^\Delta),$$

where $(l^{\alpha'})$ is a convected parametrization. If (l^α) is another parametrization (non-convected) then, from the unique representation of the surface, we have

$$\mathbf{x} = \boldsymbol{\varphi}(\tau, l^\alpha) = \boldsymbol{\varphi}(\tau, l^\alpha(\tau, l^{\alpha'})) = \boldsymbol{\varphi}'(\tau, l^{\alpha'})$$

and

$$u_n \mathbf{n}_\tau \equiv \mathbf{u}'_\tau = \frac{\partial \boldsymbol{\varphi}'}{\partial \tau} = \left.\frac{\partial \boldsymbol{\varphi}}{\partial \tau}\right|_{l^\alpha = \text{const}} + \left.\frac{\partial \boldsymbol{\varphi}}{\partial l^\alpha}\frac{\partial l^\alpha}{\partial \tau}\right|_{l^{\alpha'} = \text{const}}.$$

Hence

$$c^\alpha(\tau, l^\alpha) = g^{\alpha\beta}\frac{\partial \boldsymbol{\varphi}}{\partial l^\beta}\cdot \mathbf{c}_\tau = -\frac{\partial l^\alpha}{\partial \tau}(\tau, l^{\alpha'}),$$

where (cf. (1.13))

$$\mathbf{c}_\tau = \frac{\partial \boldsymbol{\varphi}}{\partial \tau}(\tau, l^\alpha).$$

It follows that

$$V^\alpha \equiv \left.\frac{\partial l^\alpha}{\partial \tau}\right|_{L^\Delta = \text{const}} = \left.\frac{\partial l^\alpha}{\partial \tau}\right|_{l^{\alpha'} = \text{const}} + \frac{\partial l^\alpha}{\partial l^{\beta'}}\frac{\partial \hat{l}^{\beta'}}{\partial \tau} = -c^\alpha + \frac{\partial l^\alpha}{\partial l^{\beta'}}V^{\beta'}.$$

We have therefore obtained the connection

$$V^\alpha + c^\alpha = \frac{\partial l^\alpha}{\partial l^{\beta'}}V^{\beta'}$$

between the velocity components V^α and $V^{\alpha'}$ of a point momentarily lying on a singular surface with respect to an arbitrary and a convected parametrization respectively. $\square$

Clearly, it is only the sum $V^\alpha + c^\alpha$ which forms the components of a surface vector, that is a vector satisfying the transformation law

$$V^{\alpha''} + c^{\alpha''} = \frac{\partial l^{\alpha''}}{\partial l^\alpha}(V^\alpha + c^\alpha)$$

under a change of surface parameters (l^α) to $(l^{\alpha''})$. This conclusion will be used later. Moreover, since $(V^\alpha + c^\alpha)\boldsymbol{\varphi}_{,\alpha} = \tilde{V}^\alpha \boldsymbol{\varphi}_{,\alpha} =: \tilde{\mathbf{V}}$ in (10.15), it is not difficult to show that $\dot{\mathbf{V}} = \frac{\delta}{\delta t}\mathbf{V} + (\mathrm{grad}_{\mathscr{S}}\mathbf{V})\tilde{\mathbf{V}}$ with $\mathrm{grad}_{\mathscr{S}}(\cdot) := \frac{\partial}{\partial l^\alpha}(\cdot)\otimes\boldsymbol{\varphi}^\alpha$, where $\dot{\mathbf{V}}$ has been calculated in Example 10.1.

10.2.2 *The law of balance of momentum*

The momentum $\mathbf{F}_t(\mathscr{P})$ of a subbody $\mathscr{P}$ at time t in the motion $\boldsymbol{\chi}_t$ is given by

$$\Psi_t \equiv \mathbf{F}_t(\mathscr{P}) = \int_{\boldsymbol{\chi}_t(\mathscr{P}) - \mathscr{S}(t)} \varrho \mathbf{v}\, dv + \int_{\mathscr{S}(t)} \varrho_{\mathscr{S}} \mathbf{V}\, da. \tag{10.23}$$

For the production of momentum we assume, following $(8.5)_2$, the decomposition

$$P(\mathbf{F}_t(\mathscr{P})) = \int_{\boldsymbol{\chi}_t(\mathscr{P}) - \mathscr{S}(t)} \mathbf{p}_{\mathbf{F}}\, dv + \int_{\mathscr{S}(t)} \mathbf{p}_{\mathbf{F}\mathscr{S}}\, da, \tag{10.24}$$

where $\mathbf{p}_{\mathbf{F}}$ and $\mathbf{p}_{\mathbf{F}\mathscr{S}}$ denote the vector momentum densities.

For the efflux of momentum we write

$$W(\mathbf{F}_t(\mathscr{P})) = -\int_{\partial\boldsymbol{\chi}_t(\mathscr{P}) - \mathscr{S}(t)} \mathbf{t}\, da - \int_{\partial\mathscr{S}(t)} \mathbf{t}_{\mathscr{L}}\, dl. \tag{10.25}$$

The vector $\mathbf{t}$ characterizes the load acting on the boundary of the subbody $\mathscr{P}$ in the motion $\boldsymbol{\chi}_t$ and $\mathbf{t}_{\mathscr{L}}$ represents the curvilinear load. According to Hypothesis 8.3 (the flux principle) there exist tensor functions $\mathbf{T}$ and $\mathbf{S}$ such that

$$\mathbf{t} = \mathbf{T}\mathbf{m}, \quad \mathbf{t}_{\mathscr{L}} = \mathbf{S}\tilde{\mathbf{n}}, \tag{10.26}$$

where $\tilde{\mathbf{n}}$ denotes a vector in the tangent plane to $\mathscr{S}(t)$ which is also the outward unit normal to the curve $\partial\mathscr{S}(t)$.

The vector $\mathbf{t}$ is called the **stress vector** acting on an oriented element of the surface of $\mathscr{P}$ with unit outward normal vector $\mathbf{m}$. The vector $\mathbf{t}_{\mathscr{L}}$ is a field of curvilinear force with dimension force per unit length of the curve $\partial\mathscr{S}(t)$, and it is produced by that part of the surface $\mathscr{S}_{\mathscr{B}}(t)$ towards which $\tilde{\mathbf{n}}$ is directed acting on the remaining part of the surface, namely $\mathscr{S}(t)$. The tensor $\mathbf{T}$ is called the **Cauchy stress tensor**, while we shall refer to $\mathbf{S}$ as the **surface stress** (or **surface tension**) **tensor**.

In index notation equations (10.26) take the forms

$$t^k = T^{ki}m_i, \qquad t^k_{\mathscr{L}} = S^{k\alpha}\tilde{n}_\alpha. \tag{10.26a}$$

Finally, for the supply of momentum we adopt the integral representation

$$R(\mathbf{F}_t(\mathscr{P})) = \int\limits_{\chi_t(\mathscr{P})-\mathscr{S}(t)} \varrho\mathbf{b}\,dv + \int\limits_{\mathscr{S}(t)} \varrho_{\mathscr{S}}\mathbf{b}_{\mathscr{S}}\,da, \tag{10.27}$$

where $\varrho\mathbf{b}$ is the body force density and $\varrho_{\mathscr{S}}\mathbf{b}_{\mathscr{S}}$ the surface force density. The vector $\mathbf{b}$ has dimension force per unit mass, while $\mathbf{b}_{\mathscr{S}}$ is a force per unit surface mass.

Substituting the representations (10.23)–(10.27) into equation (8.7) we obtain the global balance law for linear momentum. Its local counterpart may be formulated (see Corollaries 8.2 and 8.3) as follows:

Proposition 10.2

For points of the region $\chi_t(\mathscr{P})-\mathscr{S}(t)$ the momentum balance equation is equivalent to

$$\frac{\partial}{\partial t}(\varrho\mathbf{v}) + \operatorname{div}(\varrho\mathbf{v}\otimes\mathbf{v}-\mathbf{T}) = \mathbf{p}_F + \varrho\mathbf{b}, \tag{10.28}$$

where all densities are treated as functions of t and $\mathbf{x}$. For points of the singular surface $\mathscr{S}(t)$, $t \in I$, the momentum balance equation is equivalent to

$$\frac{\partial}{\partial t}(\varrho_{\mathscr{S}}V^k) + (\varrho_{\mathscr{S}}V^kV^\alpha - S^{k\alpha})_{;\alpha} + \varrho_{\mathscr{S}}V^k(c^\alpha_{;\alpha} - 2u_nK_m)$$
$$= [\![\varrho v^k(v_n-u_n)]\!] - [\![T^{ki}]\!]n_i + p^k_{F\mathscr{S}} + \varrho_{\mathscr{S}}b^k_{\mathscr{S}}, \tag{10.29}$$

in which all functions depend on t and l^α. □

To see the whole picture we recall that V^k are the components of the velocity $\mathbf{V}$ given by (10.15), v^k are the components of the velocity $\mathbf{v}$ (see (7.7)), and $p^k_{F\mathscr{S}}$ and $b^k_{\mathscr{S}}$ are the components of the densities $p_{F\mathscr{S}}$ and $\mathbf{b}_{\mathscr{S}}$ respectively.

A brief examination of the balance equations (10.6) and (10.29) shows that, because of the parametrization of the moving surface, each contains an invariant time derivative which measures the rate of change along the normal trajectory of the surface.

We recall that it is the displacement derivative introduced by Thomas (1957) that constitutes the particular time derivative which coincides with the partial time derivative when the moving surface is given in a convected parametrization; in an arbitrary parametrization it is given by the formula (2.7). The two balance equations may therefore be written in the forms

$$\frac{\delta}{\delta t}\varrho_{\mathscr{S}}+\left(\varrho_{\mathscr{S}}(V^{\alpha}+c^{\alpha})+w^{\alpha}_{\varrho\mathscr{S}}\right)_{;\alpha}-2u_n\varrho_{\mathscr{S}}K_m$$
$$=[\![\varrho(v_n-u_n)]\!]+[\![w_{\varrho}]\!]+p_{\varrho\mathscr{S}}+r_{\varrho\mathscr{S}}, \tag{10.6$'$}$$

$$\frac{\delta}{\delta t}(\varrho_{\mathscr{S}}V^k)+\left(\varrho_{\mathscr{S}}V^k(V^{\alpha}+c^{\alpha})-S^{k\alpha}\right)_{;\alpha}-2u_n\varrho_{\mathscr{S}}V^kK_m$$
$$=[\![\varrho v^k(v_n-u_n)]\!]-[\![T^{ki}]\!]n_i+p^k_{\mathrm{F}\mathscr{S}}+\varrho_{\mathscr{S}}b^k_{\mathscr{S}}. \tag{10.29$'$}$$

A particular case of the local momentum balance equation is given by

Corollary 10.3

If $\mathbf{p}_{\mathrm{F}}=\mathbf{0}$ in the region $\boldsymbol{\chi}_t(\mathscr{P})^+\cup\boldsymbol{\chi}_t(\mathscr{P})^-$ then (10.28) reduces to

$$\frac{\partial}{\partial t}(\varrho\mathbf{v})+\operatorname{div}(\varrho\mathbf{v}\otimes\mathbf{v}-\mathbf{T})=\varrho\mathbf{b}.\ \square \tag{10.30}$$

Carrying out the differentiations in (10.30), we obtain

$$\left(\frac{\partial\varrho}{\partial t}+\operatorname{div}(\varrho\mathbf{v})\right)\mathbf{v}+\varrho\left(\frac{\partial\mathbf{v}}{\partial t}+\mathbf{v}\cdot\operatorname{grad}\mathbf{v}\right)=\operatorname{div}\mathbf{T}+\varrho\mathbf{b}.$$

In view of (7.5), (7.7) and $(10.8)_1$ we may deduce

Corollary 10.4

If the mass $M(\mathscr{P})$ is conserved in the motion $\boldsymbol{\chi}_t$ and $\mathbf{p}_{\mathrm{F}}=\mathbf{0}$ then the momentum balance equation in the region $\boldsymbol{\chi}_t(\mathscr{P})-\mathscr{S}(t)$ takes one of the two equivalent forms

$$\varrho\dot{\mathbf{v}}=\operatorname{div}\mathbf{T}+\varrho\mathbf{b},\qquad \varrho\ddot{\boldsymbol{\chi}}=\operatorname{div}\mathbf{T}+\varrho\mathbf{b}. \tag{10.31}$$

The second equation in (10.31) is called Cauchy's first law of motion. For the balance law on the surface $\mathscr{S}(t)$ we have

Corollary 10.5

If the mass $M(\mathscr{P})$ is conserved in the whole region $\boldsymbol{\chi}_t(\mathscr{P})$ and $\mathbf{p}_{\mathrm{F}\mathscr{S}}=\mathbf{0}$, then (10.29) becomes

$$\varrho_{\mathscr{S}}\dot{\mathbf{V}}^k=S^{k\alpha}_{;\alpha}+\varrho_{\mathscr{S}}b^k_{\mathscr{S}}+[\![\varrho(v_n-u_n)(v^k-V^k)]\!]-[\![T^{ki}]\!]n_i. \tag{10.32}$$

Proof

Let $\dot{\mathbf{V}}$ denote the acceleration of a surface point, and let $\dot{V}^k$ denote the components of $\dot{\mathbf{V}}$. Then

$$\dot{V}^k:=\left.\frac{\partial V^k}{\partial t}\right|_{L^{\Delta}=\mathrm{const}}=\left.\frac{\partial V^k}{\partial t}\right|_{l^{\alpha}=\mathrm{const}}+V^k_{;\alpha}V^{\alpha}\equiv\frac{\delta V^k}{\delta t}+V^k_{,\alpha}(V^{\alpha}+c^{\alpha}).$$

Performing the differentiations in (10.29) and regrouping the terms we obtain

$$\left\{\frac{\delta\varrho_{\mathscr{S}}}{\delta t}+[\varrho_{\mathscr{S}}(V^{\alpha}+c^{\alpha})]_{;\alpha}-2\varrho_{\mathscr{S}}u_n K_m\right\}V^k+\varrho_{\mathscr{S}}\dot{V}^k$$
$$= S^{k\alpha}_{;\alpha}+\varrho_{\mathscr{S}}b^k_{\mathscr{S}}+[\![\varrho v^k(v_n-u_n)]\!]-[\![T^{ki}]\!]n_i+p^k_{\mathrm{F}\mathscr{S}}.$$

On use of (10.6) and the identity

$$[\![\varrho(v_n-u_n)]\!]V^k = [\![\varrho V^k(v_n-u_n)]\!]$$

the above equation may be rearranged as

$$\varrho_{\mathscr{S}}\dot{V}^k = S^{k\alpha}_{;\alpha}+\varrho_{\mathscr{S}}b^k_{\mathscr{S}}-[\![T^{ki}]\!]n_i+[\![\varrho(v^k-V^k)(v_n-u_n)]\!]$$
$$+(w^{\alpha}_{\varrho\mathscr{S};\alpha}-[\![w^i_{\varrho}]\!]n_i-p_{\varrho\mathscr{S}}-r_{\varrho\mathscr{S}})V^k+p^k_{\mathrm{F}\mathscr{S}}. \tag{10.33}$$

Since mass is conserved and $p^k_{\mathrm{F}\mathscr{S}} = 0$ equation (10.33) becomes (10.32). □

Note that if we define a generalized surface force by

$$\varrho_{\mathscr{S}}\tilde{\mathbf{b}}_{\mathscr{S}} := [\![\varrho(v_n-u_n)(\mathbf{v}-\mathbf{V})]\!]-[\![\mathbf{T}]\!]\mathbf{n}+\varrho_{\mathscr{S}}\mathbf{b}_{\mathscr{S}},$$

equation (10.32) takes the form

$$\varrho_{\mathscr{S}}\dot{\mathbf{V}} = \mathrm{div}_{\mathscr{S}}\mathbf{S}+\varrho_{\mathscr{S}}\tilde{\mathbf{b}}_{\mathscr{S}}, \tag{10.34}$$

this holding on $\mathscr{S}(t)$. This is very similar to $(10.31)_1$, which holds for points in the region $\boldsymbol{\chi}_t(\mathscr{P})-\mathscr{S}(t)$. Here $\mathrm{div}_{\mathscr{S}}$ is the divergence operator on the surface. It has the property

$$\mathrm{div}_{\mathscr{S}}(\mathbf{S}^{\mathrm{T}}\mathbf{k}) = (\mathrm{div}_{\mathscr{S}}\mathbf{S})\cdot\mathbf{k} \equiv k_i S^{i\alpha}_{;\alpha}$$

for any vector $\mathbf{k}$ in $\mathscr{E}^3$ with covariant components k_i.

In the remaining part of this subsection we shall discuss the case of a moving material surface $\{\mathscr{S}(t)\}_{t\in I}$, so that $u_n = \mathbf{V}\cdot\mathbf{n} = v_n$.

Corollary 10.6

If a moving surface $\{\mathscr{S}(t)\}_{t\in I}$ is a material surface and the assumptions of Corollary 10.5 are satisfied then the local balance equations of mass and momentum are respectively

$$\frac{\delta\varrho_{\mathscr{S}}}{\delta t}+\{\varrho_{\mathscr{S}}(V^{\alpha}+c^{\alpha})\}_{;\alpha}-2u_n\varrho_{\mathscr{S}}K_m = 0,$$
$$\varrho_{\mathscr{S}}\dot{V}^k = S^{k\alpha}_{;\alpha}+\varrho_{\mathscr{S}}b^k_{\mathscr{S}}-[\![T^{ki}]\!]n_i. \tag{10.35}$$

In direct notation the latter equation may be rewritten as

$$\varrho_{\mathscr{S}}\dot{\mathbf{V}} = \mathrm{div}_{\mathscr{S}}\mathbf{S}+\varrho_{\mathscr{S}}\mathbf{b}_{\mathscr{S}}-[\![\mathbf{T}]\!]\mathbf{n}.\ □$$

The proof of the above follows by setting $u_n-v_n = 0$ in (10.8) and (10.33).

A particular case of a material surface is the boundary of a subbody $\mathscr{P}$ characterized by

$$\mathscr{S}(t) = \hat{\boldsymbol{\chi}}_t(\mathscr{S}_{\mathscr{P}}) \subset \partial\boldsymbol{\chi}_t(\mathscr{P}), \quad \mathscr{S}_{\mathscr{P}} = \partial\boldsymbol{\varkappa}_0(\mathscr{P})\cap\partial\boldsymbol{\varkappa}_0(\mathscr{B}).$$

Then the region $\chi_t(\mathscr{P})^-$ coincides with $\chi_t(\mathscr{P})$, and $\chi_t(\mathscr{P})^+$ is the empty set. Let the natural extension of any function on $\chi_t(\mathscr{P})^+$ vanish. Then, for the stress jump, we obtain

$$[\![\mathbf{T}]\!] = \mathbf{T}|_{\partial\chi_t(\mathscr{P})}.$$

Hence, instead of $(10.35)_2$, the equation

$$\varrho_{\mathscr{S}}\ddot{\chi} = \mathrm{div}_{\mathscr{S}}\mathbf{S}+\varrho_{\mathscr{S}}\mathbf{b}_{\mathscr{S}}-\mathbf{Tn} \tag{10.36}$$

holds on the boundary.

Corollary 10.7

If a subbody $\mathscr{P}$ is bounded by a material surface $\mathscr{S}_{\mathscr{P}}$ which has a non-vanishing density $\varrho_{\mathscr{S}}$ and surface stress $\mathbf{S}$, then the local balance of momentum equation has the form (10.36) provided the assumptions of Corollary 10.5 are satisfied. □

We now return for a while to the general situation governed by (10.28) and (10.29). If the mass $M_t(\mathscr{P})$ and momentum $\mathbf{F}_t(\mathscr{P})$ are conserved in the region $\chi_t(\mathscr{P})$ during the motion χ_t which began at the instant $t = 0$ with $J = \det\mathbf{F}(0, \mathbf{x}) = 1$, then the relations

$$\varrho\dot{\mathbf{v}}+J^{-1}m_\varrho\mathbf{v} = \mathrm{div}\mathbf{T}+\varrho\mathbf{b}+\mathbf{p}_{\mathbf{F}},$$
$$[\![\varrho\mathbf{v}(v_n-u_n)-\mathbf{Tn}]\!] = \mathbf{m}_{\mathbf{F}}$$

correspond to (10.31) and (10.32) respectively, where

$$(\varrho J)^{\cdot} = m_\varrho, \quad [\![\varrho(v_n-u_n)]\!] = m_{\varrho\mathscr{S}},$$
$$\mathbf{m}_{\mathbf{F}} = (m_{\varrho\mathscr{S}}+p_{\varrho\mathscr{S}}+r_{\varrho\mathscr{S}}-\mathrm{div}_{\mathscr{S}}\mathbf{w}_{\varrho\mathscr{S}})\mathbf{V}+\varrho_{\mathscr{S}}\dot{\mathbf{V}}-\mathrm{div}_{\mathscr{S}}\mathbf{S}-\varrho_{\mathscr{S}}\mathbf{b}_{\mathscr{S}}-\mathbf{p}_{\mathbf{F}\mathscr{S}},$$

and

$$m_{\varrho\mathscr{S}} := \frac{\delta}{\delta t}\varrho_{\mathscr{S}}+\{\varrho_{\mathscr{S}}(V^\alpha+c^\alpha)\}_{;\alpha}-2u_n\varrho_{\mathscr{S}}K_m+\mathrm{div}_{\mathscr{S}}\mathbf{w}_{\varrho\mathscr{S}}-p_{\varrho\mathscr{S}}-r_{\varrho\mathscr{S}}.$$

Note that here the mass flux $\mathbf{w}_\varrho$ has been set to zero since we are considering the case in which mass is conserved within the subbody $\mathscr{P}$. The condition for conservation of linear momentum within $\chi_t(\mathscr{P})$ is

$$\int_{\chi_t(\mathscr{P})-\mathscr{S}(t)}\mathbf{p}_{\mathbf{F}}\,dv+\int_{\mathscr{S}(t)\cap\chi_t(\mathscr{P})}\mathbf{m}_{\mathbf{F}}\,da = \mathbf{0}.$$

As in the case of the balance of mass (10.14) the above relation may be regarded as the non-local formulation of the momentum balance if $\mathscr{P}$ is replaced by the whole body $\mathscr{B}$.

10.3 Balance of Moment of Momentum

Let $\mathbf{r}$ denote the position vector $\mathbf{x}-\mathbf{z}$ of the placement field $\mathbf{x} = \chi_t$ with respect to a fixed point $\mathbf{z}$ in the region $\chi_t(\mathscr{P})$. As in the previous sections we specify the terms appearing in (8.1) and (8.5), Ψ now being the moment of momentum tensor $\mathbf{M}(\mathscr{P})$ of a subbody $\mathscr{P}$. In the equations below $\wedge$ denotes the exterior tensor product, defined by

$$\mathbf{a}\wedge\mathbf{b} = \mathbf{a}\otimes\mathbf{b}-\mathbf{b}\otimes\mathbf{a}.$$

We have

$$\begin{aligned}
\Psi_t \equiv \mathbf{M}_t(\mathscr{P}) &= \int_{\chi_t(\mathscr{P})-\mathscr{S}(t)} \mathbf{r}\wedge\varrho\mathbf{v}\,dv + \int_{\mathscr{S}(t)} \mathbf{r}\wedge\varrho_{\mathscr{S}}\mathbf{V}\,da,\\
P(\mathbf{M}_t(\mathscr{P})) &= \int_{\chi_t(\mathscr{P})-\mathscr{S}(t)} \mathbf{r}\wedge\mathbf{p}_{\mathbf{F}}\,dv + \int_{\mathscr{S}(t)} \mathbf{r}\wedge\mathbf{p}_{\mathbf{F}\mathscr{S}}\,da,\\
W(\mathbf{M}_t(\mathscr{P})) &= -\int_{\partial\chi_t(\mathscr{P})-\mathscr{S}(t)} \mathbf{r}\wedge\mathbf{t}\,da - \int_{\partial\mathscr{S}(t)} \mathbf{r}\wedge\mathbf{t}_{\mathscr{L}}\,dl,\\
R(\mathbf{M}_t(\mathscr{P})) &= \int_{\chi_t(\mathscr{P})-\mathscr{S}(t)} \mathbf{r}\wedge\varrho\mathbf{b}\,dv + \int_{\mathscr{S}(t)} \mathbf{r}\wedge\varrho_{\mathscr{S}}\mathbf{b}_{\mathscr{S}}\,da.
\end{aligned} \tag{10.37}$$

In these representations we do not account for the existence of moments acting on $\mathscr{P}$ other than the rotational momentum arising from the momentum $\mathbf{F}(\mathscr{P})$. The theory therefore deals only with so-called **non-polar** continua.

Before formulating the balance law for the moment of momentum we examine the representation for $W(\mathbf{M}_t(\mathscr{P}))$. First, we may write

$$\mathbf{r}\wedge\mathbf{t} \equiv \mathbf{r}\wedge\mathbf{Tm} = \mathbf{r}\otimes\mathbf{Tm}-\mathbf{Tm}\otimes\mathbf{r} = \{\mathbf{r}\otimes\mathbf{T}-(\mathbf{r}\otimes\mathbf{T})^{(1\overset{\mathsf{T}}{,}2)}\}\mathbf{m}.$$

Similarly

$$\mathbf{r}\wedge\mathbf{t}_{\mathscr{L}} \equiv \mathbf{r}\wedge\mathbf{S}\tilde{\mathbf{n}} = \{\mathbf{r}\otimes\mathbf{S}-(\mathbf{r}\otimes\mathbf{S})^{(1\overset{\mathsf{T}}{,}2)}\}\tilde{\mathbf{n}}.$$

In the above identities the symbol $(1\overset{\mathsf{T}}{,}2)$ denotes transposition of the first and second indices of the third-order tensors $\mathbf{r}\otimes\mathbf{T}$ and $\mathbf{r}\otimes\mathbf{S}$.

Substituting the representation (10.37) into the local form of the balance law, namely (8.12), we obtain

$$\frac{\partial}{\partial t}(\mathbf{r}\wedge\varrho\mathbf{v})+\operatorname{div}\{(\mathbf{r}\wedge\varrho\mathbf{v})\otimes\mathbf{v}-\mathbf{r}\otimes\mathbf{T}+(\mathbf{r}\otimes\mathbf{T})^{(1\overset{\mathsf{T}}{,}2)}\} = \mathbf{r}\wedge\mathbf{p}_{\mathbf{F}}+\mathbf{r}\wedge\varrho\mathbf{b}.$$

In the differentiation of the second term we make use of the identities

$$\begin{aligned}
&\operatorname{div}(\mathbf{r}\otimes\mathbf{T}) = \mathbf{r}\otimes\operatorname{div}\mathbf{T}+(\operatorname{grad}\mathbf{r})\mathbf{T}^{\mathsf{T}},\\
&\operatorname{div}\{(\mathbf{r}\otimes\mathbf{T})^{(1\overset{\mathsf{T}}{,}2)}\} = \{\operatorname{div}(\mathbf{r}\otimes\mathbf{T})\}^{\mathsf{T}},\\
&\operatorname{div}\{(\mathbf{r}\wedge\varrho\mathbf{v})\otimes\mathbf{v}\} = \mathbf{r}\wedge\operatorname{div}(\varrho\mathbf{v}\otimes\mathbf{v}),
\end{aligned}$$

where $\operatorname{grad}\mathbf{r}$ should be replaced by the identity tensor $\mathbf{1}$.

Using the above together with (10.28) we obtain finally

$$\mathbf{T}^{\mathsf{T}} = \mathbf{T}. \tag{10.38}$$

This is called Cauchy's second law of motion. In the notation (10.37) the local balance law (8.16) on the p-material surface takes the form

$$\begin{aligned}\mathbf{r}\wedge\varrho_{\mathscr{S}}\dot{\mathbf{V}} &= \operatorname{div}_{\mathscr{S}}(\mathbf{r}\wedge\mathbf{S})+\mathbf{r}\wedge\varrho_{\mathscr{S}}\mathbf{b}_{\mathscr{S}}+\mathbf{r}\wedge\mathbf{p}_{F\mathscr{S}}\\ &\quad+\mathbf{r}\wedge[\![\varrho(v_n-u_n)(\mathbf{v}-\mathbf{V})]\!]-\mathbf{r}\wedge[\![\mathbf{T}]\!]\mathbf{n}.\end{aligned}$$

If mass is conserved, application of (10.29) and (10.33) reduces the above to

$$(\operatorname{grad}_{\mathscr{S}}\mathbf{r})\mathbf{S}^{\mathsf{T}} = \mathbf{S}(\operatorname{grad}_{\mathscr{S}}\mathbf{r})^{\mathsf{T}}, \tag{10.39}$$

where

$$\operatorname{grad}_{\mathscr{S}}\mathbf{r} = \boldsymbol{\varphi}_{,\alpha}\otimes\boldsymbol{\varphi}^{\alpha},$$

with the surface described by (1.1).

Assume now that the surface stress $\mathbf{S}$ is represented by

$$\mathbf{S} = \mathbf{n}\otimes\mathbf{S}_{(n)}+\boldsymbol{\varphi}_{,\beta}\otimes\mathbf{S}^{\beta}, \tag{10.40}$$

where $\mathbf{S}_{(n)}$ and $\mathbf{S}^{\beta}$ form a set of three vectors tangential to $\mathscr{S}(t)$. In index notation (10.40) may be written

$$S^{k\alpha} = n^k S^{\alpha}_{(n)}+\varphi^k_{,\beta}S^{\alpha\beta}.$$

Substituting (10.40) into (10.39) enables us to deduce for the p-material surface

$$S^{\alpha}_{(n)} = 0, \qquad S^{\alpha\beta} = S^{\beta\alpha}. \tag{10.41}$$

Proposition 10.3

If the balance laws of mass and linear momentum hold then the law of balance of moment of momentum is equivalent to (10.38) together with (10.41). □

10.4 Balance of Energy

We begin with the representation for the energy $E_t(\mathscr{P})$ of a subbody $\mathscr{P}$ in the motion χ_t, $t\in I$, together with the energy production, efflux and supply from (8.5)

$$\begin{aligned}\Psi_t \equiv E_t(\mathscr{P}) &= \int_{\chi_t(\mathscr{P})-\mathscr{S}(t)} \varrho(e+\tfrac{1}{2}\mathbf{v}\cdot\mathbf{v})\,dv\\ &\quad+\int_{\mathscr{S}(t)} \varrho_{\mathscr{S}}(e_{\mathscr{S}}+\tfrac{1}{2}\mathbf{V}\cdot\mathbf{V})\,da,\\ P(E_t(\mathscr{P})) &= \int_{\chi_t(\mathscr{P})-\mathscr{S}(t)} p_e\,dv+\int_{\mathscr{S}(t)} p_{e\mathscr{S}}\,da,\\ W(E_t(\mathscr{P})) &= \int_{\partial\chi_t(\mathscr{P})-\partial\mathscr{S}(t)} (q-\mathbf{t}\cdot\mathbf{v})\,da+\int_{\partial\mathscr{S}(t)} (q_{\mathscr{L}}-\mathbf{t}_{\mathscr{L}}\cdot\mathbf{V})\,dl,\\ R(E_t(\mathscr{P})) &= \int_{\chi_t(\mathscr{P})-\mathscr{S}(t)} \varrho(r_e+\mathbf{b}\cdot\mathbf{v})\,dv\\ &\quad+\int_{\mathscr{S}(t)} \varrho_{\mathscr{S}}(r_{e\mathscr{S}}+\mathbf{b}_{\mathscr{S}}\cdot\mathbf{V})\,da.\end{aligned} \tag{10.42}$$

Here, e is the density of internal energy per unit mass, $e_{\mathscr{S}}$ per unit surface mass; p_e is the volumetric energy production with the region $\boldsymbol{\chi}_t(\mathscr{P})-\mathscr{S}(t)$ and $p_{\mathscr{S}}$ is the surface energy production on $\mathscr{S}(t)$; q and $q_{\mathscr{L}}$ respectively are the densities of energy and surface energy flux transmitted by the boundaries $\partial\boldsymbol{\chi}_t(\mathscr{P})-\mathscr{S}(t)$ and $\partial\mathscr{S}(t)$; r_e and $r_{e\mathscr{S}}$ are the densities of the volumetric and surface energy sources respectively; $\frac{1}{2}\varrho\mathbf{v}\cdot\mathbf{v}$ is the density of kinetic energy and $\frac{1}{2}\varrho_{\mathscr{S}}\mathbf{V}\cdot\mathbf{V}$ the density of surface kinetic energy; $\mathbf{t}\cdot\mathbf{v}$ is the density of mechanical energy flux over the boundary $\partial\boldsymbol{\chi}_t(\mathscr{P})-\mathscr{S}(t)$ and $\mathbf{t}_{\mathscr{L}}\cdot\mathbf{V}$ the density of mechanical surface energy flux transmitted by the boundary $\partial\mathscr{S}(t)$ of $\mathscr{S}(t)$; $\varrho\mathbf{b}\cdot\mathbf{v}$ and $\varrho_{\mathscr{S}}\mathbf{b}_{\mathscr{S}}\cdot\mathbf{V}$ are the power densities of the volumetric and surface forces, $\varrho\mathbf{b}$ and $\varrho_{\mathscr{S}}\mathbf{b}_{\mathscr{S}}$, respectively.

In accordance with Hypothsis 8.3 we adopt the representations

$$q = \mathbf{q}\cdot\mathbf{m} \equiv q^k m_k, \qquad q_{\mathscr{L}} = \mathbf{q}_{\mathscr{S}}\cdot\tilde{\mathbf{n}} \equiv q^{\alpha}_{\mathscr{S}}\tilde{n}_{\alpha},$$

and it follows that

$$q-\mathbf{t}\cdot\mathbf{v} = (\mathbf{q}-\mathbf{v}\mathbf{T})\cdot\mathbf{m}, \qquad q_{\mathscr{L}}-\mathbf{t}_{\mathscr{L}}\cdot\mathbf{V} = (\mathbf{q}_{\mathscr{S}}-\mathbf{V}\mathbf{S})\cdot\tilde{\mathbf{n}}.$$

In local form the energy balance law may be written as (see (8.12) and (8.17))

$$\begin{aligned}
&\frac{\partial}{\partial t}\{\varrho(e+\tfrac{1}{2}\mathbf{v}\cdot\mathbf{v})\}+\operatorname{div}\{\varrho(e+\tfrac{1}{2}\mathbf{v}\cdot\mathbf{v})\mathbf{v}+\mathbf{q}-\mathbf{v}\mathbf{T}\}\\
&= p_e+\varrho(r_e+\mathbf{b}\cdot\mathbf{v}) \quad \text{in } \boldsymbol{\chi}_t(\mathscr{P})-\mathscr{S}(t),\\
&p_{e\mathscr{S}}+\varrho_{\mathscr{S}}(r_{e\mathscr{S}}+\mathbf{b}_{\mathscr{S}}\cdot\mathbf{V})+[\![\varrho(e+\tfrac{1}{2}\mathbf{v}\cdot\mathbf{v})(v_n-u_n)+(\mathbf{q}-\mathbf{v}\mathbf{T})\cdot\mathbf{n}]\!]\\
&= \frac{\delta}{\delta t}\{\varrho_{\mathscr{S}}(e_{\mathscr{S}}+\tfrac{1}{2}\mathbf{V}\cdot\mathbf{V})\}+\{\varrho_{\mathscr{S}}(e_{\mathscr{S}}+\tfrac{1}{2}\mathbf{V}\cdot\mathbf{V})(V^{\alpha}+c^{\alpha})\\
&\quad+q^{\alpha}_{\mathscr{S}}-V^k S^{\alpha}_k\}_{;\alpha}-\varrho_{\mathscr{S}}(e_{\mathscr{S}}+\tfrac{1}{2}\mathbf{V}\cdot\mathbf{V})2u_nK_m \quad \text{on } \mathscr{S}(t).
\end{aligned} \tag{10.43}$$

On use of the previous balance laws we may deduce the following from (10.43):

Proposition 10.4

If the equations of momentum and moment of momentum balance, (10.28) and (10.38), are satisfied for points of the region $\boldsymbol{\chi}_t(\mathscr{P})-\mathscr{S}(t)$, then the energy balance law $(10.43)_1$ is equivalent to either one of the following two equations:

$$\begin{aligned}
&\left\{\frac{\partial\varrho}{\partial t}+\operatorname{div}(\varrho\mathbf{v})\right\}(e-\tfrac{1}{2}\mathbf{v}\cdot\mathbf{v})+\varrho\left(\frac{\partial e}{\partial t}+\mathbf{v}\cdot\operatorname{grad}e\right)\\
&= \operatorname{tr}\{(\operatorname{grad}\mathbf{v})\mathbf{T}^{\mathrm{T}}\}-\operatorname{div}\mathbf{q}+\varrho r_e-\mathbf{p}_{\mathbf{F}}\cdot\mathbf{v}+p_e,\\
&(p_{\varrho}+r_{\varrho}-\operatorname{div}\mathbf{w}_{\varrho})(e-\tfrac{1}{2}\mathbf{v}\cdot\mathbf{v})+\varrho\left(\frac{\partial e}{\partial t}+\mathbf{v}\cdot\operatorname{grad}e\right)\\
&= \operatorname{tr}\{(\operatorname{grad}\mathbf{v})\mathbf{T}^{\mathrm{T}}\}-\operatorname{div}\mathbf{q}+\varrho r_e-\mathbf{p}_{\mathbf{F}}\cdot\mathbf{v}+p_e. \;\square
\end{aligned} \tag{10.44}$$

Note that if mass is conserved then the first term in each equation vanishes.

We now examine the final two terms in each equation in (10.44). If it is assumed that the energy production p_e in $\chi_t(\mathscr{P})-\mathscr{S}(t)$ arises solely from the production of momentum $\mathbf{p_F}$ then

$$p_e = \mathbf{p_F}\cdot\mathbf{v}. \tag{10.45}$$

This leads to

Corollary 10.8

If the momentum and moment of momentum balance equations, (10.28) and (10.38), are satisfied for points in the region $\chi_t(\mathscr{P})-\mathscr{S}(t)$, and the mass conservation equation $(10.8)_1$ holds, then the energy balance equation takes the form

$$\varrho\dot{e} = \mathrm{tr}\,\{(\mathbf{grad\,v})\mathbf{T}^{\mathrm{T}}\}-\mathrm{div}\,\mathbf{q}+\varrho r_e \tag{10.46}$$

if the identity (10.45) is satisfied. The symbol $\dot{e}$ denotes the derivative $\partial e/\partial t|_{\mathbf{X}=\mathrm{const}} = \partial e/\partial t|_{\mathbf{x}=\mathrm{const}}+\mathbf{v}\cdot\mathbf{grad}\,e$. □

It should be pointed out that the simultaneous vanishing of the mass source r_ϱ, the production p_ϱ and the flux w_ϱ is one of the conditions ensuring that mass is conserved (see the remark following Corollary 10.2).

It remains to derive a reduced form of energy balance for the surface. With the time derivative $\dot{e}_{\mathscr{S}}$ defined by

$$\dot{e}_{\mathscr{S}} \equiv \left.\frac{\partial e_{\mathscr{S}}}{\partial t}\right|_{L^{\Delta}=\mathrm{const}} = \left.\frac{\partial e_{\mathscr{S}}}{\partial t}\right|_{l^{\alpha}=\mathrm{const}}+e_{\mathscr{S},\alpha}V^{\alpha} = \frac{\delta e_{\mathscr{S}}}{\delta t}+e_{\mathscr{S},\alpha}(V^{\alpha}+c^{\alpha})$$

we have

Proposition 10.5

If the balance laws of mass, momentum and moment of momentum are satisfied, then the local form of the energy balance law on the surface $\mathscr{S}(t)$ is

$$\begin{aligned}
&\varrho_{\mathscr{S}}\dot{e}_{\mathscr{S}}+\mathrm{div}_{\mathscr{S}}\mathbf{q}_{\mathscr{S}}-\mathrm{tr}\,\{(\mathbf{grad}_{\mathscr{S}}\mathbf{V})\mathbf{S}^{\mathrm{T}}\}-\varrho_{\mathscr{S}}r_{\varrho\mathscr{S}}\\
&= [\![\varrho\,\{e+\tfrac{1}{2}(\mathbf{v}-\mathbf{V})\cdot(\mathbf{v}-\mathbf{V})\}(v_n-u_n)]\!]+[\![\mathbf{q}-(\mathbf{v}-\mathbf{V})\mathbf{T}]\!]\cdot\mathbf{n}\\
&\quad-[\![\varrho e_{\mathscr{S}}(v_n-u_n)]\!]+p_{e\mathscr{S}}-\mathbf{p}_{\mathbf{F}\mathscr{S}}\cdot\mathbf{V}\\
&\quad+(\tfrac{1}{2}\mathbf{V}\cdot\mathbf{V}-e_{\mathscr{S}})(p_{\varrho\mathscr{S}}+r_{\varrho\mathscr{S}}+[\![\mathbf{w}_\varrho]\!]\cdot\mathbf{n}-\mathrm{div}_{\mathscr{S}}\mathbf{w}_{\varrho\mathscr{S}}).
\end{aligned} \tag{10.47}$$

Proof

Performing the differentiation in $(10.43)_2$ and regrouping the terms, we obtain

$$\begin{aligned}
&\left\{\frac{\delta}{\delta t}\varrho_{\mathscr{S}}+\left(\varrho_{\mathscr{S}}(V^{\alpha}+c^{\alpha})\right)_{;\alpha}-2u_n\varrho_{\mathscr{S}}K_m\right\}(e_{\mathscr{S}}+\tfrac{1}{2}\mathbf{V}\cdot\mathbf{V})\\
&+\varrho_{\mathscr{S}}\left\{\frac{\delta e_{\mathscr{S}}}{\delta t}+e_{\mathscr{S},\alpha}(V^{\alpha}+c^{\alpha})\right\}+\mathbf{V}\cdot(\varrho_{\mathscr{S}}\dot{\mathbf{V}}-\mathrm{div}_{\mathscr{S}}\mathbf{S}-\varrho_{\mathscr{S}}\mathbf{b}_{\mathscr{S}})\\
&+q^{\alpha}_{\mathscr{S};\alpha}-V^{k}_{;\alpha}S^{\alpha}_{k} = [\![\varrho(e+\tfrac{1}{2}\mathbf{v}\cdot\mathbf{v})(v_n-u_n)]\!]+[\![\mathbf{q}-\mathbf{v}\mathbf{T}]\!]\cdot\mathbf{n}+p_{e\mathscr{S}}+\varrho r_{e\mathscr{S}}.
\end{aligned}$$

On the use of (10.6) and (10.33) this reduces to the required form. □

Corollary 10.9

If the assumptions of Proposition 10.5 hold and, moreover, mass is conserved, then the surface energy balance equation becomes

$$\begin{aligned}&\varrho_{\mathscr{S}}\dot{e}_{\mathscr{S}}+\operatorname{div}_{\mathscr{S}}\mathbf{q}_{\mathscr{S}}-\operatorname{tr}\{(\operatorname{grad}_{\mathscr{S}}\mathbf{V})\mathbf{S}^{\mathsf{T}}\}-\varrho_{\mathscr{S}}r_{e\mathscr{S}}\\&=[\![\varrho\{e-e_{\mathscr{S}}+\tfrac{1}{2}(\mathbf{v}-\mathbf{V})\cdot(\mathbf{v}-\mathbf{V})\}(v_n-u_n)]\!]\\&\quad+[\![\mathbf{q}-(\mathbf{v}-\mathbf{V})\mathbf{T}]\!]\cdot\mathbf{n}+p_{e\mathscr{S}}-\mathbf{p}_{F\mathscr{S}}\cdot\mathbf{V}.\ \square\end{aligned}\tag{10.48}$$

As in the discussion of the balance law for points of $\boldsymbol{\chi}_t(\mathscr{P})-\mathscr{S}(t)$ we may demand that the surface energy production arises from the momentum production on the surface, so that

$$p_{e\mathscr{S}}=\mathbf{p}_{F\mathscr{S}}\cdot\mathbf{V}.\tag{10.49}$$

Then the last two terms in (10.48) drop out. The resulting equation is called the principle of energy balance for the surface.

It should be stressed that the conditions (10.45) and (10.49) are only assumptions, although based on practical observations. However, it remains possible that in certain physical phenomena factors other than the momentum production will contribute to the energy production.

Finally, we consider the case of a material surface $\{\mathscr{S}(t)\}_{t\in I}$. Then

$$u_n=\mathbf{V}\cdot\mathbf{n}=\mathbf{v}\cdot\mathbf{n}$$

and, moreover, the velocity $\mathbf{V}$ of surface points is equal to $\dot{\boldsymbol{\varphi}}$ (see (10.19)). If (10.49) holds the energy balance equation (10.48) simplies to

$$\begin{aligned}&\varrho_{\mathscr{S}}\dot{e}_{\mathscr{S}}+\operatorname{div}_{\mathscr{S}}\mathbf{q}_{\mathscr{S}}-\operatorname{tr}\{(\operatorname{grad}_{\mathscr{S}}\mathbf{V})\mathbf{S}^{\mathsf{T}}\}-\varrho_{\mathscr{S}}r_{e\mathscr{S}}\\&=[\![\mathbf{q}-(\mathbf{v}-\dot{\boldsymbol{\varphi}})\mathbf{T}]\!]\cdot\mathbf{n}.\end{aligned}\tag{10.50}$$

When the material surface is the boundary of a subbody $\mathscr{P}$ we have*

$$\partial\boldsymbol{\varkappa}_0(\mathscr{P})\supset\mathscr{S}_{\mathscr{P}},\qquad\mathscr{S}(t)=\hat{\boldsymbol{\chi}}_t(\mathscr{S}_{\mathscr{P}})=\partial\boldsymbol{\chi}_t(\mathscr{P})\cap\partial\boldsymbol{\chi}_t(\mathscr{B}),$$

and $\boldsymbol{\chi}_t(\mathscr{P})-\mathscr{S}(t)$ coincides with $\boldsymbol{\chi}_t(\mathscr{P})$. In parallel with the discussion of the equation of momentum balance for a boundary (Section 10.2), we may write

$$[\![\mathbf{q}-(\mathbf{v}-\dot{\boldsymbol{\varphi}})\mathbf{T}]\!]=\mathbf{q}|_{\partial\boldsymbol{\chi}_t(\mathscr{P})}$$

in view of (10.21).

Instead of (10.50) we now have

$$\varrho_{\mathscr{S}}\dot{e}_{\mathscr{S}}+\operatorname{div}_{\mathscr{S}}\mathbf{q}_{\mathscr{S}}-\operatorname{tr}\{(\operatorname{grad}_{\mathscr{S}}\mathbf{V})\mathbf{S}^{\mathsf{T}}\}-\varrho_{\mathscr{S}}r_{e\mathscr{S}}=\mathbf{q}\cdot\mathbf{n}\tag{10.51}$$

on the boundary $\partial\boldsymbol{\chi}_t(\mathscr{P})$, where $\mathbf{V}=\dot{\boldsymbol{\chi}}$.

The above results are covered by

---

* As in Section 10.3, $\hat{\boldsymbol{\chi}}$ needs to be extended to the set $\boldsymbol{\varkappa}_0(\mathscr{P})\cup\partial\boldsymbol{\varkappa}\,(\mathscr{P})$.

Corollary 10.10

If $\{\mathscr{S}(t)\}_{t\in I}$ is a material surface and the conditions of Corollary 10.9 are satisfied, then the surface energy balance equation has the form (10.50). Moreover, if the surface is the image of the boundary of a subbody $\mathscr{P}$ in the motion $\hat{\chi}_t$, then equation (10.51) is the energy balance equation for the boundary. □

We now return to the general situation governed by (10.43). If the mass $M_t(\mathscr{P})$, momentum $\mathbf{F}_t(\mathscr{P})$ and energy $E_t(\mathscr{P})$ are conserved in the region $\chi_t(\mathscr{P})$ during the motion χ_t which began at time $t = 0$ with $\det \mathbf{F}(0, \mathbf{x}) = 1$, then the following relations holds:

$$\varrho\dot{\varepsilon}+J^{-1}m_\varrho(e-\tfrac{1}{2}\mathbf{v}\cdot\mathbf{v})+\mathbf{p}_\mathrm{F}\cdot\mathbf{v}$$
$$= \mathrm{tr}\,\{(\mathrm{grad}\,\mathbf{v})\mathbf{T}^\mathrm{T}\}-\mathrm{div}\,\mathbf{q}+\varrho r_e+p_e \quad \text{in } \chi_t(\mathscr{P})-\mathscr{S}(t),$$
$$[\![\varrho(e+\tfrac{1}{2}\mathbf{v}\cdot\mathbf{v})(v_n-u_n)+(\mathbf{q}-\mathbf{v}\mathbf{T})\cdot\mathbf{n}]\!] = m_e \quad \text{on } \mathscr{S}(t)\cap\chi_t(\mathscr{P}),$$

where

$$m_e := (e_\mathscr{S}-\tfrac{1}{2}\mathbf{V}\cdot\mathbf{V})(m_{\varrho\mathscr{S}}+p_{\varrho\mathscr{S}}+r_{\varrho\mathscr{S}}-\mathrm{div}_\mathscr{S}\mathbf{w}_{\varrho\mathscr{S}})+\mathbf{m}_\mathrm{F}\cdot\mathbf{V}$$
$$+\varrho_\mathscr{S}\dot{e}_\mathscr{S}+\mathrm{div}_\mathscr{S}\mathbf{q}_\mathscr{S}-\mathrm{tr}\,\{(\mathrm{grad}_\mathscr{S}\mathbf{V})\mathbf{S}^\mathrm{T}\}+\mathbf{p}_{\mathrm{F}\mathscr{S}}\cdot\mathbf{V}-p_{e\mathscr{S}}-\varrho_\mathscr{S}r_{e\mathscr{S}}.$$

Energy is conserved in $\chi_t(\mathscr{P})$ as a whole if

$$\int_{\chi_t(\mathscr{P})-\mathscr{S}(t)} p_e\,\mathrm{d}v+\int_{\mathscr{S}(t)\cap\chi_t(\mathscr{P})} m_e\,\mathrm{d}a = 0.$$

If $\mathscr{P}$ is replaced by the whole body $\mathscr{B}$ in this equation, and p_e and m_e do not vanish simultaneously, then the last three equations above can be treated as statements of the non-local theory in which there is a net transfer of energy from one subbody to the other but balance of energy holds for the body as a whole.

In the special case in which $\varrho_\mathscr{S} = 0$ on $\mathscr{S}(t)$ we obtain

$$m_{\varrho\mathscr{S}}+p_{\varrho\mathscr{S}}+r_{\varrho\mathscr{S}}-\mathrm{div}_\mathscr{S}\mathbf{w}_{\varrho\mathscr{S}} = 0$$

and hence

$$\mathbf{m}_\mathrm{F} = -\mathrm{div}_\mathscr{S}\mathbf{S}-\mathbf{p}_{\mathrm{F}\mathscr{S}}$$

and

$$m_e = \mathrm{div}_\mathscr{S}(\mathbf{q}_\mathscr{S}-\mathbf{V}\mathbf{S})-p_{e\mathscr{S}}.$$

We can interpret this result as follows: if there is no surface concentration of mass then the only effects manifest on the surface are due to the surface tension and the surface heat flux provided $\mathbf{p}_{\mathrm{F}\mathscr{S}}$ and $p_{e\mathscr{S}}$ are zero.

We now consider under what conditions the energy balance equation may be used to obtain the local forms of the mass, momentum and moment of momentum balance equations. It does not seem possible that the Noether theorem of classical mechanics can be applied here; nor can it to the results of Green and Rivlin (1964). In the absence of production and surface effects, however, Green and Rivlin obtained the above-mentioned balance equations in their classical forms by making use of the invariance of the local and global energy equations under Euclidean transformations. In the present approach the arbitrariness of the terms defining the production of thermodynamic quantities does not allows us to repeat the argument used by Green and Rivlin. But if we adopt explicit relations between p_ϱ, p_e and $\mathbf{p_F}$ we may obtain results similar to those mentioned above (see Kosiński, 1983, 1986).

10.5 Entropy Balance

It is not our aim to discuss the existence of entropy for a material continuum. It should be pointed out, however, that there are numerous proofs of the existence of entropy for particular classes of material media (see Section 10.6 for references).

In order to achieve the required result we shall assume the existence of the following thermodynamic quantities for a subbody $\mathscr{P}$:

the entropy $N_t(\mathscr{P})$ with densities $\varrho\eta$ and $\varrho_{\mathscr{S}}\eta_{\mathscr{S}}$,

$$\Psi_t \equiv N_t(\mathscr{P}) = \int_{\chi_t(\mathscr{P})-\mathscr{S}(t)} \varrho\eta \,\mathrm{d}v + \int_{\mathscr{S}(t)} \varrho_{\mathscr{S}}\eta_{\mathscr{S}} \,\mathrm{d}a;$$

the entropy production $P(N_t(\mathscr{P}))$ with densities p_η and $p_{\eta\mathscr{S}}$,

$$P(N_t(\mathscr{P})) = \int_{\chi_t(\mathscr{P})-\mathscr{S}(t)} p_\eta \,\mathrm{d}v + \int_{\mathscr{S}(t)} p_{\eta\mathscr{S}} \,\mathrm{d}a; \tag{10.52}$$

the entropy flux $W(N_t(\mathscr{P}))$ with densities m_η and $m_{\mathscr{L}}$,

$$W(N_t(\mathscr{P})) = \int_{\partial\chi_t(\mathscr{P})-\partial\mathscr{S}(t)} m_\eta \,\mathrm{d}a + \int_{\partial\mathscr{S}(t)} m_{\mathscr{L}} \,\mathrm{d}l;$$

and the entropy sources $R(N_t(\mathscr{P}))$ with densities ξ and $\xi_{\mathscr{S}}$,

$$R(N_t(\mathscr{P})) = \int_{\chi_t(\mathscr{P})-\mathscr{S}(t)} \xi \,\mathrm{d}v + \int_{\mathscr{S}(t)} \xi_{\mathscr{S}} \,\mathrm{d}a.$$

For the flux densities we assume the representations

$$m_\eta = \mathbf{m}_\eta \cdot \mathbf{m}, \qquad m_{\mathscr{L}} = \mathbf{m}_{\eta\mathscr{S}} \cdot \tilde{\mathbf{n}}.$$

From (8.12) and (8.16) we deduce that the local form of the entropy balance law may be written

$$\frac{\partial}{\partial t}(\varrho\eta) + \mathrm{div}(\varrho\eta\mathbf{v} + \mathbf{m}_\eta) = p_\eta + \xi \quad \text{in } \chi_t(\mathscr{P}) - \mathscr{S}(t),$$

$$\frac{\delta}{\delta t}(\varrho_{\mathscr{S}}\eta_{\mathscr{S}})+\{\varrho_{\mathscr{S}}\eta_{\mathscr{S}}(V^{\alpha}+c^{\alpha})+m^{\alpha}_{\eta\mathscr{S}}\}_{;\alpha}-2\varrho_{\mathscr{S}}\eta_{\mathscr{S}}u_n K_m$$
$$= [\![\varrho\eta(v_n-u_n)]\!]+[\![\mathbf{m}_\eta]\!]\cdot\mathbf{n}+p_{\eta\mathscr{S}}+\xi_{\mathscr{S}} \quad \text{on } \mathscr{S}(t).$$

The quantities in the first equation are treated as functions of t and $\mathbf{x}$ and those in the second equation as functions of t and l^{α}.

We make use of the notations

$$\dot{\eta} := \left.\frac{\partial\eta}{\partial t}\right|_{\mathbf{X}=\text{const}} = \left.\frac{\partial\eta}{\partial t}\right|_{\mathbf{x}=\text{const}} + \mathbf{v}\cdot\operatorname{grad}\eta,$$

$$\dot{\eta}_{\mathscr{S}} := \left.\frac{\partial\eta_{\mathscr{S}}}{\partial t}\right|_{L^{A}=\text{const}} = \left.\frac{\partial\eta_{\mathscr{S}}}{\partial t}\right|_{l^{\alpha}=\text{const}} + \eta_{\mathscr{S},\alpha}V^{\alpha}.$$

Proposition 10.6

The entropy balance law is equivalent to the equations

$$\left\{\frac{\partial\varrho}{\partial t}+\operatorname{div}(\varrho\mathbf{v})\right\}\eta+\varrho\dot{\eta}+\operatorname{div}\mathbf{m}_\eta = p_\eta+\xi \quad \text{in } \chi_t(\mathscr{P})-\mathscr{S}(t),$$

$$\left\{\frac{\delta\varrho_{\mathscr{S}}}{\delta t}+((\varrho_{\mathscr{S}}(V^{\alpha}+c^{\alpha}))_{;\alpha}-2\varrho_{\mathscr{S}}u_n K_m\right\}\eta_{\mathscr{S}}+\varrho_{\mathscr{S}}\dot{\eta}_{\mathscr{S}} \tag{10.53}$$
$$+\operatorname{div}_{\mathscr{S}}\mathbf{m}_{\eta\mathscr{S}} = [\![\varrho\eta(v_n-u_n)]\!]+[\![\mathbf{m}_\eta]\!]\cdot\mathbf{n}$$
$$+p_{\eta\mathscr{S}}+\xi_{\mathscr{S}} \quad \text{on } \mathscr{S}(t). \ \square$$

Use of the mass balance law then enables us to obtain

$$\varrho\dot{\eta}+\operatorname{div}\mathbf{m}_\eta = p_\eta+\xi-\eta(p_\varrho+r_\varrho-\operatorname{div}\mathbf{w}_\varrho),$$
$$\varrho_{\mathscr{S}}\dot{\eta}_{\mathscr{S}}+\operatorname{div}_{\mathscr{S}}\mathbf{m}_{\eta\mathscr{S}} = [\![\varrho(\eta-\eta_{\mathscr{S}})(v_n-u_n)]\!]+[\![\mathbf{m}_\eta-\eta_{\mathscr{S}}\mathbf{w}_\varrho]\!]\cdot\mathbf{n}$$
$$+p_{\eta\mathscr{S}}+\xi_{\mathscr{S}}-\eta_{\mathscr{S}}(p_{\varrho\mathscr{S}}+r_{\varrho\mathscr{S}}-\operatorname{div}_{\mathscr{S}}\mathbf{w}_{\varrho\mathscr{S}}).$$

Taking into account equations (10.5), (10.6) and (10.8) we obtain

Corollary 10.11

If mass is conserved then balance of entropy implies that

$$\varrho\dot{\eta}+\operatorname{div}\mathbf{m}_\eta = p_\eta+\xi \quad \text{in } \chi_t(\mathscr{P})-\mathscr{S}(t),$$
$$\varrho_{\mathscr{S}}\dot{\eta}_{\mathscr{S}}+\operatorname{div}_{\mathscr{S}}\mathbf{m}_{\eta\mathscr{S}} = [\![\varrho(\eta-\eta_{\mathscr{S}})(v_n-u_n)]\!]+[\![\mathbf{m}_\eta]\!]\cdot\mathbf{n} \tag{10.54}$$
$$+p_{\eta\mathscr{S}}+\xi_{\mathscr{S}} \quad \text{on } \mathscr{S}(t). \ \square$$

In the case of a material surface we have (see Corollaries 10.7 and 10.10)

Corollary 10.12

If $\{\mathscr{S}(t)\}_{t\in I}$ is a moving material surface and mass is conserved, then the equation of entropy balance on the surface is

$$\varrho_{\mathscr{S}}\dot{\eta}_{\mathscr{S}}+\operatorname{div}\mathbf{m}_{\eta\mathscr{S}} = [\![\mathbf{m}_\eta]\!]\cdot\mathbf{n}+p_{\eta\mathscr{S}}+\xi_{\mathscr{S}}. \tag{10.55}$$

Moreover, if the surface is the image of the boundary $\mathscr{S}_{\mathscr{P}}$ of a subbody $\mathscr{P}$ in the motion $\boldsymbol{\chi}_t$ the entropy balance equation becomes

$$\varrho_{\mathscr{S}}\dot{\eta}_{\mathscr{S}}+\operatorname{div}\mathbf{m}_{\eta\mathscr{S}}-\mathbf{m}_\eta\cdot\mathbf{n} = p_{\eta\mathscr{S}}+\xi_{\mathscr{S}}. \ \square \tag{10.56}$$

Finally, we consider the situation in which mass $M_t(\mathscr{P})$, momentum $\mathbf{F}_t(\mathscr{P})$, energy $E_t(\mathscr{P})$ and entropy $N_t(\mathscr{P})$ are conserved in the region $\boldsymbol{\chi}_t(\mathscr{P})$. Then the appropriate equations are

$$\varrho\dot{\eta}+J^{-1}m_\varrho\eta+\operatorname{div}\mathbf{m}_\eta = p_\eta+\xi \quad \text{in } \boldsymbol{\chi}_t(\mathscr{P})-\mathscr{S}(t),$$
$$[\![\varrho\eta(v_n-u_n)+\mathbf{m}_\eta\cdot\mathbf{n}]\!]+p_{\eta\mathscr{S}} = w_{\eta\mathscr{S}} \quad \text{on } \mathscr{S}(t)\cap\boldsymbol{\chi}_t(\mathscr{P}),$$

where

$$w_{\eta\mathscr{S}} := \eta_{\mathscr{S}}(m_{\varrho\mathscr{S}}+p_{\varrho\mathscr{S}}+r_{\varrho\mathscr{S}}-\operatorname{div}_{\mathscr{S}}\mathbf{w}_{\varrho\mathscr{S}})+\varrho_{\mathscr{S}}\dot{\eta}_{\mathscr{S}}+\operatorname{div}_{\mathscr{S}}\mathbf{m}_{\eta\mathscr{S}}-\xi_{\mathscr{S}}.$$

The case of entropy balance is rather subtle since the entropy increase within the volume $\boldsymbol{\chi}_t(\mathscr{P})$ may not be equilibrated by the supply terms and the efflux across the boundary. For reversible processes, and those reversible "in the large", that is within $\boldsymbol{\chi}_t(\mathscr{P})$, one can require that

$$\int\limits_{\boldsymbol{\chi}_t(\mathscr{P})-\mathscr{S}(t)} p_\eta\,\mathrm{d}v = 0.$$

However, the entropy balance law for the singular surface is of a more complex nature since it is sometimes the entropy and its non-vanishing concentrated production which characterize the surface. For shock waves in an elastic medium or in an ideal non-viscous fluid, where by definition both media are non-dissipative so that there is no production of entropy in any smooth thermodynamic process, production of entropy appears on the shock. Then the surface representing the shock wave becomes a singular surface in the sense that some thermodynamic quantities possess jump discontinuities while others possess concentrations on it.

Consequently, one can imagine, and one can even find, physical situations in which the integral

$$\int\limits_{\boldsymbol{\chi}_t(\mathscr{P})-\mathscr{S}(t)} p_\eta\,\mathrm{d}v$$

vanishes together with the integrand, but

$$\int\limits_{\mathscr{S}(t)} p_{\eta\mathscr{S}}\,\mathrm{d}a > 0$$

with $w_{\eta\mathscr{S}} = 0$ on $\mathscr{S}(t)$.

Finally, we point out that the entropy balance law is not independent of the other balance laws but arises as a consequence of the other laws coupled with constitutive relations—which describe the material of the body—and the entropy production inequality (see Kosiński, 1983).

10.5.1 Entropy balance in rational thermodynamics

We now consider the specific forms of the entropy flux and entropy sources which are often, though not always, used in the thermodynamics of continuous media (see Section 10.6). It is assumed that there exists a positive field function $\theta(t)$, $t \in I$, in the region $\boldsymbol{\chi}_t(\mathscr{P})$ such that the entropy flux $\mathbf{m}_\eta$ and the entropy source ξ may be expressed in the form

$$\mathbf{m}_\eta = \mathbf{q}\theta^{-1}, \qquad \xi = \varrho r_e \theta^{-1}. \tag{10.57}$$

These equations state that the volumetric source of energy r_e and the heat flux density are directly proportional to the entropy source ξ and the entropy flux density $\mathbf{m}_\eta$ respectively, and that the common coefficient of proportionality for $\varrho r_e/\xi$ and q/m_η is the absolute temperature θ. Similarly, we assumed the existence of a surface temperature $\theta_{\mathscr{S}}(t)$ on a moving surface $\{\mathscr{S}(t)\}_{t\in I}$, and this allows us to determine the surface flux $\mathbf{m}_{\eta\mathscr{S}}$ and surface source $\xi_{\mathscr{S}}$ as follows:

$$\mathbf{m}_{\eta\mathscr{S}} = \mathbf{q}_{\mathscr{S}}\theta_{\mathscr{S}}^{-1}, \qquad \xi_{\mathscr{S}} = \varrho_{\mathscr{S}} r_{e\mathscr{S}}\theta_{\mathscr{S}}^{-1}. \tag{10.58}$$

As a consequence of these definitions we obtain the following result from the balance law (10.54).

Lemma 10.1

If the energy balance equations (10.46) and (10.48) and the condition (10.49) are satisfied, then the entropy balance equations (10.54) coupled with (10.57) and (10.58) yield

$$\begin{aligned}
&\varrho\theta\dot{\eta} + \operatorname{tr}\{(\operatorname{grad}\mathbf{v})\mathbf{T}^{\mathrm{T}}\} - \varrho\dot{e} - \theta^{-1}\mathbf{q}\cdot\operatorname{grad}\theta = \theta p_\eta \quad \text{in } \boldsymbol{\chi}_t(\mathscr{P}) - \mathscr{S}(t),\\
&\varrho_{\mathscr{S}}\theta_{\mathscr{S}}\dot{\eta}_{\mathscr{S}} + \operatorname{tr}\{(\operatorname{grad}_{\mathscr{S}}\mathbf{V})\mathbf{S}^{\mathrm{T}}\} - \varrho_{\mathscr{S}}\dot{e}_{\mathscr{S}} - \theta^{-1}\mathbf{q}_{\mathscr{S}}\cdot\operatorname{grad}_{\mathscr{S}}\theta_{\mathscr{S}}\\
&= [\![\varrho\{\theta_{\mathscr{S}}(\eta - \eta_{\mathscr{S}}) - (e - e_{\mathscr{S}}) - \tfrac{1}{2}(\mathbf{v}-\mathbf{V})\cdot(\mathbf{v}-\mathbf{V})\}(v_n - u_n)]\!]\\
&\quad - \left[\!\!\left[\mathbf{q}\left(1 - \frac{\theta_{\mathscr{S}}}{\theta}\right) - (\mathbf{v}-\mathbf{V})\mathbf{T}\right]\!\!\right]\cdot\mathbf{n} + \theta_{\mathscr{S}} p_{\eta\mathscr{S}} \quad \text{on } \mathscr{S}(t).
\end{aligned} \tag{10.59}$$

Proof

We determine the term $\operatorname{div}\mathbf{q}$ from the law of energy conservation (10.46):

$$\operatorname{div}\mathbf{q} = \operatorname{tr}\{(\operatorname{grad}\mathbf{v})\mathbf{T}^{\mathrm{T}}\} - \varrho\dot{e} + \varrho r_e.$$

By use of (10.57), equation (10.54) becomes

$$\varrho\dot{\eta} + \theta^{-1}\operatorname{div}\mathbf{q} - \theta^{-2}\mathbf{q}\cdot\operatorname{grad}\theta = p_\eta + \theta^{-1}\varrho r_e.$$

On substitution of the expression for $\operatorname{div}\mathbf{q}$ into this we obtain $(10.59)_1$. Equation $(10.59)_2$ is obtained in a similar way. Determining $\operatorname{div}\mathbf{q}_{\mathscr{S}}$ from (10.48), using the assumption (10.49), and inserting the resulting term into $(10.54)_2$ we obtain $(10.59)_2$. □

By analogy with the results of Corollaries (10.7) (10.10) and (10.12), we can write down the entropy balance equation $(10.59)_2$ for a material surface $\mathscr{S}(t)$. Thus

$$\varrho_{\mathscr{S}}\theta_{\mathscr{S}}\dot{\eta}_{\mathscr{S}}+\operatorname{tr}\{(\operatorname{grad}_{\mathscr{S}}\mathbf{V})\mathbf{S}^{\mathsf{T}}\}-\varrho_{\mathscr{S}}\dot{e}_{\mathscr{S}}-\theta^{-1}\mathbf{q}_{\mathscr{S}}\cdot\operatorname{grad}_{\mathscr{S}}\theta_{\mathscr{S}}$$

$$=\theta_{\mathscr{S}}p_{\eta\mathscr{S}}-\left[\!\left[\mathbf{q}\left(1-\frac{\theta_{\mathscr{S}}}{\theta}\right)-(\mathbf{v}-\dot{\boldsymbol{\varphi}})\mathbf{T}\right]\!\right]\cdot\mathbf{n}, \tag{10.60}$$

where $\dot{\boldsymbol{\varphi}}$ is given by (10.19).

If $\mathscr{S}(t)$ is the boundary $\partial\boldsymbol{\chi}_t(\mathscr{P})\cap\partial\boldsymbol{\chi}_t(\mathscr{B})$, then the local law of entropy balance is equivalent to

$$\varrho_{\mathscr{S}}\theta_{\mathscr{S}}\dot{\eta}_{\mathscr{S}}+\operatorname{tr}\{(\operatorname{grad}_{\mathscr{S}}\mathbf{V})\mathbf{S}^{\mathsf{T}}\}-\varrho_{\mathscr{S}}\dot{e}_{\mathscr{S}}-\theta_{\mathscr{S}}^{-1}\mathbf{q}_{\mathscr{S}}\cdot\operatorname{grad}_{\mathscr{S}}\theta_{\mathscr{S}}$$

$$=\left(\frac{\theta_{\mathscr{S}}}{\theta}-1\right)\mathbf{q}\cdot\mathbf{n}+\theta_{\mathscr{S}}p_{\eta\mathscr{S}}, \tag{10.61}$$

where $\mathbf{V}=\dot{\boldsymbol{\chi}}|_{\partial\boldsymbol{\chi}_t(\mathscr{P})}$.

Note that when $\theta_{\mathscr{S}}=\theta$ the entropy production $p_{\eta\mathscr{S}}$ on the surface is not counterbalanced by the heat outflow $\mathbf{q}\cdot\mathbf{n}$ through the surface.

10.6 Bibliographical and Historical Notes

Various types of surface effect require the use of the balance laws as formulated in this chapter. For example, the adsorption process (see Section 10.2.1) provides a possible illustration of the theory.

If we examine in more detail the liquid/gas interaction we find that the cohesive forces between molecules are unbalanced at the boundary between phases and that the resultant force acting on the surface particles from the liquid side exceeds that acting from the gas side. The surface particles of the liquid are then drawn into the liquid phase, whose surface area tends to diminish. The force normal to the surface of the liquid is accompanied by a tangential force which opposes any surface expansion. This latter force, per unit length (of the surface cross-section), is a measure of the surface tension. Since the surface of the liquid tends to contract, its expansion requires work to be done to overcome the surface tension. This work is measured by the product of the surface tension and the increase in surface area and performance of the work causes a change in the surface energy.

Besides the surface phenomena discussed above, there is a wide class of problems which require that interfacial elasticity, residual surface tension (tension which prevails in the absence of strain) and surface distribution of mass density (different from the interior density) are accounted for. These surface effects derive from the disturbance of homogeneity in the neighbourhood of the boundary of a body undergoing some technological process such as coating a metal with a thin layer of a different metal.

The local form of the mass balance equation (10.5) with mass source and outflow but without mass production was derived by Truesdell and Toupin (1960, p. 527). The equation of mass balance on the surface without the terms corresponding to the source and production of mass was introduced by Moeckel (1974) for a surface given in a convected parametrization.

The Stokes–Christoffel condition was obtained in the discussion of the behaviour of mass in a shock wave (Stokes, 1848; Christoffel, 1877). Its derivation from the balance equations is due to Kotchine (1926); see Truesdell and Toupin (1960).

The local form of momentum balance (10.28) and (10.29), without the source and production of momentum, was obtained by Moeckel (1974). Cauchy's first law of motion (10.31) was found by Cauchy (1827). The reduced form of momentum balance (10.33) on the surface, given in a general parametrization, was derived for the first time in the Polish edition.

In the author's opinion attention should be paid to the equation of motion (10.36) for a material surface which is also the boundary of a body. This result was obtained by Gurtin and Murdoch (1975) in their development of a continuum theory with material surfaces (see also Murdoch, 1976). It seems that this equation may be regarded as common both to the theory of the motion of singular surfaces and to non-linear shell theory.

Classically a shell is treated as a three-dimensional medium contained between two surfaces. All forces and moments affecting a shell, and its balance equations, are derived from the general theory of three-dimensional continua (see Woźniak, 1966, or Naghdi, 1976, in the latter of which the Cosserat equations are obtained from the three-dimensional equations). However, there is another point of view in which a shell is regarded as a surface, and new balance laws have to be formulated without recourse to the three-dimensional theory. The problem was treated more widely by Truesdell and Toupin (1960, pp. 556–560) who derived an equation similar to (10.36). For this reason some of the results given here may be regarded as an introduction to the theory of shells geometrically represented by surfaces.

The consequences of the local principle of balance of momentum are derived in Section 10.3 in the form of the restriction (10.38) and (10.41) for a deformable continuum with a p-material surface.

In view of the possibility of applying the results of the theory of singular surfaces to shell theory it would seem advisable to incorporate moment effects not resulting directly from momentum into the moment of momentum balance law. This is not done here, however.

An original way of deriving balance laws with surface effects was introduced by Wilmański (1977). He includes surface sources in the form of integrals with surface distributions extending over the region containing the singular surface. Unfortunately, the balance equations on the surface obtained by Wilmański have not been developed to the point of showing the influence of singular surface curvature K_m. In Wilmański (1975, 1977) the generalized Kotchine condition for singular surfaces has been obtained, and the invariance of the balance equations under Galilean transformations was discussed in the latter of these two papers.

Returning to the contents of this chapter, we note that the symmetry condition (10.41) satisfied by the surface stress tensor was given by Moeckel (1974). In the continuum theory with material surfaces a similar result may be found in the paper by Gurtin and Murdoch (1975).

The local form (10.43) of the balance of energy equation without the energy source and production terms was given by Moeckel (1974), but the reduced form (10.44) has not been derived before (see Ghez, 1966).

The principle of energy balance (10.46) is the first law of thermodynamics and is most frequently dealt with in problems of thermomechanics (see Truesdell and Toupin, 1960, p. 609).

The reduced form (10.47) of the energy balance equation for a surface has not been derived previously. Equation (10.51) is an originally-derived equation of energy balance for the surface of a deformable continuum, and should be regarded as an extension of the results of Gurtin and Murdoch (1975) taking into account thermal effects (see Murdoch, 1976).

An entropy balance law similar to equations (10.53) but with some simplifying assumptions is contained in Moeckel (1974). The reduced forms (10.53) and (10.54) derived here are original (see the Polish edition), as is (10.56) for the boundary of a continuum. The entropy balance law was first introduced by Eckart (1940) and Tolman and Fine (1948). In rational thermodynamics an entropy balance was proposed by Green and Naghdi (1977).

The representations (10.57) for the entropy flux and the source of entropy are classical (see Meixner and Reik, 1959).

In the axiomatic formulation of thermodynamics presented by Gurtin and Williams (1967) there are two temperatures as derived quantities instead of one absolute temperature: one is associated with heat condition and the other with radiation. They proved that in many circumstances but not always, these two temperatures coincide (see Gurtin, 1976).

Fisher and Leitman (1970) presented a law of energy balance for the surface of a rigid conductor, taking into account the existence of radiational and conductive temperatures.

The most general treatment of entropy flux may be found in the paper by Müller (1971), who did not assume in advance any connection with thermal energy flux. Only after suggesting certain constitutive equations did he prove that for thermo-elastic materials and simple fluids there must exist a certain linear relationship between these fluxes, with the coefficient of proportionality depending on the temperature and its time rate of change. Müller's derivation was connected with his attempt to formulate a thermodynamic theory of continua in which thermal disturbances (waves) propagate with finite speed (see also Müller, 1973, and Liu, 1972).

The reader interested in the proof of the existence of entropy may find, in addition to the classical work of Carathéodory (1909), certain partial results in the following papers: Valanis (1971), Day (1969, 1972a, 1975, 1976), Nemat-Nasser (1973), Wilmański (1974), Šilhavy (1977, 1978, 1980, 1982), Coleman and Owen (1974, 1977), Bree and Beevers (1979), Willems (1972), Day and Šilhavy (1977) and Coleman *et al.* (1981).

Axiomatic approaches to the thermodynamics of continua, dealing with special balance equations, may be found in the following papers: Gurtin and Williams (1967), Fisher and Leitman (1968), Truesdell (1969), Williams (1971, 1972), Day (1972), Perzyna (1974, 1978), Gurtin (1973), Noll (1973), Wang and Truesdell (1973), Wilmański (1974, 1975), Uziembło (1974), Elżanowski (1975) and Eringen (1975).

The study of surface effects on the boundaries of bodies was initiated by Gibbs (1928) who employed the term "dividing surface" in his description of phase changes. Further development of the thermodynamics of an interface was provided by Guggenheim (1957); see also Buff (1960). In Gibbs' model with the hypothetical dividing surface the surface energy may depend on the principal curvatures of the surface. But subsequently Gibbs argued that it is always possible to choose a dividing surface in such a way that the surface energy is independent of the mean curvature. In his idealization, thermodynamic functions, determined on the dividing surface, are homogeneous, the surface tension is isotropic and the surface energy is independent of the mean curvature. The Gibbs model is not adequate for the description of irreversible and non-stationary phenomena.

Guggenheim (1957) takes this model further and argues that surface tension has the same value for a curved surface as it has for a plane surface provided that the thickness of the interface is relatively small and we concentrate on the equilibrium state. It is worth mentioning here a result of Moeckel (1974) in which the independence of constitutive functions for all surface quantities (not only for surface tension) is a proved consequence of the entropy production inequality if we consider the thermodynamic processes for material interfacial surfaces.

Ościk, in his monograph (1982), investigated the physical aspects of adsorption and analysed in detail the existing hypotheses and descriptions of surface phenomena of this type. Unfortunately, in his theoretical study of the thermodynamics of adsorption, Ościk does not in principle go beyond the thermodynamics of Guggenheim (1957) and Manual (1972). The apparatus that he presents is firmly embedded in the assumption that the surface of an adsorbant (in our terminology, a singular surface) is homogeneous in the sense that there is no need to introduce surface fields, in particular surface distributions of energy. However, most adsorbant surfaces are, in fact, energetically heterogeneous (Pierotti and Thomas, 1971).

Over the last 25 years investigations of the effects of surface heterogeneity of solids adsorbing the gaseous phase have undergone a very rapid and stormy development. It seems that within the framework of the theory presented here we can provide another, richly-described phenomenological model of such phenomena.

The first attempt to describe precisely the motion of a deformable surface of a Newtonian fluid and to derive the mass and momentum balance laws was in an article by Scriven (1960). This was also included in an extended version of the book by Aris (1962). Slattery (1967) has also described the motion of an interfacial surface together with that of the adjacent interfacial regions. Papers by Slattery (1964) and Ghez (1966) have extended the results obtained by Scriven (see also Napolitano, 1977, 1978, 1979; Lindsay and Straughan, 1979; Fergola and Romano, 1983; and Romano, 1982).

The phenomenon of surface tension in solid bodies has been of great interest to physicists and chemists for quite a long time, and much has been written on its role in phase transitions, crystallization and the propagation of ultrasonic waves. Some references to such work are: Semenchenko (1957), Herring (1953), Greguss (1960), Adamson (1967) and Orowan (1970). Orowan's article discusses the contributions of T. Young, P. S. Laplace, J. W. Gibbs and others.

Particular problems concerning surface tension have been discussed, for example, in papers by Wesołowski (1968), Blinowski (1970, 1971), Gurtin *et al.* (1976a, b); see also Cohen and De Silva (1966), Cohen and Suh (1970), and De Silva and Whitman (1971).

Attention should also be paid to an article by Blinowski (1971) in which the dependence of surface tension on surface curvature is discussed in detail. It also contains proposals for particular constitutive relations together with their experimental motivation (see also Mindlin, 1965; Blinowski, 1970; Blinowski and Trzęsowski, 1981; and Murdoch, 1979).

Since electromagnetic effects have not been included explicitly the reader interested in the additional balance laws for magnetic and electric fields may consult Truesdell and Toupin (1960) or McCarthy and Tiersten (1978).

11 THE ENTROPY PRODUCTION INEQUALITY

So far we have examined five balance laws. The first three of these, namely the balance laws of mass, momentum and moment of momentum, are concerned with the motion of a body and the forces and moments influencing that motion. Together these laws form the axioms of dynamics.

Although many continuum problems can be solved by use of the laws of dynamics, coupled with the constitutive relations for the material under consideration, complete results can be obtained only if the laws of thermodynamics are satisfied. Thermodynamics is concerned not only with motion but also with heat and temperature. The energy balance law is the first law of thermodynamics. The second law of thermodynamics incorporates the entropy balance law, but also entails the entropy production inequality

$$P\big(N_t(\mathscr{P})\big) \geqslant 0 \quad \text{for any } \mathscr{P} \subset \mathscr{B}. \tag{11.1}$$

As for the laws discussed in Section 10 the entropy production inequality must be satisfied for every subbody $\mathscr{P}$ in an arbitrary motion $\boldsymbol{\chi}_t$, $t \in I$, at an arbitrary instant of time t in I and for an arbitrary part $\mathscr{S}(t)$ of the surface $\mathscr{S}_{\mathscr{B}}(t)$ dividing the area $\boldsymbol{\chi}_t(\mathscr{P})$ into two parts $\boldsymbol{\chi}_t(\mathscr{P})^+$ and $\boldsymbol{\chi}_t(\mathscr{P})^-$ with the property (7.21).

Proposition 11.1

The entropy production inequality (11.1) for a subbody $\mathscr{P}$ in the motion $\boldsymbol{\chi}_t$, $t \in I$, is equivalent to the inequalities

$$\begin{aligned} p_\eta &= \frac{\partial}{\partial t}(\varrho\eta) + \operatorname{div}(\varrho\eta\mathbf{v} + \mathbf{m}_\eta) - \xi \geqslant 0 \quad \text{in } \boldsymbol{\chi}_t(\mathscr{P}) - \mathscr{S}(t), \\ p_{\eta\mathscr{S}} &= \frac{\delta}{\delta t}(\varrho_{\mathscr{S}}\eta_{\mathscr{S}}) + \big(\varrho_{\mathscr{S}}\eta_{\mathscr{S}}(V^\alpha + c^\alpha) + m^\alpha_{\eta\mathscr{S}}\big)_{;\alpha} - 2\varrho_{\mathscr{S}}\eta_{\mathscr{S}}u_n K_m \\ &\quad - [\![\varrho\eta(v_n - u_n)]\!] - [\![\mathbf{m}_\eta]\!] \cdot \mathbf{n} - \xi_{\mathscr{S}} \geqslant 0 \quad \text{on } \mathscr{S}(t), \end{aligned} \tag{11.2}$$

which, on use of the mass balance equations (10.5) and (10.6), may be written

$$\begin{aligned} &\varrho\dot{\eta} + \operatorname{div}\mathbf{m}_\eta + \eta(p_\varrho + r_\varrho - \operatorname{div}\mathbf{w}_\varrho) - \xi \geqslant 0 \quad \text{in } \boldsymbol{\chi}_t(\mathscr{P}) - \mathscr{S}(t), \\ &\varrho_{\mathscr{S}}\dot{\eta}_{\mathscr{S}} + \operatorname{div}_{\mathscr{S}}\mathbf{m}_{\eta\mathscr{S}} + \eta_{\mathscr{S}}(p_{\varrho\mathscr{S}} + r_{\varrho\mathscr{S}} - \operatorname{div}_{\mathscr{S}}\mathbf{w}_{\varrho\mathscr{S}}) \\ &- [\![\varrho(\eta - \eta_{\mathscr{S}})(v_n - u_n)]\!] - [\![\mathbf{m}_\eta - \eta_{\mathscr{S}}\mathbf{w}_\varrho]\!] \cdot \mathbf{n} - \xi_{\mathscr{S}} \geqslant 0 \end{aligned} \tag{11.3}$$

on $\mathscr{S}(t)$. □

The following result is a consequence of Corollaries 10.11 and 10.12.

Corollary 11.1

If mass is conserved then the local entropy production inequality is equivalent to

$$\begin{aligned}&\varrho\dot{\eta}+\operatorname{div}\mathbf{m}_\eta-\xi\geqslant 0\quad\text{in }\boldsymbol{\chi}_t(\mathscr{P})-\mathscr{S}(t),\\&\varrho_{\mathscr{S}}\dot{\eta}_{\mathscr{S}}+\operatorname{div}_{\mathscr{S}}\mathbf{m}_{\eta\mathscr{S}}-[\![\varrho(\eta-\eta_{\mathscr{S}})(v_n-u_n)]\!]-[\![\mathbf{m}_{\eta\mathscr{S}}]\!]\cdot\mathbf{n}\\&-\xi_{\mathscr{S}}\geqslant 0\quad\text{on }\mathscr{S}(t).\end{aligned}\tag{11.4}$$

If $\mathscr{S}(t)$ is a material surface then $(11.4)_2$ is replaced by

$$\varrho_{\mathscr{S}}\dot{\eta}_{\mathscr{S}}+\operatorname{div}_{\mathscr{S}}\mathbf{m}_{\eta\mathscr{S}}-[\![\mathbf{m}_\eta]\!]\cdot\mathbf{n}-\xi_{\mathscr{S}}\geqslant 0\quad\text{on }\mathscr{S}(t).\tag{11.5}$$

Moreover, if $\mathscr{S}(t)$ is the boundary $\partial\boldsymbol{\chi}_t(\mathscr{P})\cap\partial\boldsymbol{\chi}_t(\mathscr{B})$ then (11.5) becomes

$$\varrho_{\mathscr{S}}\dot{\eta}_{\mathscr{S}}+\operatorname{div}_{\mathscr{S}}\mathbf{m}_{\eta\mathscr{S}}-\mathbf{m}_\eta\cdot\mathbf{n}-\xi_{\mathscr{S}}\geqslant 0.\ \square\tag{11.6}$$

11.1 The Clausius–Duhem Inequality

Finally, we discuss a particular form of the entropy production inequality $(11.4)_1$, namely

$$\varrho\dot{\eta}+\operatorname{div}(\theta^{-1}\mathbf{q})-\theta^{-1}\varrho r_e\geqslant 0,\tag{11.7}$$

which is a result of assumptions (10.57).

Corollary 11.2

If the flux and source of entropy are given by (10.57) and (10.58) then the local entropy production inequality becomes the Clausius–Duhem inequality of rational thermodynamics, namely

$$\varrho\theta\dot{\eta}+\operatorname{tr}\{(\operatorname{grad}\mathbf{v})\mathbf{T}^{\mathsf{T}}\}-\varrho\dot{e}-\theta^{-1}\mathbf{q}\cdot\operatorname{grad}\theta\geqslant 0\quad\text{in }\boldsymbol{\chi}_t(\mathscr{P})-\mathscr{S}(t),\tag{11.8}$$

while inequality $(11.4)_2$ may be rewritten as

$$\begin{aligned}&\varrho_{\mathscr{S}}\theta_{\mathscr{S}}\dot{\eta}_{\mathscr{S}}+\operatorname{tr}\{(\operatorname{grad}_{\mathscr{S}}\mathbf{V})\mathbf{S}^{\mathsf{T}}\}-\varrho_{\mathscr{S}}\dot{e}_{\mathscr{S}}-\mathbf{q}_{\mathscr{S}}\cdot\operatorname{grad}_{\mathscr{S}}(\ln\theta_{\mathscr{S}})\\&-[\![\varrho\{\theta_{\mathscr{S}}(\eta-\eta_{\mathscr{S}})-(e-e_{\mathscr{S}})-\tfrac{1}{2}(\mathbf{v}-\mathbf{V})\cdot(\mathbf{v}-\mathbf{V})\}(v_n-u_n)]\!]\\&+\left[\!\!\left[\left(1-\frac{\theta_{\mathscr{S}}}{\theta}\right)\mathbf{q}-(\mathbf{v}-\mathbf{V})\mathbf{T}\right]\!\!\right]\cdot\mathbf{n}\geqslant 0\quad\text{on }\mathscr{S}(t).\ \square\end{aligned}\tag{11.9}$$

The inequality (11.9) is a particularly interesting generalization of the Clausius–Duhem inequality to singular surfaces.

If $\mathscr{S}(t)$ is a material surface then (11.9) becomes

$$\begin{aligned}&\varrho_{\mathscr{S}}\theta_{\mathscr{S}}\dot{\eta}_{\mathscr{S}}+\operatorname{tr}\{(\operatorname{grad}_{\mathscr{S}}\mathbf{V})\mathbf{S}^{\mathsf{T}}\}-\varrho_{\mathscr{S}}\dot{e}_{\mathscr{S}}-\mathbf{q}_{\mathscr{S}}\cdot\operatorname{grad}_{\mathscr{S}}(\ln\theta_{\mathscr{S}})\\&+\left[\!\!\left[\left(1-\frac{\theta_{\mathscr{S}}}{\theta}\right)\mathbf{q}-(\mathbf{v}-\dot{\boldsymbol{\varphi}})\mathbf{T}\right]\!\!\right]\cdot\mathbf{n}\geqslant 0.\end{aligned}\tag{11.10}$$

For the case in which $\mathcal{S}(t)$ is the boundary $\partial\chi_t(\mathcal{P})\cap\partial\chi_t(\mathcal{B})$ we have

$$\varrho_{\mathcal{S}}\theta_{\mathcal{S}}\dot{\eta}_{\mathcal{S}}+\operatorname{tr}\{(\operatorname{grad}_{\mathcal{S}}\mathbf{V})\mathbf{S}^{\mathsf{T}}\}-\varrho_{\mathcal{S}}\dot{e}_{\mathcal{S}}-\mathbf{q}_{\mathcal{S}}\cdot\operatorname{grad}_{\mathcal{S}}(\ln\theta_{\mathcal{S}})$$

$$+\left(1-\frac{\theta_{\mathcal{S}}}{\theta}\right)\mathbf{q}\cdot\mathbf{n}\geqslant 0. \qquad (11.11)$$

11.2 Bibliographical and Historical Notes

The second law of thermodynamics has a rich and controversial history. A characteristic feature of the development of thermodynamics so far is the lack of a single universal formulation of the second law. The second law itself quantifies the physical observation that every change in energy and every flux of mechanical energy may be transformed into heat or heat flux but not vice versa. It has given rise to numerous mathematical expressions, a list of which would fill a dozen or more pages.

Until some ten or so years ago there were no more than three main approaches to the thermodynamics of continuous media. It is in the form of the second law that the three approaches have differed. First, there is the classical Onsager thermodynamics of irreversible processes (see De Groot and Mazur, 1962; Gumiński, 1962; and Wiśniewski, Staniszewski and Szymanik, 1976) embodying the hypothesis of local state (see Kestin, 1966). Second is the thermodynamics of passive processes (see Meixner, 1966a, b) and third, rational thermodynamics (see Coleman and Noll, 1963; and Truesdell, 1969).

Over the past fifteen years many new interpretations of the second law have appeared, along with many new approaches to the thermodynamics of irreversible processes. In particular, we should mention here the Polish achievements in axiomatic, neo-classical thermodynamics, whose foundations are laid in the book by Wilmański (1974a). A broader review of the present state of thermodynamics can be found in the books by Perzyna (1978) and Hanyga (1986) as well as Lawenda (1979) and the review papers by Hutter (1977), Müller (1979) and Ruggeri (1983). The rational approach is currently that most widely accepted and used. The mathematical theory in which the second law is written in the form of the Clausius–Duhem inequality (11.7) has proved the most useful in practice and for the last twenty two years the majority of papers have adopted this theory.*

Returning to the contents of the present chapter, we note that the way in which the second law of thermodynamics is presented here is similar to that in Müller (1967, 1971a, b). In the writer's opinion it contains two char-

* In 1985 exactly twenty two years had passed since publication of the article by Coleman and Noll (1963) which is generally regarded as the beginning of rational thermodynamics.

acteristic features. First, it expresses precisely the physical sense of the second law; second, it is sufficiently general for appropriate specification of the source and flux of entropy to lead to rational thermodynamics and the Clausius–Duhem inequality, as a particular case.

The generality of the forms of the balance of entropy law, (10.53) and (10.54), and the entropy production inequalities, (11.2)–(11.4), in contrast to the more limited forms, (10.59) and (11.7), in rational thermodynamics, has an advantage over the latter in that there is no contradiction with existing theories. Indeed, as proved by Müller (1971) and Truesdell (1969), the Clausius–Duhem inequality (11.7) is not acceptable in the kinetic theory of gases because of the form of $(10.57)_1$ postulated in rational thermodynamics. Moreover, a number of papers based on so-called semi-nonlocal effects have appeared; in these an extended irreversible thermodynamics dealing with general entropy fluxes and sources is used (see Müller, 1967, 1979; Kranyš, 1972; Israel, 1976; Ruggeri, 1983; Carrassi and Morro, 1972). A slightly different view of the extra entropy flux is presented by Kosiński (1983); see also Green and Laws (1972), Gyarmati (1977), Dafermos and Kosiński (1984) and Kosiński (1985).

12 THE BASIC RELATIONS WITHOUT SURFACE EFFECTS

In the previous parts of this chapter we have dealt with derivations of the balance laws and the entropy production inequality for the general case of a deformable continuum. In these derivations we took account of surface effects in the form of concentrations of sources, fluxes and productions of the dynamic and thermodynamic quantities involved. In deriving the conservation of mass equation for a singular surface in Section 10.1 we obtained the Stokes–Christoffel condition (10.9) for the case of vanishing surface mass density.

At this point our aim is to investigate the remaining balance equations together with the entropy production inequality when the surface quantities vanish. Thus, instead of the general representations (8.5) in Hypothesis 8.2, we have

$$\begin{aligned} \Psi_t &= \int_{\chi_t(\mathscr{P})} \psi \, \mathrm{d}v, & W(\Psi_t) &= \int_{\partial\chi_t(\mathscr{P})} w \, \mathrm{d}a, \\ P(\Psi_t) &= \int_{\chi_t(\mathscr{P})} p \, \mathrm{d}v, & R(\Psi_t) &= \int_{\chi_t(\mathscr{P})} r \, \mathrm{d}v \end{aligned} \tag{12.1}$$

for every subbody $\mathscr{P}$, motion χ_t and instant $t \in I$.

The mass conservation equation $(10.7)_2$ reduces to

$$[\![\varrho(v_n - u_n)]\!] = 0, \tag{12.2}$$

and the momentum conservation equation (10.29) to

$$[\![\varrho\mathbf{v}(v_n - u_n) - \mathbf{T}\mathbf{n}]\!] = \mathbf{0}, \tag{12.3}$$

which, for a material surface, becomes

$$[\![\mathbf{T}]\!]\mathbf{n} = \mathbf{0}, \tag{12.4}$$

specializing $(10.35)_2$. The energy balance equation $(10.43)_2$, (10.47) or (10.48) takes the form

$$[\![\varrho(e + \tfrac{1}{2}\mathbf{v}\cdot\mathbf{v})(v_n - u_n)]\!] + [\![\mathbf{q} - \mathbf{v}\mathbf{T}]\!]\cdot\mathbf{n} = 0, \tag{12.5}$$

and for a material surface either (10.50) or (12.5) yields

$$[\![\mathbf{q} - \mathbf{v}\mathbf{T}]\!]\cdot\mathbf{n} = 0. \tag{12.6}$$

From the entropy balance equation $(10.53)_2$, or $(10.54)_2$, we obtain

$$[\![\varrho\eta(v_n - u_n) + \mathbf{m}_\eta\cdot\mathbf{n}]\!] = 0, \tag{12.7}$$

and its specialization to rational thermodynamics is

$$[\![\varrho\eta(v_n - u_n) + \theta^{-1}\mathbf{q}\cdot\mathbf{n}]\!] = 0 \tag{12.8}$$

from $(10.59)_2$. For a material surface (10.55) and (10.60) yield

$$[\![\mathbf{m}_\eta]\!]\cdot\mathbf{n} = 0, \qquad [\![\theta^{-1}\mathbf{q}]\!]\cdot\mathbf{n} = 0. \tag{12.9}$$

The entropy production inequality in the form (11.2) or $(11.4)_2$ reduces to

$$[\![\varrho\eta(u_n - v_n) - \mathbf{m}_\eta\cdot\mathbf{n}]\!] \geqslant 0, \tag{12.10}$$

or, with the assumption $(10.57)_1$, to

$$[\![\varrho\eta(u_n - v_n) - \theta^{-1}\mathbf{q}\cdot\mathbf{n}]\!] \geqslant 0. \tag{12.11}$$

For a material surface (12.10) and (12.11) respectively become

$$-[\![\mathbf{m}_\eta]\!]\cdot\mathbf{n} \geqslant 0, \qquad -[\![\theta^{-1}\mathbf{q}]\!]\cdot\mathbf{n} \geqslant 0. \tag{12.12}$$

Between equations (12.7)–(12.9) and inequalities (12.10)–(12.12) there is apparently a contradiction whose source is the local entropy production inequality $(11.2)_2$. The derivation above is based on the assumption that the surface production of entropy $p_{\eta\mathscr{S}}$, appearing on the left-hand side of $(11.2)_2$, disappears. This implies the simultaneous requirements

$$p_{\eta\mathscr{S}} = 0 \quad \text{and} \quad p_{\eta\mathscr{S}} \geqslant 0. \tag{12.13}$$

Clearly, if $(12.13)_1$ is satisfied then $(12.13)_2$ follows automatically.

Chapter 5

Singular surfaces of the motion

In this chapter we introduce shock and acceleration waves, which appear as surface singularities of certain derivatives of the motion $\boldsymbol{\chi}_t$. Attention is confined to the kinematics described by $\boldsymbol{\chi}_t$, thermodynamic considerations being set aside temporarily. The basic properties of shock waves, vortex sheets and acceleration waves will be examined independently of the material properties of the medium in motion.

13 SINGULAR SURFACES OF ORDER ONE

Let $\{\mathscr{S}_{\mathscr{B}}(t)\}_{t \in I}$ be a moving surface in the Euclidean space $\mathscr{E}^3$. Assume that in $\mathscr{E}^3$ and in the same time interval I a topological motion $\hat{\boldsymbol{\chi}}_t$, $t \in I$, of a deformable body $\mathscr{B}$ takes place with respect to some reference placement $\boldsymbol{\varkappa}_0$ (see Section 9.1). In addition to

$$\mathscr{S}_{\mathscr{B}}(t) \subset \boldsymbol{\chi}_t(\mathscr{B}) \quad \text{for } t \in I \tag{13.1}$$

we assume that the surface $\{\mathscr{S}_{\mathscr{B}}(t)\}_{t \in I}$ has the properties (7.19)–(7.22) with $\mathscr{P}$ replaced by $\mathscr{B}$.

If the parametric representation of the surface is of the form

$$\mathbf{x} = \boldsymbol{\varphi}(t, l^1, l^2), \quad t \in I, \ (l^1, l^2) \in \mathscr{U} \tag{13.2}$$

then there exists an implicit representation (recall (1.20))

$$g(t, \mathbf{x}) = 0 \tag{13.3}$$

with the property

$$g(t, \boldsymbol{\varphi}) \equiv 0, \quad t \in I.$$

Into the reference placement $\boldsymbol{\varkappa}_0$ we now introduce the family of surfaces $\{\Sigma_{\mathscr{B}}(t)\}_{t \in I}$ with the representation

$$G(t, \mathbf{X}) = 0 \tag{13.4}$$

defined by

$$G(t, \mathbf{X}) \equiv g(t, \hat{\boldsymbol{\chi}}_t(\mathbf{X})). \tag{13.5}$$

We choose the parameters $l^{\Delta}, \Delta = 1, 2$, so that

$$\Sigma_{\mathscr{B}}(t): \mathbf{X} = \mathbf{P}(t, l^{\Delta}), \quad t \in I, \ (l^{\Delta}) \in \mathscr{U}, \tag{13.6}$$

$G(t, \mathbf{P}) \equiv 0$, and the function $\mathbf{P}$, with non-vanishing Jacobian, is continuously differentiable. The family $\{\Sigma_{\mathscr{B}}(t)\}_{t \in I}$ forms a moving surface in the reference placement $\boldsymbol{\varkappa}_0$ of the body $\mathscr{B}$ (see Figure 13.1) and for every t the surface $\Sigma_{\mathscr{B}}(t)$ is the image of $\mathscr{S}_{\mathscr{B}}(t)$ in the placement $\boldsymbol{\varkappa}_0(\mathscr{B})$.

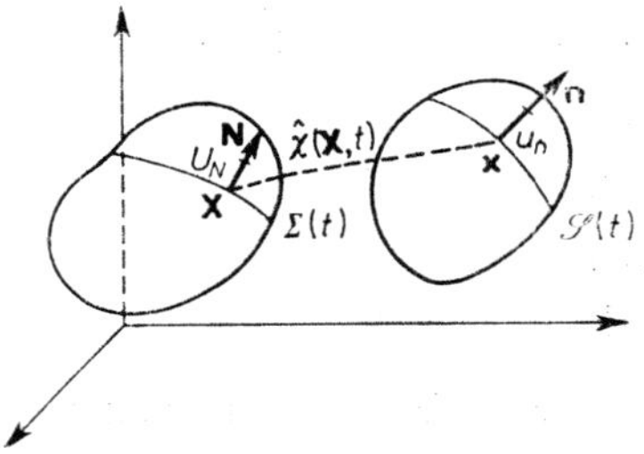

Fig. 13.1. Material representation of a singular surface $\Sigma(t)$ and its spatial representation $\mathscr{S}(t)$

It is worth pointing out here the duality between the representations of the singular surfaces $\mathscr{S}(t)$ and $\Sigma(t)$, their normal vectors and their normal velocity components given below. Since a reference placement and a placement corresponding to the current instant t are both used in continuum mechanics this leads to a dual treatment of the dynamic and thermodynamic quantities used in the description of the continuum. In certain situations the quantities are regarded as functions of time in the region $\boldsymbol{\varkappa}_0(\mathscr{B})$; in others we choose the spatial description with functions of time defined in the placement $\boldsymbol{\chi}$, that is in the region $\bigcup_{t \in I} \boldsymbol{\chi}_t(\mathscr{B})$. Both approaches are equally admissible, the particular circumstances determining which is the most convenient to use.

Thus, if the motion $\hat{\boldsymbol{\chi}}_t$, $t \in I$, is defined on $\boldsymbol{\varkappa}_0(\mathscr{B})$ then the discontinuities of the displacement tensor $\mathbf{F}$ and the velocity field $\dot{\boldsymbol{\chi}}$ are determined from the geometry of the moving singular surface $\{\Sigma_{\mathscr{B}}(t)\}_{t \in I}$.

By reference to equations (1.22) and (1.23) we define the unit normal $\mathbf{N}$ to the surface $\Sigma_{\mathscr{B}}(t)$ as

$$\mathbf{N} = \frac{\operatorname{Grad} G(t, \mathbf{X})}{|\operatorname{Grad} G|_3} \tag{13.7}$$

and the normal component of the surface velocity as

$$U_N = -\frac{\partial G/\partial t}{|\mathrm{Grad}\, G|_3}. \tag{13.8}$$

The moving surface $\{\Sigma_{\mathscr{B}}(t)\}_{t\in I}$ is uniquely determined by the moving surface $\{\mathscr{S}_{\mathscr{B}}(t)\}_{t\in I}$ because the motion $\hat{\boldsymbol{\chi}}_t$, $t \in I$, is a continuous family of homeomorphisms of open regions of $\mathscr{E}^3$. This means that if $\hat{\boldsymbol{\chi}}_t$ is differentiable then the unit normal $\mathbf{N}$ and the velocity component U_N are given by (13.7) and (13.8) since*

$$\mathrm{Grad}\, G(t, \mathbf{X}) = \{\mathrm{grad}\, g(t, \mathbf{x})\}\mathbf{F},$$

$$|\mathrm{Grad}\, G| = |(\mathrm{grad}\, g)\mathbf{F}| = |\mathrm{grad}\, g||\mathbf{nF}|,$$

where $\mathbf{F} = \mathrm{Grad}\, \hat{\boldsymbol{\chi}}(t, \mathbf{X})$.

Using equations (1.22), (1.23), (13.7) and (13.8) we obtain

$$\mathbf{N} = \mathbf{nF}\frac{|\mathrm{grad}\, g|}{|\mathrm{Grad}\, G|} = \frac{\mathbf{nF}}{|\mathbf{nF}|} = \frac{\mathbf{F}^{\mathrm{T}}\mathbf{n}}{|\mathbf{F}^{\mathrm{T}}\mathbf{n}|}. \tag{13.9}$$

The unit normal $\mathbf{n}$ to the surface $\mathscr{S}(t)$ is related to $\mathbf{N}$ by

$$\mathbf{n} = \mathbf{NF}^{-1}\frac{|\mathrm{Grad}\, G|}{|\mathrm{grad}\, g|} = \frac{\mathbf{N}\overset{-1}{\mathbf{F}}}{|\mathbf{N}\overset{-1}{\mathbf{F}}|} = \frac{\overset{-1}{\mathbf{F}^{\mathrm{T}}}\mathbf{N}}{|\overset{-1}{\mathbf{F}^{\mathrm{T}}}\mathbf{N}|}. \tag{13.10}$$

The normal components of velocities u_n and U_N of the surfaces $\mathscr{S}(t)$ and $\Sigma(t)$ respectively are connected by

$$U_N = (\mathbf{c} - \dot{\boldsymbol{\chi}}) \cdot \mathbf{n}\frac{|\mathrm{grad}\, g|}{|\mathrm{Grad}\, G|} = (u_n - v_n)\frac{1}{|\mathbf{F}^{\mathrm{T}}\mathbf{n}|} = (u_n - v_n)|\overset{-1}{\mathbf{F}^{\mathrm{T}}}\mathbf{n}|. \tag{13.11}$$

In terms of the local speed U the above result may be written

$$U_N = \frac{U}{|\mathbf{F}^{T}\mathbf{n}|} = U|\overset{-1}{\mathbf{F}^{\mathrm{T}}}\mathbf{n}|. \tag{13.12}$$

Under the assumption that $UU_N \neq 0$ equations (13.10)–(13.12) lead to

$$\mathbf{n} = \frac{U}{U_N}\overset{-1}{\mathbf{F}^{\mathrm{T}}}\mathbf{N}, \qquad \mathbf{N} = \frac{U_N}{U}\mathbf{F}^{\mathrm{T}}\mathbf{N}. \tag{13.13}$$

We now consider the case in which $\{\Sigma(t)\}_{t\in I}$ is a singular surface of order one for the motion $\hat{\boldsymbol{\chi}}_t$, $t \in I$. Then, according to Definition 5.3, $\hat{\boldsymbol{\chi}}_t$,

---

* We use the standard notation $\mathrm{grad}\, g = \nabla g$, $\mathrm{Grad}\, G = \nabla G$, where ∇ is the gradient operator, g is a function of $\mathbf{x}$ and G is a function of $\mathbf{X}$. Henceforth, we omit the subscript 3 from $|\cdot|_3$ and the letter $\mathscr{B}$ labelling the surface.

$t \in I$, is a continuous function on $\hat{\mathcal{N}} \equiv I \times \varkappa_0(\mathscr{B})$, while its first partial derivatives are continuous only on $\hat{\mathcal{N}}^+ \cup \hat{\mathscr{S}}$ and $\hat{\mathcal{N}}^- \cup \hat{\mathscr{S}}$, where

$$\hat{\mathscr{S}} = \bigcup_{t \in I} \{t\} \times \Sigma(t). \tag{13.14}$$

On the surface $\hat{\mathscr{S}}$ there is a jump discontinuity in the derivatives of the function $\hat{\boldsymbol{\chi}}\colon \hat{\mathcal{N}} \to \mathscr{E}^3$, and at least one of

$$[\![\dot{\boldsymbol{\chi}}]\!] \neq \mathbf{0} \quad \text{or} \quad [\![\mathbf{F}]\!] \neq \mathbf{0}$$

holds.

If f is a function for which $\hat{\mathscr{S}}$ is a surface of jump discontinuity and f_+ and f_- are smooth extensions of the functions $f\|_{\hat{\mathcal{N}}^+ \cup \hat{\mathscr{S}}}$ and $f\|_{\hat{\mathcal{N}}^- \cup \hat{\mathscr{S}}}$, then let $(f)^+$ and $(f)^-$ denote the restrictions of the extensions to the surface $\hat{\mathscr{S}}$, that is

$$(f)^+ \equiv f_+|_{\hat{\mathscr{S}}}, \quad (f)^- \equiv f_-|_{\hat{\mathscr{S}}}. \tag{13.15}$$

From the uniqueness of the normals $\mathbf{N}$ and $\mathbf{n}$ and the velocity components U_N and u_n we deduce

$$\begin{gathered}
\frac{(\mathbf{F}^\mathsf{T})^-\mathbf{n}}{|(\mathbf{F}^\mathsf{T})^-\mathbf{n}|} = \frac{(\mathbf{F}^\mathsf{T})^+\mathbf{n}}{|(\mathbf{F}^\mathsf{T})^+\mathbf{n}|}, \qquad \frac{U_N}{(U)^-}(\mathbf{F}^\mathsf{T})^-\mathbf{n} = \frac{U_N}{(U)^+}(\mathbf{F}^\mathsf{T})^+\mathbf{n}, \\
\frac{(\overset{-1}{\mathbf{F}^\mathsf{T}})^-\mathbf{N}}{|(\overset{-1}{\mathbf{F}^\mathsf{T}})^-\mathbf{N}|} = \frac{(\overset{-1}{\mathbf{F}^\mathsf{T}})^+\mathbf{N}}{|(\overset{-1}{\mathbf{F}^\mathsf{T}})^+\mathbf{N}|}, \qquad \frac{(U)^-}{U_N}(\overset{-1}{\mathbf{F}^\mathsf{T}})^-\mathbf{N} = \frac{(U)^+}{U_N}(\overset{-1}{\mathbf{F}^\mathsf{T}})^+\mathbf{N}, \\
\frac{(U)^-}{|(\mathbf{F}^\mathsf{T})^-\mathbf{n}|} = \frac{(U)^+}{|(\mathbf{F}^\mathsf{T})^+\mathbf{n}|}, \qquad (U)^-|(\overset{-1}{\mathbf{F}^\mathsf{T}})^-\mathbf{N}| = (U)^+|(\overset{-1}{\mathbf{F}^\mathsf{T}})^+\mathbf{N}|.
\end{gathered} \tag{13.16}$$

In view of (5.14) the above may be written as

$$\begin{gathered}
\left[\!\!\left[\frac{\mathbf{F}^\mathsf{T}}{|\mathbf{F}^\mathsf{T}\mathbf{n}|}\right]\!\!\right]\mathbf{n} = \mathbf{0}, \qquad \left[\!\!\left[\frac{\overset{-1}{\mathbf{F}^\mathsf{T}}}{|\overset{-1}{\mathbf{F}^\mathsf{T}}\mathbf{N}|}\right]\!\!\right]\mathbf{N} = \mathbf{0}, \\
\left[\!\!\left[\frac{U}{|\mathbf{F}^\mathsf{T}\mathbf{n}|}\right]\!\!\right] = 0, \qquad \left[\!\!\left[\frac{\mathbf{F}^\mathsf{T}}{U}\right]\!\!\right]\mathbf{n} = \mathbf{0}, \\
[\![U\overset{-1}{\mathbf{F}^\mathsf{T}}]\!]\mathbf{N} = \mathbf{0}, \qquad [\![U|\overset{-1}{\mathbf{F}^\mathsf{T}}\mathbf{N}|]\!] = 0,
\end{gathered} \tag{13.17}$$

the latter four being valid provided $UU_N \neq 0$.

For the derivatives of $\hat{\boldsymbol{\chi}}$ the compatibility conditions (6.10) become

$$[\![\mathbf{F}]\!] = \left[\!\!\left[\frac{\partial\hat{\boldsymbol{\chi}}}{\partial N}\right]\!\!\right] \otimes \mathbf{N}, \qquad [\![\dot{\boldsymbol{\chi}}]\!] = -U_N\left[\!\!\left[\frac{\partial\hat{\boldsymbol{\chi}}}{\partial N}\right]\!\!\right]. \tag{13.18}$$

In the notation introduced in Section 2 (see (2.3)), we have

$$\frac{\partial\hat{\boldsymbol{\chi}}}{\partial N} \equiv \mathbf{N} \cdot \mathrm{Grad}\,\hat{\boldsymbol{\chi}} = \mathbf{F}\mathbf{N}.$$

For the singular surface $\hat{\mathscr{S}}$ the displacement derivatives $\delta/\delta t$ has the form

$$\frac{\delta}{\delta t}(\cdot) = \frac{\partial}{\partial t}(\cdot)+U_N \mathbf{Grad}\,(\cdot)\mathbf{N} = \frac{\partial}{\partial t}(\cdot)+U_N\frac{\partial(\cdot)}{\partial N}, \quad (13.19)$$

by (2.10).

Application of the compatibility condition $(6.12)_3$ to the second derivatives of $\hat{\chi}$ yields

$$2U_N\frac{\delta}{\delta t}\left[\!\left[\frac{\partial\hat{\chi}}{\partial N}\right]\!\right]+\left[\!\left[\frac{\partial\hat{\chi}}{\partial N}\right]\!\right]\frac{\delta U_N}{\delta t} = U_N^2\left[\!\left[\frac{\partial^2\hat{\chi}}{\partial N^2}\right]\!\right]-[\![\ddot{\chi}]\!]. \quad (13.20)$$

Assuming that $U_N > 0$ we may simplify (13.20) and write it as one or other of the following:

$$\begin{aligned} 2\sqrt{U_N}\,\frac{\delta}{\delta t}\left(\sqrt{U_N}\left[\!\left[\frac{\partial\hat{\chi}}{\partial N}\right]\!\right]\right) &= U_N^2\left[\!\left[\frac{\partial^2\hat{\chi}}{\partial N^2}\right]\!\right]-[\![\ddot{\chi}]\!],\\ 2\sqrt{U_N}\,\frac{\delta}{\delta t}\left(\frac{\left[\!\left[\frac{\partial\hat{\chi}}{\partial N}\right]\!\right]}{\sqrt{U_N}}\right) &= [\![\ddot{\chi}]\!]-U_N^2\left[\frac{\partial^2\hat{\chi}}{\partial N^2}\right]. \end{aligned} \quad (13.21)$$

In each of these we have adopted the identity

$$(\mathbf{Grad}\,\mathbf{Grad}\,\hat{\chi})\cdot(\mathbf{N}\otimes\mathbf{N}) \equiv \frac{\partial^2\hat{\chi}}{\partial N^2} = (\mathbf{Grad}\,\mathbf{F})\cdot(\mathbf{N}\otimes\mathbf{N})$$

in accordance with the notation introduced in Section 3.

Equations (13.21) govern the evolution of the jump in the normal derivative of $\hat{\chi}$ and the jump in the velocity along the normal trajectory. In the theory of one-dimensional waves the equation corresponding to (13.21) is called the **amplitude equation** of the shock wave.

13.1 Absolutely Material Surfaces

We return briefly to the compatibility conditions (13.18). Clearly, the vector

$$\mathbf{s} \equiv \left[\!\left[\frac{\partial\hat{\chi}}{\partial N}\right]\!\right] \quad (13.22)$$

is the vital quantity in the definition of a singular surface of order one for the motion $\hat{\chi}$. We shall refer to it as the **characteristic vector** or, following Truesdell and Toupin (1960), the **singularity vector** of the surface $\hat{\mathscr{S}}$. If $\mathbf{s}$ is zero then the jumps of both derivatives of $\hat{\chi}$ automatically vanish. We observe that continuity of the velocity field $\dot{\chi}$ across $\Sigma(t)$ does not necessarily imply the continuity of the displacement tensor $\mathbf{F}$ since $\mathbf{s} \neq \mathbf{0}$ may be accompanied by $U_N = 0$. This situation is characterized by the following definition:

Definition 13.1

A singular surface $\hat{\mathscr{S}}$ is called an **absolutely material surface** if the normal component $U_N(t)$ of its velocity of propagation vanishes for $t \in I$.

For such surfaces we formulate the following proposition:

Proposition 13.1

On a singular absolutely material surface of order one only the displacement tensor **F** suffers a jump discontinuity, while the velocity field $\dot{\boldsymbol{\chi}}$ is continuous, that is

$$[\![\mathbf{F}]\!] = \left[\!\!\left[\frac{\partial \hat{\boldsymbol{\chi}}}{\partial N} \right]\!\!\right] \otimes \mathbf{N} = \mathbf{s} \otimes \mathbf{N}, \qquad [\![\dot{\boldsymbol{\chi}}]\!] = \mathbf{0}. \tag{13.23}$$

Moreover, the acceleration field is continuous on $\hat{\mathscr{S}}$.

Proof

Equations (13.23) have been established above. The validity of the second property follows from the compatibility conditions (6.12), (6.17) and (6.19); see also equation (13.21). □

The above result may be formulated as a sufficient condition for a surface to be an absolutely material surface:

Corollary 13.1

If the velocity $\dot{\boldsymbol{\chi}}$ is continuous then the surface $\hat{\mathscr{S}}$ is an absolutely material surface. □

13.2 Shock Waves

We now return to an arbitrary singular surface of order one. From Section 8.1 the local speed U of a propagating surface is defined by

$$U = u_n - v_n = u_n - \dot{\boldsymbol{\chi}} \cdot \mathbf{n}. \tag{13.24}$$

For a surface of discontinuity of the velocity field the quantities $(U)^+$ and $(U)^-$ are given by

$$(U)^+ = u_n - (\dot{\boldsymbol{\chi}})^+ \cdot \mathbf{n}, \qquad (U)^- = u_n - (\dot{\boldsymbol{\chi}})^- \cdot \mathbf{n}.$$

If the velocity field is discontinuous then $(U)^+ \neq (U)^-$, although one of these quantities may become zero. We therefore say that a singular surface of order one is a material surface with respect to the region $\hat{\mathscr{N}}^+$ (respectively $\hat{\mathscr{N}}^-$) if $(U)^+ = 0$ (respectively $(U)^- = 0$). This definition justifies the inclusion of the word "absolutely" in Defition 13.1 since only a surface which is a material surface for both regions is an absolutely material surface.

Since u_n is the normal component of the displacement velocity of the surface $\mathscr{S}(t)$ it follows that in the general case

$$[\![U]\!] = -[\![\dot{\boldsymbol{\chi}}]\!]\cdot\mathbf{n}. \tag{13.25}$$

In view of $(13.18)_2$ and (13.22) we obtain the basic equality

$$[\![U]\!] = U_N\mathbf{s}\cdot\mathbf{n} \tag{13.26}$$

which is necessary for the introduction of the concept of a shock wave.

Definition 13.2

A singular surface of order one on which the local speed of propagation U suffers a discontinuity, that is $[\![U]\!] \neq 0$, is called a **shock wave.**

As a direct conclusion from Proposition 13.1 and Corollary 13.1 we have:

Proposition 13.2

A shock wave cannot be an absolutely material surface; at most, it can be a one-sided material surface, that is a material surface in relation to either $\hat{\mathscr{N}}^+$ or $\hat{\mathscr{N}}^-$. On a shock wave the normal component $\dot{\boldsymbol{\chi}}\cdot\mathbf{n}$ of the velocity field $\dot{\boldsymbol{\chi}}$ suffers a discontinuity and the characteristic vector $\mathbf{s}$ is not orthogonal to the vector $\mathbf{n}$ normal to the surface $\mathscr{S}(t)$, $t \in I$. □

The following result is characteristic for shock waves:

Corollary 13.2

If the characteristic vector $\mathbf{s}$ is parallel to the normal vector $\mathbf{n}$ then the limit values $(\mathbf{F})^+$ and $(\mathbf{F})^-$ of the displacement tensor differ by a pure extension.

Proof

In view of Definition 13.2 and equations (13.23) the difference between the tensors $(\mathbf{F})^-$ and $(\mathbf{F})^+$ may be expressed in the form

$$(\mathbf{F})^- - (\mathbf{F})^+ = \mathbf{s}\otimes\mathbf{N}.$$

From $(13.16)_1$ it follows that

$$(\mathbf{F})^- = (\mathbf{F})^+ + \mathbf{s}\otimes\mathbf{n}(\mathbf{F})^+\frac{1}{|\mathbf{n}(\mathbf{F})^+|} = \left(\mathbf{1} + \frac{\mathbf{s}\otimes\mathbf{n}}{|(\mathbf{F}^{\mathrm{T}})^+\mathbf{n}|}\right)(\mathbf{F})^+,$$

or, more compactly,

$$(\mathbf{F})^- = \mathbf{A}(\mathbf{F})^+, \quad \text{where } \mathbf{A} = \mathbf{1} + \frac{\mathbf{s}\otimes\mathbf{n}}{|(\mathbf{F}^{\mathrm{T}})^+\mathbf{n}|}. \tag{13.27}$$

To complete the proof we require certain facts from continuum mechanics. The well-known polar decomposition theorem allows us to write the displacement tensor $\mathbf{F}$ as

$$\mathbf{F} = \mathbf{RU},$$

where **R** is called the rotation tensor and **U** the right stretch (or extension) tensor. Recalling (9.9) we see that the right Cauchy–Green tensor **C** may be written as $\mathbf{C} = \mathbf{U}^2$. The rotational tensor is orthogonal and **U** is symmetric, so that $\mathbf{R}^{-1} = \mathbf{R}^{\mathsf{T}}$ and $\mathbf{U}^{\mathsf{T}} = \mathbf{U}$. If **R** equals the identity tensor then $\mathbf{F}^{\mathsf{T}} = \mathbf{F}$.

Returning to the proof of the corollary we see from (13.27) that

$$\mathbf{A}-\mathbf{A}^{\mathsf{T}} = \frac{1}{|\mathbf{n}(\mathbf{F})^+|}\,(\mathbf{s}\otimes\mathbf{n}-\mathbf{n}\otimes\mathbf{s}).$$

Since **s** is parallel to **n** it follows that $\mathbf{s}\otimes\mathbf{n} = \mathbf{n}\otimes\mathbf{s}$ and hence that **A** is symmetric. Thus, if **A** is interpreted as a displacement tensor it represents a pure extension with $\mathbf{R} = \mathbf{1}$. □

Corollary 13.3

If the assumption of Corollary 13.2 is satified and the motion in one of the regions separated by the surface $\hat{\mathscr{S}}$ is irrotational then $\hat{\mathscr{S}}$ is at most a surface of constant discontinuity (independent of l^α) of the velocity field potential.

Proof

Assume that the motion is irrotational in the region $\hat{\mathscr{N}}^+$ so that there exists a velocity potential V such that

$$\mathbf{v} = -\operatorname{grad} V \quad \text{on} \quad \bigcup_{t\in I}\{t\}\times\boldsymbol{\chi}_t(\mathscr{B})^+.$$

If the shock wave is a surface of discontinuity of V then from (6.5) we obtain

$$[\![\mathbf{v}]\!] = -\left[\!\!\left[\frac{\partial V}{\partial n}\right]\!\!\right]\mathbf{n}-[\![V]\!]_{,\alpha}\,\boldsymbol{\varphi}^\alpha. \tag{13.28}$$

The assumption of Corollary 13.2 implies that $[\![\mathbf{v}]\!]$ is parallel to **n**, so that $[\![\mathbf{v}]\!]\cdot\boldsymbol{\varphi}_{,\beta} = 0$ for $\beta = 1, 2$. Equation (13.28) then gives

$$0 = [\![\mathbf{v}]\!]\cdot\boldsymbol{\varphi}_{,\beta} = -[\![V]\!]_{,\beta}, \qquad \beta = 1, 2.$$

This means that either $[\![V]\!] = 0$ or $[\![V]\!]$ is independent of the curvilinear coordinates l^1, l^2 of the surface $\mathscr{S}(t)$, $t\in I$, that is $[\![V]\!]$ is constant. □

We now examine the properties of the vorticity vector on a shock wave. As before we consider the situation in which the characteristic vector **s** is parallel to the normal. Such a situation arises, for example, in connection with materials whose constitutive law is of the form $\mathbf{T} = -p\mathbf{1}$, where **T** is the Cauchy stress tensor and p the hydrostatic pressure, as for a barotropic fluid.

Lemma 13.1

If the assumption of Corollary 13.2 is satisfied then the jumps of the components of the vorticity vector $\mathbf{w} = \operatorname{curl}\mathbf{v}$ satisfy

$$
\begin{aligned}
[\![w_n]\!] &\equiv 2\boldsymbol{\varphi}_{,1}\cdot[\![\mathbf{W}]\!]\boldsymbol{\varphi}_{,2} = 0,\\
[\![w_\alpha]\!] &\equiv 2\boldsymbol{\varphi}_{,\alpha}\cdot[\![\mathbf{W}]\!]\mathbf{n} = -\frac{1}{u_n}\left((u_n a)_{,\alpha}+\left[\!\!\left[\frac{\partial \mathbf{v}}{\partial t}\right]\!\!\right]\cdot\boldsymbol{\varphi}_{,\alpha}\right),
\end{aligned}
\tag{13.29}
$$

where $\mathbf{W}$ is the spin tensor defined by (9.18) and a is the projection of $[\![\mathbf{v}]\!]$ on the normal $\mathbf{n}$.

Proof

By use of (6.5) we obtain

$$[\![\mathbf{L}]\!] \equiv [\![\operatorname{grad}\mathbf{v}]\!] = [\![\mathbf{v}]\!]_{,\alpha}\otimes\boldsymbol{\varphi}^{\alpha}+\left[\!\!\left[\frac{\partial \mathbf{v}}{\partial n}\right]\!\!\right]\otimes\mathbf{n}$$

and hence

$$\boldsymbol{\varphi}_{,1}\cdot[\![\mathbf{L}-\mathbf{L}^{\mathrm{T}}]\!]\boldsymbol{\varphi}_{,2} = [\![\mathbf{v}]\!]_{,2}\cdot\boldsymbol{\varphi}_{,1}-[\![\mathbf{v}]\!]_{,1}\cdot\boldsymbol{\varphi}_{,2}.$$

Since $\mathbf{s}$ is parallel to $\mathbf{n}$ we have

$$[\![\mathbf{v}]\!] = a\mathbf{n},\qquad a\equiv[\![\mathbf{v}]\!]\cdot\mathbf{n} = -\mathbf{s}\cdot\mathbf{n}U_N.$$

On differentiation of $[\![\mathbf{v}]\!]$ we obtain

$$[\![\mathbf{v}]\!]_{,\gamma}\cdot\boldsymbol{\varphi}_{,\delta} = (a\mathbf{n})_{,\gamma}\cdot\boldsymbol{\varphi}_{,\delta} = a\mathbf{n}_{,\gamma}\cdot\boldsymbol{\varphi}_{,\delta} = -ab_{\gamma\delta},$$

use having been made of (1.6). From the symmetry of $b_{\gamma\delta}$ we deduce that

$$[\![w_n]\!] = a(b_{12}-b_{21}) = 0.$$

Next we have

$$2\boldsymbol{\varphi}_{,\alpha}\cdot[\![\mathbf{W}]\!]\mathbf{n} = \boldsymbol{\varphi}_{,\alpha}\cdot\left[\!\!\left[\frac{\partial \mathbf{v}}{\partial n}\right]\!\!\right]-\mathbf{n}\cdot[\![\mathbf{v}]\!]_{,\alpha}.$$

The second term on the right-hand side reduces to

$$\mathbf{n}\cdot[\![\mathbf{v}]\!]_{,\alpha} = \mathbf{n}\cdot(a\mathbf{n})_{,\alpha} = a_{,\alpha}.$$

Hence

$$[\![w_\alpha]\!] = \tilde{c}_\alpha - a_{,\alpha},$$

where we have introduced the notation

$$\tilde{c}_\alpha \equiv \boldsymbol{\varphi}_{,\alpha}\cdot\left[\!\!\left[\frac{\partial \mathbf{v}}{\partial n}\right]\!\!\right].$$

Applying (6.7) to $\mathbf{v}$, using (2.15) and taking the scalar product of the result with $\boldsymbol{\varphi}_{,\alpha}$ we obtain

$$\left[\!\!\left[\frac{\partial \mathbf{v}}{\partial t}\right]\!\!\right]\cdot\boldsymbol{\varphi}_{,\alpha} = -au_{n,\alpha}-u_n\tilde{c}_\alpha.$$

It follows that

$$[\![w_\alpha]\!] = -\frac{1}{u_n}\left\{au_{n,\alpha} + \left[\!\!\left[\frac{\partial \mathbf{v}}{\partial t}\right]\!\!\right]\cdot \boldsymbol{\varphi}_{,\alpha}\right\} - a_{,\alpha},$$

which can be rearranged in the required form $(13.29)_2$. □

It can be seen from the above result that if $\partial\mathbf{v}/\partial t = \mathbf{0}$ and $u_{n,\alpha} = 0$ then $[\![w_\alpha]\!] = -a_{,\alpha}$.

13.3 Vortex Sheets

Definiton 13.3

A singular non-absolutely material surface of order one whose local speed of propagation U is continuous is called a **vortex sheet** or a surface of contact discontinuity.

We note the consistent use of assumption (b) from Definition 9.2 which requires that every displacement of a body $\mathscr{B}$ be a homeomorphism. According to this condition all discontinuities of $\hat{\boldsymbol{\chi}}_t$, $t \in I$, and all non-uniqueness of the inverse of $\hat{\boldsymbol{\chi}}_t$ are rejected. This means that we do not consider singular surfaces of order zero with respect to the motion or its inverse. We remark, however, that a slip surface, in the sense of Helmholtz (1858), is an example of a singular surface of order zero on which the inverse motion $\mathbf{X} = \hat{\boldsymbol{\chi}}_t^{-1}(\mathbf{x})$ is discontinuous; such a surface is an absolutely material surface.

The adequacy of the definition of a vortex sheet is expressed by the following result:

Proposition 13.3

The characteristic vector $\mathbf{s}$ of a vortex sheet $\hat{\mathscr{S}}$ is orthogonal to the normal $\mathbf{n}$, that is

$$\mathbf{s}\cdot\mathbf{n} = 0. \tag{13.30}$$

If the motion in one of the regions, $\hat{\mathscr{N}}^+$ or $\hat{\mathscr{N}}^-$, separated by $\hat{\mathscr{S}}$ is irrotational then $\hat{\mathscr{S}}$ changes the laminarity of the velocity field in the other region (that is $\hat{\mathscr{N}}^-$ or $\hat{\mathscr{N}}^+$); in particular circumstances $\hat{\mathscr{S}}$ may initiate rotational motion.

On $\hat{\mathscr{S}}$ the vectors $\overset{-1}{\mathbf{F}}{}^{\mathrm{T}}\mathbf{N}$ and $\mathbf{F}^{\mathrm{T}}\mathbf{n}$ are continuous, so that

$$[\![\overset{-1}{\mathbf{F}}{}^{\mathrm{T}}]\!]\mathbf{N} = [\![\mathbf{F}^{\mathrm{T}}]\!]\mathbf{n} = \mathbf{0}. \tag{13.31}$$

Proof

The first property

$$[\![\mathbf{v}]\!]\cdot\mathbf{n} = \mathbf{s}\cdot\mathbf{n} = 0 \tag{13.30$_a$}$$

is obvious. For the proof of the second we assume that the motion is irrotational in the region $\hat{\mathscr{N}}^+$, that is the velocity field $\mathbf{v}(t, \mathbf{x})$, $t \in I$, $\mathbf{x} \in \boldsymbol{\chi}_t(\mathscr{B})^+$, is laminar; hence there exists a velocity potential V such that $\mathbf{v} = -\operatorname{grad} V$.

Assuming that $\hat{\mathscr{S}}$ does not change the laminarity of the velocity field, so that V is continuous across the surface $\mathscr{S}$ (which is the spatial image of $\hat{\mathscr{S}}$ in the set $\mathscr{N}^*$) or at most suffers a constant jump discontinuity, and recalling (13.28), it follows that

$$[\![\mathbf{v}]\!] = -[\![\operatorname{grad} V]\!] = -\left[\!\!\left[\frac{\partial V}{\partial n}\right]\!\!\right]\mathbf{n}. \tag{13.32}$$

In view of $(13.30)_a$ it follows that

$$\left[\!\!\left[\frac{\partial V}{\partial n}\right]\!\!\right] = 0.$$

On reference to (13.32) we see that this contradicts the assumption that $[\![\dot{\boldsymbol{\chi}}]\!] \neq \mathbf{0}$. Hence the motion in the region $\mathscr{N}^-$ cannot be laminar with continuous potential V. In other words, either the motion is laminar in $\mathscr{N}^-$ with $\mathscr{S}$ carrying a non-constant discontinuity in V,** or through $\mathscr{S}$ the irrotational motion in $\mathscr{N}^+$ becomes rotational in $\mathscr{N}^-$.

To prove the last property we return to (13.17); if $UU_N \neq 0$ then continuity of U implies (13.31). The proof of (13.31) for the case of $UU_N = 0$ is given in Example 13.2 below. □

Example 13.1

We shall prove that the connections between the normal vectors $\mathbf{N}$ and $\mathbf{n}$ may be expressed in the forms

$$\mathbf{N} = \frac{\mathbf{F}^{\mathrm{T}}\mathbf{n}}{\sqrt{\mathbf{n}\cdot\mathbf{B}\mathbf{n}}}, \qquad \mathbf{n} = \frac{\overset{-1}{\mathbf{F}}{}^{\mathrm{T}}\mathbf{N}}{\sqrt{\mathbf{N}\cdot\overset{-1}{\mathbf{C}}\mathbf{N}}}$$

where $\mathbf{B}$ is the left Cauchy–Green tensor defined by $\mathbf{B} = \mathbf{F}\mathbf{F}^{\mathrm{T}}$ and $\mathbf{C}$ is the right Cauchy–Green tensor given by (9.9).

We note that for any differentiable function $\hat{f}$ defined on $\hat{\mathscr{N}}$ and f defined on $\bigcup_{t\in I}\{t\}\times\boldsymbol{\chi}_t(\mathscr{B})$ such that

$$f(t, \hat{\boldsymbol{\chi}}_t(\mathbf{X})) = \hat{f}(t, \mathbf{X}), \qquad (t, \mathbf{X}) \in \hat{\mathscr{N}}$$

we have

$$(\operatorname{grad} f)\mathbf{F} = \operatorname{Grad}\hat{f}.$$

* See equation (14.1).

** In which case (13.28) applies rather than (13.32).

Hence

$$(\mathrm{Grad}\hat{f}) \cdot (\mathrm{Grad}\hat{f}) = \{(\mathrm{grad}f)\mathbf{F}\} \cdot (\mathbf{F}^{\mathrm{T}}\mathrm{grad}f) = (\mathrm{grad}f) \cdot (\mathbf{B}\,\mathrm{grad}f)$$

and similarly

$$(\mathrm{grad}f) \cdot (\mathrm{grad}f) = \{(\mathrm{Grad}\hat{f})\overset{-1}{\mathbf{F}}\} \cdot \{(\mathrm{Grad}\hat{f})\overset{-1}{\mathbf{F}}\}$$
$$= (\mathrm{Grad}\hat{f}) \cdot (\overset{-1}{\mathbf{C}}\,\mathrm{Grad}\hat{f}).$$

With f and $\hat{f}$ replaced by g and G respectively in the above, equations (13.9) and (13.10) yield the required formulae for the connections between $\mathbf{N}$ and $\mathbf{n}$. □

Example 13.2

Here we prove the validity of (13.31) under the assumption $UU_N = 0$. From Definition 13.3 we obtain (13.30). Hence

$$[\![\mathbf{F}^{\mathrm{T}}\mathbf{n}]\!] = \mathbf{n}[\![\mathbf{F}]\!] = \mathbf{n}(\mathbf{s}\otimes\mathbf{N}) = (\mathbf{n}\cdot\mathbf{s})\mathbf{N} = \mathbf{0}.$$

It follows that $[\![|\mathbf{F}^{\mathrm{T}}\mathbf{n}|]\!] = \mathbf{0}$ and hence that $[\![\mathbf{n}\cdot(\mathbf{B}\mathbf{n})]\!] = 0$.

In order to prove $(13.31)_2$ we make use of the transformation

$$[\![\overset{-1}{\mathbf{F}}{}^{\mathrm{T}}\mathbf{N}]\!] = \left[\!\!\left[\overset{-1}{\mathbf{F}}{}^{\mathrm{T}}\mathbf{F}^{\mathrm{T}}\frac{\mathbf{n}}{\sqrt{\mathbf{n}\cdot(\mathbf{B}\mathbf{n})}}\right]\!\!\right] = \frac{[\![\mathbf{n}]\!]}{\sqrt{\mathbf{n}\cdot(\mathbf{B}\mathbf{n})}} = \mathbf{0}. \quad □$$

The following example provides an additional property of the vortex sheet $\mathscr{S}$:

Example 13.3

If the velocity field $\mathbf{v}$ is such that the functions $\mathbf{v}||_{\mathscr{N}^-\cup\mathscr{S}}$ and $\mathbf{v}||_{\mathscr{N}^+\cup\mathscr{S}}$ are continuously differentiable then the vorticity vector $\mathbf{w} = \mathrm{curl}\,\mathbf{v}$ satisfies the integral equality

$$\int_{\chi_t(\mathscr{B})} \mathbf{w}\,dv = \int_{\partial(\chi_t(\mathscr{B}))^+ - \mathscr{S}(t)} \mathbf{m}\times\mathbf{v}_+ da + \int_{\partial(\chi_t(\mathscr{B}))^- - \mathscr{S}(t)} \mathbf{m}\times\mathbf{v}_- da$$
$$+ \int_{\mathscr{S}(t)} \mathbf{n}\times[\![\mathbf{v}]\!]\,da,$$

where $\mathbf{m}$ is the normal vector to the boundary of $\chi_t(\mathscr{B})$ and $\mathbf{v}_+$ and $\mathbf{v}_-$ are smooth extensions of the functions $\mathbf{v}||_{\mathscr{N}^+\cup\mathscr{S}}$ and $\mathbf{v}||_{\mathscr{N}^-\cup\mathscr{S}}$ respectively.

To prove this we use the Gauss–Ostrogradski identity

$$\int_{\mathscr{D}} \mathrm{curl}\,\mathbf{a}\,dv = \int_{\partial\mathscr{D}} \mathbf{m}\times\mathbf{a}\,da,$$

which is valid for any vector field $\mathbf{a}$ differentiable in the region $\mathscr{D}$ with sufficiently smooth boundary $\partial\mathscr{D}$. We write

$$\int_{\chi_t(\mathscr{B})} \mathbf{w}\, dv = \int_{\chi_t(\mathscr{B})^+} \operatorname{curl} \mathbf{v}\, dv + \int_{\chi_t(\mathscr{B})^-} \operatorname{curl} \mathbf{v}\, dv = \int_{\partial(\chi_t(\mathscr{B}))^+} \mathbf{m} \times \mathbf{v}_+ da$$

$$+ \int_{\partial(\chi_t(\mathscr{B}))^-} \mathbf{m} \times \mathbf{v}_- da = \int_{\partial(\chi_t(\mathscr{B}))^+ - \mathscr{S}(t)} \mathbf{m} \times \mathbf{v}_+ da + \int_{\mathscr{S}(t)} \mathbf{m} \times \mathbf{v}_+ da$$

$$+ \int_{\partial(\chi_t(\mathscr{B}))^- - \mathscr{S}(t)} \mathbf{m} \times \mathbf{v}_- da + \int_{\mathscr{S}(t)} \mathbf{m} \times \mathbf{v}_- da$$

$$= \int_{\partial(\chi_t(\mathscr{B}))^+ - \mathscr{S}(t)} \mathbf{m} \times \mathbf{v}_+ da + \int_{\partial(\chi_t(\mathscr{B}))^- - \mathscr{S}(t)} \mathbf{m} \times \mathbf{v}_- da + \int_{\mathscr{S}(t)} \mathbf{n} \times [\![\mathbf{v}]\!] da.$$

The integral equality is valid for any singular surface $\mathscr{S}$ of order one. If $\mathscr{S}$ is a vortex sheet then the final integrand is non-zero; on the other hand the integrand vanishes if $\mathscr{S}$ is a shock wave since $[\![\mathbf{v}]\!]$ is parallel to $\mathbf{n}$. □

The following result is characteristic for a vortex sheet (recall Corollary 13.2):

Corollary 13.4

The limit values $(\mathbf{F})^+$ and $(\mathbf{F})^-$ of the displacement tensor $\mathbf{F}$ on the vortex sheet $\mathscr{S}$ have different rotation tensors.

Proof

Equation (13.27) is valid in the present circumstances but, from (13.30), it follows that $\mathbf{s} \otimes \mathbf{n} \neq \mathbf{n} \otimes \mathbf{s}$ for a vortex sheet and hence that $\mathbf{A}$ is not symmetric. This means that the polar decomposition of $\mathbf{A}$, regarded as a displacement tensor, has the form $\mathbf{A} = \mathbf{R}_A \mathbf{U}_A$ with $\mathbf{R}_A \neq \mathbf{1}$. Thus we have proved that the tensors $(\mathbf{F})^+$ and $(\mathbf{F})^-$ have different rotational tensors. □

For a vortex sheet the analogue of Lemma 13.1 is the following result:

Lemma 13.2

On a vortex sheet discontinuities in the components of the vorticity vector satisfy

$$\begin{aligned} [\![w_n]\!] &\equiv 2\boldsymbol{\varphi}_{,1} \cdot [\![\mathbf{W}]\!] \boldsymbol{\varphi}_{,2} = a_{1;2} - a_{2;1}, \\ [\![w_\alpha]\!] &\equiv 2\boldsymbol{\varphi}_{,\alpha} \cdot [\![\mathbf{W}]\!] \mathbf{n} = -\frac{1}{u_n} \left\{ \left[\!\!\left[\frac{\partial \mathbf{v}}{\partial t} \right]\!\!\right] \cdot \boldsymbol{\varphi}_{,\alpha} - \frac{\delta a_\alpha}{\delta t} + a_\beta c^\beta_{,\alpha} \right], \end{aligned} \tag{13.33}$$

where we have used the notation

$$a_\alpha \equiv [\![\mathbf{v}]\!] \cdot \boldsymbol{\varphi}_{,\alpha}.$$

Proof

From the proof of Lemma 13.1 we have

$$[\![w_n]\!] = [\![\mathbf{v}]\!]_{,2} \cdot \boldsymbol{\varphi}_{,1} - [\![\mathbf{v}]\!]_{,1} \cdot \boldsymbol{\varphi}_{,2}.$$

On a vortex sheet $[\![\mathbf{v}]\!] = a^{\alpha}\boldsymbol{\varphi}_{,\alpha}$, and since

$$[\![\mathbf{v}]\!]_{,\gamma}\cdot\boldsymbol{\varphi}_{,\delta} = a^{\alpha}_{;\gamma}g_{\alpha\delta}$$

we obtain

$$[\![w_n]\!] = a^{\alpha}_{;2}g_{\alpha 1} - a^{\alpha}_{;1}g_{\alpha 2} = a_{1;2} - a_{2;1}$$

as required.

Next, we note that

$$[\![w_{\alpha}]\!] = \boldsymbol{\varphi}_{,\alpha}\cdot\left[\!\!\left[\frac{\partial\mathbf{v}}{\partial n}\right]\!\!\right] - \mathbf{n}\cdot[\![\mathbf{v}]\!]_{,\alpha}.$$

Successive calculations result in

$$\left[\!\!\left[\frac{\partial\mathbf{v}}{\partial t}\right]\!\!\right]\cdot\boldsymbol{\varphi}_{,\alpha} = \boldsymbol{\varphi}_{,\alpha}\cdot\frac{\delta}{\delta t}[\![\mathbf{v}]\!] - u_n\tilde{c}_{\alpha} = \frac{\delta a_{\alpha}}{\delta t} - a_{\beta}(c^{\beta}_{,\alpha} - u_n b^{\beta}_{\alpha}) - u_n\tilde{c}_{\alpha},$$

$$\mathbf{n}\cdot[\![\mathbf{v}]\!]_{,\alpha} = \mathbf{n}\cdot(a^{\delta}\boldsymbol{\varphi}_{,\delta})_{,\alpha} = a^{\delta}b_{\delta\alpha} = a_{\beta}b^{\beta}_{\alpha}.$$

Finally, in view of the formula

$$\tilde{c}_{\alpha} = -\frac{1}{u_n}\left\{\left[\!\!\left[\frac{\partial\mathbf{v}}{\partial t}\right]\!\!\right]\cdot\boldsymbol{\varphi}_{,\alpha} - \frac{\delta a_{\alpha}}{\delta t} + a_{\beta}(c^{\beta}_{,\alpha} - u_n b^{\beta}_{\alpha})\right\},$$

we obtain $(13.33)_2$. Note, that here $\dfrac{\delta a_{\alpha}}{\delta t} = \dfrac{\partial a_{\alpha}}{\partial t} - a_{\alpha,\delta}c^{\delta}$. However, we have

$$\frac{\mathrm{d}a_{\alpha}}{\delta t} - a_{\beta}c^{\beta}_{,\alpha} = \frac{\partial a_{\alpha}}{\partial t} - a_{\alpha;\beta}c^{\beta} - a_{\beta}c^{\beta}_{;\alpha}.\ \square$$

13.4 Bibliographical Notes

The definition of an absolutely material surface introduced in Section 13.1 is original. A definition of a shock wave, identical to that given is Section 13.2, may be found in Truesdell and Toupin (1960).

In differentiating surfaces with normal and tangential discontinuities we have followed work in hydrodynamics (see Landau and Lifshitz, 1959, and Kočin, 1965).

Extensive discussion of the properties of shock waves and vortex sheets may be found in Truesdell and Toupin (1960); see also Kočin (1926, 1965).

Continuity conditions for $J\overset{-1}{\mathbf{F}}{}^{\mathrm{T}}\mathbf{N}$ and $J\overset{-1}{\mathbf{F}}{}^{\mathrm{T}}\mathbf{n}$, similar to (13.17), were given by Eringen and Şuhubi (1974) and Chen and Wright (1975), who made use of the conservation of mass on a shock wave in the form of the Stokes–Christoffel condition (10.9).

The amplitude equation (13.21) was derived by Huilgol (1973) coupled with the equation of motion.

The properties of shock waves were investigated in detail by Truesdell and Toupin (1960, pp. 519–523), but their article does not mention the properties given in Corollary 13.2. A proposition corresponding to Corollary 13.3 can be found in the same article (p. 494).

The results of Lemma 13.1 for a barotropic fluid in steady plane flow were derived by Truesdell (1952); their generalizations were given by Lighthill (1957) and Hayes (1957). A more detailed derivation, utilizing the Stokes–Christoffel relation and the conditions ahead of the wave in the region $\mathcal{N}^+$, is given by Bowen, Chen and McCarthy (1976); see also Chen (1976).

Material vortex sheets were discussed in Truesdell and Toupin (1960, pp. 515–517). Our discussion of non-material vortex sheets is new and, in particular, the properties (13.31), Corollary 13.4 and Lemma 13.2 have not been given previously.

Because of the complexity of spatial problems of the propagation of shock waves there is little literature on the subject. Certain general properties of multi-dimensional shock waves in fluids and solids are discussed in Weyl (1949), Hill (1961), Bland (1964, 1969), Chu (1967), Currie (1972), Germain (1972), Mandel (1976), Wesołowski and Bürger (1977), Wesołowski (1982), Anile (1985), Hanyga (1985), Włodarczyk (1977), Freudenthal and Geiringer (1958); see also Olszak *et al.* (1965), Bejda (1972) and Kukudžanow (1974). Waves in rods were discussed in Cohen (1978) and Bachman and Cohen (1979) and in membranes by Braun (1983).

Extensive information concerning one-dimensional shock waves is contained in Achenbach (1973), Chen (1973), Schuler *et al.* (1973), Spence (1973), Nunziato *et al.* (1974), Eringen and Şuhubi (1974), McCarthy (1975), Kosiński (1976); see also Courant and Friedrichs (1948), Dunwoody (1972), W. Nowacki (1975a, b), W. K. Nowacki (1978) and Hanyga (1984, 1985, 1986). Not discussed here is the subject of coupled electromagnetic waves; see, however, McCarthy and O'Leary (1975); see also McCarthy (1966, 1973), Chen and McCarthy (1976). We have not touched at all the physical, better to say, experimental picture of the shock front (wave). From the physical viewpoint a shock wave is a three-dimensional layer, with its own structure, strongly depending on the physical properties of the medium. From the existing papers dealing with this problem we should mention the papers by Fiszdon *et al.* (1974, 1976, 1983) as well as the report by Płatkowski (1981), both for gases; see also Duvall (1961) for solids.

13.5 Referential Form of the Local Balance Equations on a Wave

In many problems it is convenient to use a form of the balance laws and compatibility conditions on a singular surface of the motion that relates to some reference placement. In the theory of finite deformations of solids, for instance, the actual shape of the deformed boundary is not in general known and must be determined from the solution of the field equations. However, if the undeformed shape of a body is known in some reference placement (as is usually the case), the boundary of the body in that placement is also known.

In order to obtain the form of the compatibility conditions which result from the local balance laws relative to the variables $(t, \mathbf{X})$, where $\mathbf{X} \in \varkappa_0(\mathscr{B})$, we require the basic relations between volume and surface elements, namely

$$dv = J\,dv_\varkappa, \qquad \mathbf{F}^T\mathbf{m}\,da = J\mathbf{m}_\varkappa\,da_\varkappa.$$

For an arbitrary surface density w with the representation $(8.6)_1$ we have

$$\int_{\partial\chi_t(\mathscr{P})} w\,da = \int_{\partial\chi_t(\mathscr{P})} \mathbf{w}\cdot\mathbf{m}\,da = \int_{\partial\varkappa_0(\mathscr{P})} \mathbf{w}_\varkappa\cdot\mathbf{m}_\varkappa\,da_\varkappa,$$

where $\mathbf{w}_\varkappa$ is the flux $\mathbf{w}$ shifted to the reference placement. The relationship between $\mathbf{w}_\varkappa$ and $\mathbf{w}$ is

$$\mathbf{w} = J^{-1}\mathbf{w}_\varkappa\mathbf{F}^T, \qquad \mathbf{w}_\varkappa = J\mathbf{w}(\mathbf{F}^T)^{-1}. \tag{13.34}$$

For an arbitrary volume density r, as appears in the representations (8.5), we have

$$\int_{\chi_t(\mathscr{P})} r\,dv = \int_{\varkappa_0(\mathscr{P})} r_\varkappa\,dv_\varkappa,$$

and hence

$$r_\varkappa = Jr.$$

In particular, if $r = \varrho$ then the local mass balance equation assumes the form

$$\frac{\partial}{\partial t}(\varrho J) = 0$$

in terms of the variables t and $\mathbf{X} \in \varkappa_0(\mathscr{B})$. If there exists a time t_0 at which the motion $\chi_t\colon \mathscr{B} \to \mathscr{E}^3$ passes through the reference placement, that is $\chi_{t_0}(\mathscr{B}) = \varkappa_0(\mathscr{B})$, then we obtain $\varrho J = \varrho_\varkappa$, where $\varrho_\varkappa$ is the mass density in the reference placement (recall (10.13)). Thus, in the material description, the first compatibility condition on a singular surface has the form $[\![\varrho_\varkappa U_N]\!] = 0$ or, equivalently,

$$U_N[\![\varrho_\varkappa]\!] = 0. \tag{13.35}$$

We interpret this as follows: either the distribution of mass density in the reference placement is continuous or we are dealing with a material singular surface ($U_N = 0$) with a discontinuous distribution of density $\varrho_\varkappa$.

From (12.2) and (13.35) we have the three equivalent conditions

$$\left[\!\!\left[\frac{\varrho U}{\varrho_\varkappa U_N}\right]\!\!\right] = 0, \quad [\![J^{-1}|\mathbf{F}^\mathsf{T}\mathbf{n}|]\!] = 0, \quad [\![J|\overset{-1}{\mathbf{F}}{}^\mathsf{T}\mathbf{N}|]\!] = 0, \tag{13.36}$$

use having been made of (13.12).

Next we consider the momentum conservation equation (12.3). We introduce the stress tensor $\mathbf{T}_\varkappa$ related to the Cauchy stress tensor $\mathbf{T}$ by

$$\mathbf{T}_\varkappa = J\mathbf{T}\overset{-1}{\mathbf{F}}{}^\mathsf{T}, \quad \mathbf{T} = J^{-1}\mathbf{T}_\varkappa\mathbf{F}^\mathsf{T}. \tag{13.37}$$

In terms of $\mathbf{T}_\varkappa$ called the first Piola–Kirchhoff tensor, (12.3) may be re-written as

$$[\![\varrho\mathbf{v}(v_n - u_n) - J^{-1}\mathbf{T}_\varkappa\mathbf{F}^\mathsf{T}\mathbf{n}]\!] = -[\![J^{-1}|\overset{-1}{\mathbf{F}}{}^\mathsf{T}\mathbf{N}|^{-1}(\varrho_\varkappa\mathbf{v}U_N + \mathbf{T}_\varkappa\mathbf{N})]\!] = \mathbf{0},$$

and it follows from (13.36) that this simplifies to

$$\varrho_\varkappa U_N[\![\mathbf{v}]\!] = -[\![\mathbf{T}_\varkappa]\!]\mathbf{N}$$

or, equivalently, to

$$[\![\varrho_\varkappa U_N\mathbf{v} + \mathbf{T}_\varkappa\mathbf{N}]\!] = \mathbf{0}. \tag{13.38}$$

In order to deal with the energy balance law we introduce the referential form $\mathbf{q}_\varkappa$ of the heat flux $\mathbf{q}$, related to $\mathbf{q}$ by

$$\mathbf{q} = J^{-1}\mathbf{q}_\varkappa\mathbf{F}^\mathsf{T}, \quad \mathbf{q}_\varkappa = J\mathbf{q}\overset{-1}{\mathbf{F}}{}^\mathsf{T}. \tag{13.39}$$

Equation (12.5) can then by rearranged as

$$[\![\varrho_\varkappa U_N(e + \tfrac{1}{2}\mathbf{v}\cdot\mathbf{v}) + (\mathbf{v}\mathbf{T}_\varkappa - \mathbf{q}_\varkappa)\cdot\mathbf{N}]\!] = 0$$

or

$$\varrho_\varkappa U_N[\![e + \tfrac{1}{2}\mathbf{v}\cdot\mathbf{v}]\!] = -[\![\mathbf{v}\mathbf{T}_\varkappa - \mathbf{q}_\varkappa]\!]\cdot\mathbf{N}. \tag{13.40}$$

With the notation

$$\mathbf{m}_{\eta\varkappa} = J\mathbf{m}_\eta\overset{-1}{\mathbf{F}}{}^\mathsf{T}$$

we obtain the entropy balance equations

$$\begin{aligned}&[\![\varrho_\varkappa \eta U_N - \mathbf{m}_{\eta\varkappa}\cdot\mathbf{N}]\!] = 0,\\ &[\![\varrho_\varkappa \eta U_N - \theta^{-1}\mathbf{q}_\varkappa\cdot\mathbf{N}]\!] = 0\end{aligned} \tag{13.41}$$

from (12.7) and (12.8) respectively.

The material counterparts of the entropy production inequalities (12.10) and (12.11) are

$$[\![\varrho_\varkappa \eta U_N - \mathbf{m}_{\eta\varkappa}\cdot\mathbf{N}]\!] \geqslant 0 \quad \text{and} \quad [\![\varrho_\varkappa \eta U_N - \theta^{-1}\mathbf{q}_\varkappa\cdot\mathbf{N}]\!] \geqslant 0 \tag{13.42}$$

respectively.

Under adiabatic conditions and assuming that U_N is positive, the inequality (13.42), together with (13.35), leads to

$$[\![\eta]\!] \geqslant 0,$$

which states that entropy cannot decrease across a shock wave in this case.

14 SINGULAR SURFACES OF ORDER TWO

As in Section 13 a singular surface of the motion is represented by $\hat{\mathscr{S}}$ or $\mathscr{S}$ according to whether the reference or current placements are used in its description. Thus

$$\hat{\mathscr{S}} = \bigcup_{t\in I}\{t\}\times\Sigma(t), \qquad \mathscr{S} = \bigcup_{t\in I}\{t\}\times\mathscr{S}(t). \tag{14.1}$$

In many papers $\Sigma(t)$ and $\mathscr{S}(t)$ are referred to respectively as the material and spatial representations of a singular surface (see Figure 13.1; the idea of this as well as to the Figures 7.1 and 1.1 has been taken from figures which appeared in Eringen and Şuhubi (1974)).

In the present section we shall consider the situation in which the first derivatives of $\boldsymbol{\chi}$ are continuous on their domains, but the second derivatives suffer jump discontinuities. This means that we shall deal with singular surfaces of order two for $\boldsymbol{\chi}$. Such a surface is characterized by the conditions

$$[\![\dot{\boldsymbol{\chi}}]\!] = \mathbf{0} \quad \text{and} \quad [\![\mathbf{F}]\!] = \mathbf{0}$$

together with

$$[\![\ddot{\boldsymbol{\chi}}]\!] \neq \mathbf{0} \quad \text{and/or} \quad [\![\operatorname{Grad}\mathbf{F}]\!] \neq \mathbf{0}.$$

In what follows the symbols $\ddot{\boldsymbol{\chi}}$, $\ddot{\mathbf{x}}$ and $\dot{\mathbf{v}}$ will all be used for the acceleration field. However, it should be remembered that $\ddot{\boldsymbol{\chi}}$ and $\ddot{\mathbf{x}}$ are defined on the set

$$\hat{\mathscr{N}} = I\times\boldsymbol{\varkappa}_0(\mathscr{B}), \tag{14.2}$$

while $\dot{\mathbf{v}}$, like $\mathbf{v}$, is defined on

$$\mathcal{N} = \bigcup_{t\in I} \{t\}\times \chi_t(\mathcal{B}). \tag{14.3}$$

For functions defined on $\mathcal{N}$ the dot denotes the **material time derivative.** Thus, for example,

$$\dot{\mathbf{v}}(t, \mathbf{x}) \equiv \frac{\partial}{\partial t}\mathbf{v}+\mathbf{v}\cdot \operatorname{grad}\mathbf{v} = \frac{\partial \mathbf{v}}{\partial t}+\dot{\boldsymbol{\chi}}\cdot \operatorname{grad}\mathbf{v}. \tag{14.4}$$

The compatibility conditions (6.14) for the second derivatives of $\hat{\boldsymbol{\chi}}$ are

$$[\![\operatorname{Grad}\mathbf{F}]\!] = \left[\!\!\left[\frac{\partial^2\hat{\boldsymbol{\chi}}}{\partial N^2}\right]\!\!\right]\otimes\mathbf{N}\otimes\mathbf{N},$$

$$[\![\dot{\mathbf{F}}]\!] = -U_N\left[\!\!\left[\frac{\partial^2\hat{\boldsymbol{\chi}}}{\partial N^2}\right]\!\!\right]\otimes\mathbf{N}, \tag{14.5}$$

$$[\![\ddot{\mathbf{x}}]\!] = U_N^2\left[\!\!\left[\frac{\partial^2\hat{\boldsymbol{\chi}}}{\partial N^2}\right]\!\!\right].$$

According to the notation introduced in Section 3 the vector $\partial^2\hat{\boldsymbol{\chi}}/\partial N^2$ is defined by

$$\frac{\partial^2\hat{\boldsymbol{\chi}}}{\partial N^2} = (\operatorname{Grad}\operatorname{Grad}\hat{\boldsymbol{\chi}})\cdot(\mathbf{N}\otimes\mathbf{N}).$$

When we substitute the velocity $\dot{\mathbf{x}}$ for the function f in $(6.12)_3$ and use the variables appropriate to the region $\hat{\mathcal{N}}$, the displacement derivative (13.19) enables us to write

$$[\![\ddot{\mathbf{x}}]\!] = -2U_N\frac{\delta}{\delta t}([\![\dot{\mathbf{F}}]\!]\mathbf{N})-[\![\dot{\mathbf{F}}]\!]\mathbf{N}\frac{\delta U_N}{\delta t}+U_N^2\left[\!\!\left[\frac{\partial\dot{\mathbf{F}}}{\partial N}\right]\!\!\right]\mathbf{N}. \tag{14.6}$$

Provided $U_N \neq 0$ equation (14.5) yields

$$[\![\dot{\mathbf{F}}]\!]\mathbf{N} = -U_N\left[\!\!\left[\frac{\partial^2\hat{\boldsymbol{\chi}}}{\partial N^2}\right]\!\!\right] = -U_N^{-1}[\![\ddot{\mathbf{x}}]\!]. \tag{14.7}$$

Hence,

$$\begin{aligned}
[\![\dddot{\mathbf{x}}]\!] &= 2\frac{\delta}{\delta t}[\![\ddot{\mathbf{x}}]\!]-U_N^{-1}[\![\ddot{\mathbf{x}}]\!]\frac{\delta U_N}{\delta t}+U_N^2\left[\!\!\left[\frac{\partial\dot{\mathbf{F}}}{\partial N}\right]\!\!\right]\mathbf{N},\\
[\![\operatorname{Grad}\dot{\mathbf{F}}]\!] &= \left[\!\!\left[\frac{\partial^2\dot{\boldsymbol{\chi}}}{\partial N^2}\right]\!\!\right]\otimes\mathbf{N}\otimes\mathbf{N}+U_N^{-1}\hat{b}_{\Gamma\Delta}[\![\ddot{\mathbf{x}}]\!]\otimes\mathbf{P}^\Gamma\otimes\mathbf{P}^\Delta\\
&\quad-(U_N^{-1}[\![\ddot{\mathbf{x}}]\!])_{,\Gamma}\otimes(\mathbf{N}\otimes\mathbf{P}^\Gamma+\mathbf{P}^\Gamma\otimes\mathbf{N}),
\end{aligned} \tag{14.8}$$

where $\hat{b}_{\Gamma\Delta}$ are the components of the second metric tensor of the surface $\hat{\mathcal{S}}$, and $\mathbf{P}^\Gamma = \hat{g}^{\Gamma\Delta}\mathbf{P}_{,}$, $\mathbf{P}$ being defined by (13.6).

Under the additional assumption that U_N is positive equation $(14.8)_1$ may be shortened to

$$2\sqrt{U_N}\,\frac{\delta}{\delta t}\left(\frac{[\![\dot{\mathbf{x}}]\!]}{\sqrt{U_N}}\right) = [\![\ddot{\mathbf{x}}]\!] - U_N^2\left[\!\!\left[\frac{\partial\dot{\mathbf{F}}}{\partial N}\right]\!\!\right]\mathbf{N}. \tag{14.9}$$

or, if (14.7) is used,

$$2\sqrt{U_N}\,\frac{\delta}{\delta t}\left(U_N^{1/2}[\![\dot{\mathbf{F}}]\!]\mathbf{N}\right) = U_N^2\left[\!\!\left[\frac{\partial\dot{\mathbf{F}}}{\partial N}\right]\!\!\right]\mathbf{N} - [\![\ddot{\mathbf{x}}]\!]. \tag{14.10}$$

The equation

$$2\sqrt{U_N}\,\frac{\delta}{\delta t}\left(U_N^{2/3}\left[\!\!\left[\frac{\partial^2\hat{\boldsymbol{\chi}}}{\partial N^2}\right]\!\!\right]\right) = [\![\ddot{\mathbf{x}}]\!] - U_N^2\left[\!\!\left[\frac{\partial\dot{\mathbf{F}}}{\partial N}\right]\!\!\right]\mathbf{N} \tag{14.11}$$

is equivalent to (14.9) and (14.10).

Thus we have derived the basic equations governing the evolution of jump discontinuities of the second derivatives of $\hat{\chi}$ along a normal trajectory. We should remember, however, that equations (14.9)–(14.11) have been derived under the assumption that U_N is positive. The case in which U_N is negative is governed by (14.6) and (14.7), and by

$$3U_N\left[\!\!\left[\frac{\partial^2\hat{\boldsymbol{\chi}}}{\partial N^2}\right]\!\!\right]\frac{\delta U_N}{\delta t} + 2U_N^2\,\frac{\delta}{\delta t}\left[\!\!\left[\frac{\partial^2\hat{\boldsymbol{\chi}}}{\partial N^2}\right]\!\!\right] = [\![\ddot{\mathbf{x}}]\!] - U_N^2\left[\!\!\left[\frac{\partial\dot{\mathbf{F}}}{\partial N}\right]\!\!\right]\mathbf{N}. \tag{14.12}$$

In the theory of one-dimensional waves the equation corresponding to those above is called the **amplitude equation** for acceleration waves.

14.1 Material Surfaces

In Section 13.1 we defined an absolutely material surface and in Section 13.2 a one-sided material surface. For the surface now under consideration the two definitions are indistinguishable since the velocity is continuous.

Definition 14.1

A singular surface of order two on which the normal component U_N of the speed vanishes is called a **material surface**.

If we ignore situations in which the growth of $|\mathbf{F}^{\mathrm{T}}\mathbf{n}|$ is unlimited or $|\overset{-1}{\mathbf{F}}{}^{\mathrm{T}}\mathbf{N}| = 0$ then the following proposition is valid:

Proposition 14.1

A singular surface of order two is a material surface if and only if one of the following conditions holds:

(a) the local speed vanishes, that is

$$U = 0, \tag{14.13}$$

(b) the acceleration is continuous and there is a jump discontinuity in the gradient of the displacement tensor, that is

$$[\![\ddot{\mathbf{x}}]\!] = \mathbf{0} \quad \text{and} \quad [\![\mathrm{Grad}\,\mathbf{F}]\!] \neq \mathbf{0}, \tag{14.14}$$

(c) the time derivative of the displacement tensor is continuous, and there is a jump discontinuity in the gradient of the displacement tensor, that is

$$[\![\dot{\mathbf{F}}]\!] = \mathbf{0} \quad \text{and} \quad [\![\mathrm{Grad}\,\mathbf{F}]\!] \neq \mathbf{0}. \;\square \tag{14.15}$$

The proof of this proposition is obvious. Clearly, on a material surface we have $[\![\dddot{\mathbf{x}}]\!] = \mathbf{0}$.

The material surface defined here is used in the analysis of the phenomenon of deformation instability. To illustrate this we consider a model of instability in which the notion of localized deformation is used. We say that in a material continuum there is a localization of the deformation if a continuous contact force rate field results in a discontinuous deformation gradient field, that is

$$[\![\dot{\mathbf{T}}]\!] = \mathbf{0} \quad \text{and} \quad [\![\mathrm{Grad}\,\mathbf{F}]\!] \neq \mathbf{0}.$$

With an appropriate constitutive relation this is equivalent to (14.15). According to Definition 14.1 this means that a surface of localized deformation is a material surface.

14.2 Acceleration Waves

Definition 14.2

A singular surface of order two on which the acceleration field suffers a jump discontinuity is called an **acceleration wave**.

The characteristic quantity in the analysis of acceleration waves is the non-vanishing vector $\hat{\mathbf{a}} \equiv [\![\partial^2\hat{\boldsymbol{\chi}}/\partial N^2]\!]$.

The compatibility conditions (14.5) for an acceleration wave may be expressed in terms of the normal $\mathbf{n}$ and the local speed of propagation U. Using (13.13) we obtain

$$\begin{aligned}
[\![\mathrm{Grad}\,\mathbf{F}]\!] &= \left(\frac{U_N}{U}\right)^2 \hat{\mathbf{a}} \otimes \mathbf{F}^{\mathsf{T}}\mathbf{n} \otimes \mathbf{F}^{\mathsf{T}}\mathbf{n}, \\
[\![\dot{\mathbf{F}}]\!] &= -\left(\frac{U_N}{U}\right) U_N \hat{\mathbf{a}} \otimes \mathbf{F}^{\mathsf{T}}\mathbf{n}, \\
[\![\ddot{\mathbf{x}}]\!] &= U_N^2 \hat{\mathbf{a}}.
\end{aligned} \tag{14.16}$$

Introducing the amplitude vector

$$\mathbf{a} = \left(\frac{U_N}{U}\right)^2 \hat{\mathbf{a}} \tag{14.17}$$

of the acceleration wave we can write (14.16) in the form

$$\begin{aligned} [\![\mathrm{Grad}\,\mathbf{F}]\!] &= \mathbf{a} \otimes \mathbf{F}^{\mathsf{T}}\mathbf{n} \otimes \mathbf{F}^{\mathsf{T}}\mathbf{n}, \\ [\![\dot{\mathbf{F}}]\!] &= -U\mathbf{a} \otimes \mathbf{F}^{\mathsf{T}}\mathbf{n}, \\ [\![\ddot{\mathbf{x}}]\!] &= U^2\mathbf{a}. \end{aligned} \tag{14.18}$$

The acceleration wave is called a **longitudinal wave** if

$$\mathbf{a} \times \mathbf{n} = \mathbf{0} \tag{14.19}$$

and a **transverse wave** if

$$\mathbf{a} \cdot \mathbf{n} = 0. \tag{14.20}$$

Proposition 14.2

For a longitudinal wave

$$\begin{aligned} \mathbf{a}_{,\alpha} \cdot \boldsymbol{\varphi}_{,\alpha} &= -b_{\gamma\alpha}\mathbf{a} \cdot \mathbf{n}, \\ \hat{\mathbf{a}}_{,\gamma} \cdot \boldsymbol{\varphi}_{,\alpha} &= -b_{\gamma\alpha}\hat{\mathbf{a}} \cdot \mathbf{n}, \end{aligned} \tag{14.21}$$

and for a transverse wave

$$\begin{aligned} \mathbf{a}_{,\gamma} \cdot \mathbf{n} &= -b_{\gamma}^{\alpha}\mathbf{a} \cdot \boldsymbol{\varphi}_{,\alpha}, \\ \hat{\mathbf{a}}_{,\gamma} \cdot \mathbf{n} &= -b_{\gamma}^{\alpha}\hat{\mathbf{a}} \cdot \boldsymbol{\varphi}_{,\alpha}. \end{aligned} \tag{14.22}$$

The proof follows by direct differentiation and application of the Gauss–Weingarten formulae (1.7). □

Corollary 14.1

On an acceleration wave the following equations hold:

$$\begin{aligned} &[\![\mathrm{div}\,\mathbf{v}]\!] = -U\mathbf{a} \cdot \mathbf{n}, \qquad [\![\mathrm{curl}\,\mathbf{v}]\!] = -U\mathbf{a} \times \mathbf{n}, \\ &2u_n \frac{\delta}{\delta t}\left[\!\!\left[\frac{\partial \mathbf{v}}{\partial n}\right]\!\!\right] + \left[\!\!\left[\frac{\partial \mathbf{v}}{\partial t}\right]\!\!\right]\frac{\delta u_n}{\delta t} \\ &\quad = u_n^2\left([\![\mathrm{div}\,\mathrm{grad}\,\mathbf{v}]\!] + 2K_m\left[\!\!\left[\frac{\partial \mathbf{v}}{\partial n}\right]\!\!\right]\right) - \left[\!\!\left[\frac{\partial^2 \mathbf{v}}{\partial t^2}\right]\!\!\right]. \end{aligned} \tag{14.23}$$

Proof

In the current placement we deal with functions defined on the set $\mathcal{N}$. For the velocity field $\mathbf{v}$ we have

$$\left[\!\!\left[\frac{\partial \mathbf{v}}{\partial t}\right]\!\!\right] = -u_n\left[\!\!\left[\frac{\partial \mathbf{v}}{\partial n}\right]\!\!\right], \qquad [\![\mathrm{grad}\,\mathbf{v}]\!] = \left[\!\!\left[\frac{\partial \mathbf{v}}{\partial n}\right]\!\!\right] \otimes \mathbf{n}.$$

In view of (14.4) we have

$$[\![\ddot{\mathbf{x}}]\!] = \left[\!\!\left[\frac{\partial \mathbf{v}}{\partial t}\right]\!\!\right] + [\![\operatorname{grad}\mathbf{v}]\!]\mathbf{v} = -(u_n - v_n)\left[\!\!\left[\frac{\partial \mathbf{v}}{\partial n}\right]\!\!\right].$$

Hence

$$[\![\ddot{\mathbf{x}}]\!] = -U\left[\!\!\left[\frac{\partial \mathbf{v}}{\partial n}\right]\!\!\right]. \tag{14.24}$$

Equations $(14.23)_{1,2}$ follow on use of $(14.18)_3$.

We note that if the wave is longitudinal then $[\![\partial \mathbf{v}/\partial n]\!] \times \mathbf{n} = \mathbf{0}$.

To obtain $(14.23)_3$ we apply the compatibility condition $(6.12)_3$ to $\mathbf{v}$ to give

$$\left[\!\!\left[\frac{\partial^2 \mathbf{v}}{\partial t^2}\right]\!\!\right] = -2u_n \frac{\delta}{\delta t}\left[\!\!\left[\frac{\partial \mathbf{v}}{\partial n}\right]\!\!\right] - \left[\!\!\left[\frac{\partial \mathbf{v}}{\partial n}\right]\!\!\right]\frac{\delta u_n}{\delta t} + u_n^2\left[\!\!\left[\frac{\partial^2 \mathbf{v}}{\partial n^2}\right]\!\!\right].$$

Use of (6.22) leads to

$$[\![\operatorname{div}\operatorname{grad}\mathbf{v}]\!] = \left[\!\!\left[\frac{\partial^2 \mathbf{v}}{\partial n^2}\right]\!\!\right] - 2K_m\left[\!\!\left[\frac{\partial \mathbf{v}}{\partial n}\right]\!\!\right]$$

and thence to the required result. □

From $(14.23)_{1,2}$ we see that the vorticity is continuous across a longitudinal acceleration wave, while the dilatation rate is continuous across a transverse wave. If mass is conserved then we see from (10.11) that the rate of change of the mass density is also continuous across a transverse wave. The result $(14.23)_1$ shows that there can be no longitudinal waves in an incompressible body.

To end this section we point to a possible alternative approach to the discussion of acceleration waves from that based on the displacement derivative of Thomas and the amplitude equation in the forms (14.9)–(14.12). The Thomas derivative measures the change in mechanical and thermodynamical quantities or their jumps defined on a singular surface along the normal trajectory of the moving surface $\{\mathscr{S}(t)\}_{t\in I}$.

For many problems, especially those concerned with isotropic media and waves propagating into undisturbed regions, the equation governing changes in the characteristic vector of an acceleration wave is an ordinary differential equation with respect to time t (the parameter of the normal trajectory). But this is not always the case. Very often the amplitude equation contains tangential derivatives with respect to the surface parameters (l^1, l^2), as in (14.8), in addition to the time derivative. In such cases it is not in general possible to find a solution of the equation.

The appearance of the tangential derivatives in the amplitude equation implies that the normal trajectories are not curves along which the principal disturbances connected with the acceleration wave propagate. A disturbance is principal if its rate of change with respect to the parameter of the curve (trajectory) is independent of the surface parameters.

In order to avoid tangential derivatives in the amplitude equation we have to find the non-orthogonal trajectories (curves) along which the principal disturbances propagate and replace the Thomas displacement derivative by a derivative measuring changes in the disturbance along the new trajectories.

From (2.7), (2.8) and (2.26) we can see that in the case of a convected parametrization the displacement derivative is identical to the partial derivative with respect to time. If the curve along which principal disturbances propagate is not the normal trajectory then it is necessary to use a non-convected parametrization instead. This parametrization should be chosen in such a way that the displacement velocity **c** of the moving surface defines the direction of the curve (recall (1.15)). In other words, the curves along which the principal disturbances propagate are integral curves of the field **c**.

In the theory of hyperbolic equations these curves are called **bi-characteristics**.

In Chapter 6 we shall present a formal theory of acceleration waves in which we do not assume that the principal disturbances propagate along the normal trajectories.

14.3 Bibliographical Notes

The problem of localization of the deformation in material surfaces is discussed in Rice (1976), Perzyna (1982) and Hill (1967) for elastic-plastic materials, and in Wesołowski (1977) for elastic materials (see also Drugan and Rice (1983)).

It is difficult to establish the origin of the formulae (14.8)–(14.12) for the amplitude of an acceleration wave.

Equations (14.21) and (14.22) for longitudinal and transverse waves respectively were derived by Chen (1968a), while $(14.23)_{1,2}$ were discovered by Weingarten (1901). The amplitude equation $(14.23)_3$ was given by Doria and Bowen (1970).

After the time of Hadamard modern investigations of acceleration waves were revived by the elegant paper of Ericksen (1953) on waves in incompressible isotropic elastic materials. Following the important general paper by Thomas (1957) the next major step was taken by Truesdell (1961) who, in a work of characteristic erudition, unified and extended the earlier work on wave propagation in non-linear elastic media.

The year 1964 brought further achievements, with Green (1964), using the Thomas derivative, analyzing the variation of the amplitude of plane acceleration waves propagating into a homogeneously deformed elastic material; Chu (1964) and Bland (1964) obtained simple wave solutions for simple shear waves and for simple longitudinal waves in incompressible and compressible non-heat conducting elastic media. In the following year, in the series of papers by Coleman *et al.* (1965) on the propagation of waves in simple materials with fading memory, appeared an article by Coleman and Gurtin (1965) on three-dimensional acceleration waves; see also Coleman *et al.* (1966).

Since 1965 numerous original papers and surveys dealing with the analysis of acceleration waves in elastic, visco-elastic, elastic-plastic and generally dissipative materials have been published. For non-elastic materials, however, much of the literature is concerned with one-dimensional waves.

We mention the following papers which deal with three-dimensional acceleration waves: Mandel (1962, 1976), Chadwick and Powdrill (1965), Varley and Dunwoody (1965), Green (1965), Varley (1965a, b), Gurtin and Walsh (1967), Bürger (1968), Chen (1968a, b, c), Hayes (1968), Currie and Hayes (1969), Şuhubi (1970), McCarthy (1970), Balaban, Green and Naghdi (1970), Doria and Wang (1970), Bowen and Wang (1970, 1971), Ogden (1970, 1974), Chadwick and Currie (1972), Bowen and Rankin (1973), Tokuoka (1973, 1974, 1984a, b), Tokuoka and Kusunoki (1982), Nunziato and Walsh (1975), Piau (1975), Raniecki (1976), Mihăilescu and Suliciu (1976), Ting (1977), Kosiński and Szmit (1977), Nowacki (1980), and Hanyga (1984).

For a more extensive review of the literature see Chen (1973, 1976), Achenbach (1973), McCarthy (1975) and Eringen and Şuhubi (1974). One-dimensional acceleration waves have been dealt with in Schuler, Nunziato and Walsh (1973), Nunziato *et al.* (1974) and Kosiński (1976).

Particular applications of acceleration wave propagation in different media may be found in McCarthy (1969), Chadwick and Ogden (1971), Currie and O'Leary (1978), Ting (1977), Cohen (1978), Cohen and Epstein (1979), Cohen and Tallin (1982), Cohen and Wang (1983), Ram and Pandey (1979), Parker and Seymour (1980), and Hayes (1984).

Chapter 6

Kinematic theory of acceleration waves

Hyperbolic equations describe many problems in continuum mechanics. The classical approach used in their solution is the method of characteristic curves and surfaces. This method reduces the hyperbolic partial differential equations to ordinary differential equations along curves called bi-characteristics or rays. The subsequent analysis and numerical calculation is then simplified.

In the general theory of quasi-linear hyperbolic equations it is shown that the characteristic surfaces are surfaces of discontinuity of the derivatives of the solution. The equations of motion containing acceleration as the time derivative of the particle velocity are included in the set of equations describing dynamic problems for deformable media and the characteristic surfaces are carriers of acceleration discontinuities since the particle velocity itself is one element in the solution of the equations.

The importance of the method of characteristics should be stressed in the discussion of acceleration waves in anisotropic media, for which the Thomas approach is sometimes ineffective. The reason for its ineffectiveness is that it often leads to a partial differential equation for the wave amplitude (involving both normal and surface derivatives). With the method of characteristic surfaces and rays ordinary differential equations are usually obtained for the amplitude (this is always the case when the normal component of the velocity of propagation is an isolated root of the characteristic equation). Moreover, these equations hold along the rays. If the material is isotropic and, in addition, the region ahead of the wave is in a state of uniform pure dilatation then the normal trajectories and the rays coincide. In this case the two methods are identical.

In many papers in which the Thomas method has been employed the problems of surface variations in the amplitude are neglected and the surface derivatives disappear; in consequence the amplitude equation becomes an ordinary differential equation along the normal trajectory. In this case the equation may be solved by means of quadrature. However, in most cases the equation is of Bernoulli type with variable coefficients depending on the mean and Gaussian curvatures of the singular surface as well as on the normal speed of propagation. Unfortunately, the normal trajectories are rarely known in advance although there are two typical cases where the curvatures, normal velocities and normal trajectories may be inferred. These are plane waves and waves of arbitrary curvature which propagate in homogeneous isotropic media where the region ahead of the wave is in a state of uniform dilatation.

In the ray method, on the other hand, the ordinary differential equation for the amplitude is coupled with ordinary differential equations for the normal velocity and for the rays themselves, all in terms of a single ray parameter. It is worth mentioning that if the normal speed of propagation depends only on the normal itself and not explicitly on position in space and time, then the rays, being straight lines, are not necessarily the same as the normal trajectories.

A separate ray theory, known as the method of bi-characteristics, has its origin in the method of characteristic surfaces. This theory was developed for systems of linear equations of hyperbolic type by Courant (1962). In physics ray theory is applied to geometrical optics and non-linear geometrical acoustics. In continuum mechanics, however, the theory of rays is used less frequently.

In what follows we present the main ideas and results of what is called the **kinematic theory of wave propagation**. The term originates from the papers by Lighthill and Whitham (1955) and Hayes (1970). Kinematic wave theory may be regarded as part of ray theory.

15 Dispersion Relations

Let a moving singular surface $\{\Sigma(t)\}_{t\in I}$ described by the function $G(\cdot,\cdot)$ in (13.4) and (13.5) represent an acceleration wave in a region $\hat{\mathcal{N}}$. This implies that in the considerations that follow the normal component U_N has constant sign since $U_N \neq 0$. For convenience we assume that $U_N > 0$. Moreover,* it is assumed that the G function defined on $\hat{\mathcal{N}}$ is such that the independent

* Since $\dot{G}(t, \mathbf{X})$ vanishes nowhere in the region $\hat{\mathcal{N}}$, the equation $\varphi - G(t, \mathbf{X}) = 0$ may be solved for t to yield $t = \hat{t}(\varphi, \mathbf{X})$. The time at which the acceleration wave has reached the point $\mathbf{X}$ is, accordingly, $\hat{t}(0, \mathbf{X})$.

variables $(t, \mathbf{X})$ may be replaced by new variables $(\varphi, \mathbf{X})$, where φ is the value of G. Thus

$$\varphi = G(t, \mathbf{X}), \qquad (t, \mathbf{X}) \in \hat{\mathcal{N}}. \tag{15.1}$$

The equation $\varphi = 0$ determines the singular surface $\hat{\mathcal{S}}$. In the kinematic theory of waves G is called the **phase.**

In what follows we shall write

$$\omega(t, \mathbf{X}) := -\frac{\partial G}{\partial t} \equiv -\dot{G}, \qquad \mathbf{k}(t, \mathbf{X}) := \operatorname{Grad} G. \tag{15.2}$$

The scalar ω is called the **frequency** and the vector $\mathbf{k}$ the **wave vector**, its magnitude being called the **wave number**. Note that

$$\operatorname{Grad}\omega + \frac{\partial \mathbf{k}}{\partial t} = \mathbf{0}, \qquad \mathbf{Grad\,k} - (\mathbf{Grad\,k})^{\mathsf{T}} = \mathbf{0}. \tag{15.3}$$

The basic assumption of kinematic wave theory is that there exists a dispersion relation. This relation* is written either explicitly as

$$\omega = \Omega(\mathbf{k}, t, \mathbf{X}) \tag{15.4}$$

or implicitly as

$$F(\mathbf{k}, \omega, t, \mathbf{X}) = 0 \quad \text{with} \quad \partial_\omega F \neq 0. \tag{15.5}$$

The standard procedure is to obtain the dispersion relation for each point $(\mathbf{X}, t)$ in space-time from a plane-wave analysis in a uniform medium whose properties are given by appropriate constitutive relations. Thus the kinematic theory of wave propagation is based on the assumption, valid in an asymptotic sense, that waves in a physical system (the medium) may be approximated by plane waves locally in space-time (the propagation space**). This is appropriate as far as geometrical optics is concerned, but for acceleration wave propagation in a non-linear (and possibly dissipative) medium the dispersion relation follows immediately from the so-called secular (or characteristic) equation, and it is exact. No *a priori* assumptions concerning the natural length and acceleration measures are introduced into the derivation. The only assumption is the existence of a moving surface $\varphi = 0$, called an

---

* Examples in which the dispersion relations are obtained from the field equations are given in Section 17.2.

** In certain situations (see Hayes, 1970) a distinction is made between physical space and the propagation space. If the propagation space is the entire physical space $(\mathbf{X}, t)$ the waves are termed **local.** If the waves are not local they are called **modal,** and physical space is the product of the propagation space and a cross space.

acceleration front, across which the particle acceleration changes discontinuously as the front passes, while the particle velocity remains continuous. Geometrical optics and the kinematic theory of acceleration waves are distinguished by the quantities which are transported along the rays. In geometrical optics these quantities are dependent variables, that is the solutions of the field equations; in the kinematic theory of acceleration waves, however, the jump discontinuities of the solution are under consideration and non-linear transport equations governing these jumps have to be derived.

In what follows the results are not restricted to acceleration waves (for which the dispersion relation is homogeneous of degree one; see Section 16).

It should be pointed out that the arguments of the function Ω and also of the so-called wave-propagation function F appearing in (15.4) and (15.5) are regarded as independent variables. The argument space $(\mathbf{k}, t, \mathbf{X})$ of Ω must be distinguished from the propagation space. This is done by distinguishing ω and Ω. We shall assume that Ω and F are twice continuously differentiable. Since both functions depend on $\mathbf{X}$ and t through thermodynamic and dynamic variables they are not continuously differentiable across the singular surface $\hat{\mathscr{S}}$, but they are differentiable in the regions $\hat{\mathscr{N}}^+ \cup \hat{\mathscr{S}}$ and $\hat{\mathscr{N}}^- \cup \hat{\mathscr{S}}$ if $\hat{\mathscr{S}}$ represents an acceleration front.

We write $\partial_t \Omega$ and $\partial_t F$ for the time derivatives of Ω and F respectively and $\partial_\omega F$ for the derivative of F with respect to ω. For the derivatives of Ω and F with respect to $\mathbf{k}$ and $\mathbf{X}$ we use the symbol ∇ with the appropriate subscript, that is $\nabla_{\mathbf{k}} \Omega$, $\nabla_{\mathbf{X}} \Omega$, $\nabla_{\mathbf{k}} F$, $\nabla_{\mathbf{X}} F$.

The dispersion relation is often expressed in the implicit form (15.5), especially in the case of acceleration waves, when the characteristic equation cannot be solved explicitly and there are several solution branches. Moreover, the implicit form emphasizes that there is no mathematical distinction between t and any coordinate in $\mathbf{X}$.

15.1 The Eikonal Equation

On introducing the quantities ω and $\mathbf{k}$ defined by (15.2) into the dispersion relation (15.4) or (15.5) we obtain the so-called **eikonal equation**

$$\dot{G} + \Omega(\operatorname{Grad} G, t, \mathbf{X}) = 0 \quad \text{or} \quad F(\operatorname{Grad} G, -\dot{G}, t, \mathbf{X}) = 0, \quad (15.6)$$

which is a first-order partial differential equation for the phase G. In order to solve the equation by the method of characteristics (see John, 1978; and Jeffrey and Taniuti, 1964) one has to find a solution of the following system of ordinary differential equations:

$$
\begin{aligned}
&\frac{\mathrm{d}t}{\mathrm{d}l_0} = 1,\\
&\frac{\mathrm{d}\mathbf{X}}{\mathrm{d}l_0} = \nabla_{\mathbf{k}}\Omega = -\frac{\nabla_{\mathbf{k}}F}{\partial_\omega F},\\
&\frac{\mathrm{d}\omega}{\mathrm{d}l_0} = \partial_t\Omega = -\frac{\partial_t F}{\partial_\omega F},\\
&\frac{\mathrm{d}\mathbf{k}}{\mathrm{d}l_0} = -\nabla_{\mathbf{X}}\Omega = \frac{\nabla_{\mathbf{X}}F}{\partial_\omega F},\\
&\frac{\mathrm{d}G}{\mathrm{d}l_0} = \dot{G} + (\mathrm{Grad}\,G)\cdot\nabla_{\mathbf{k}}\Omega = \dot{G} - (\mathrm{Grad}\,G)\cdot\frac{\nabla_{\mathbf{k}}F}{\partial_\omega F},
\end{aligned}
\tag{15.7}
$$

with the initial values ω_0, $\mathbf{k}_0$, G_0 given at the instant $\hat{t}(0) = t_0$ on some initial surface

$$\Sigma(t_0)\colon\ \mathbf{X}(0) = \mathbf{P}(t_0, l^1, l^2). \tag{15.8}$$

On the left-hand side of (15.7) the quantities appear as functions of the ray parameter l^0 for the given values of the Gaussian surface coordinates (l^1, l^2). Equation $(15.7)_1$ with the initial condition $\hat{t}(0) = t_0$ shows that l^0 may be identified with time t. Equations $(15.7)_{2,3}$ are called **ray equations**; in the three-dimensional case the ray equations form a system of six equations. Equation $(15.7)_2$ defines the rays themselves, that is curves originating from the initial surface $\Sigma(t_0)$ at the point $(l^1, l^2) = \mathrm{const}$ with tangent $\nabla_{\mathbf{k}}\Omega$ (not necessarily a unit vector). For a given fixed value of l^0 the solution of the ray equation represents the actual position of the surface $(\Sigma\hat{t}(l^0))$ and the distribution of the surface normal vector $\mathbf{k}$. We note that in (15.7) the particular choice of the ray parameter has been adopted (see $(15.7)_1$). If one introduces $(X^\varrho) = (t, X^1, X^2, X^3)$, $\varrho = 0, 1, 2, 3$, along with the notation

$$G_{,X^\varrho} := \frac{\partial G}{\partial X^\varrho}, \qquad \nabla_{G,X^\varrho}F := \frac{\partial F}{\partial\left(\dfrac{\partial G}{\partial X^\varrho}\right)},$$

then the eikonal equation (15.6) takes the form

$$F(G_{,X^\varrho}, X^\varrho) = 0 \tag{15.5$'$}$$

and the characteristic equations are

$$\frac{\mathrm{d}X^\varrho}{\mathrm{d}s} = \nabla_{G,X^\varrho}F, \qquad \frac{\mathrm{d}G_{,X^\varrho}}{\mathrm{d}s} = -\nabla_{X^\varrho}F \tag{15.7$'$}$$

with the initial conditions

$$X^\varrho(s)|_{s=0} = X^\varrho_0, \qquad G_{,X}^{\ \varrho}(s)|_{s=0} = (G_{,X_\varrho})_0.$$

The parameter s of the characteristics, and the ray parameter l^0 are connected by the equation

$$\frac{dl^0}{ds} = -\partial_\omega F(= \nabla_{G, X^0} F).$$

In the case of acceleration waves appearing as weak discontinuities of a hyperbolic system the rays are the characteristics of the characteristic equation and therefore the term "bi-characteristics" is in common use to describe rays.

Note that the initial conditions cannot be given arbitrarily for (15.7). They must satisfy a compatibility condition, called the strip condition (Courant, 1962, pp. 77–80), namely*

$$\begin{gathered} F(\mathbf{k}_0, \omega_0, t_0, \mathbf{X}(0)) = 0, \\ \frac{\partial G_0}{\partial l^\Delta} = \frac{\partial \mathbf{X}(0)}{\partial l^\Delta} \cdot \mathbf{k}_0, \qquad \Delta = 1, 2, \end{gathered} \tag{15.9}$$

on $\Sigma(t_0)$, with $\mathbf{k}_0$ normal to $\Sigma(t_0)$.

Since $\partial \mathbf{X}(0)/\partial l^\Delta$ are tangent to the initial surface, we see that G_0 is constant on $\Sigma(t_0)$. The solution of equations (15.6) and (15.7) is unique if the Jacobian determinant of $\mathbf{X}$ with respect to the parameters (l^0, l^1, l^2) at $l^0 = 0$ does not vanish, that is

$$j_0 := j|_{l^0=0} := \det(\nabla_l \mathbf{P})|_{l^0=0} \neq 0, \tag{15.10}$$

where $\mathbf{l} = (l^0, l^1, l^2)$. The Jacobian j plays an important role in the subsequent analysis.

The present analysis of the solution of the Cauchy problem (15.7) leads to the following observation: the geometrical representation of the solution to $(15.7)_2$ is a family of surfaces parametrized by $l^0 > 0$, where l^0 may be identified with the natural time t since the initial condition $\hat{t}(0) = t_0$ implies the relation

$$l^0 = t - t_0. \tag{15.11}$$

Without distinguishing between $\mathbf{P}(l^0 + t_0, l^1, l^2)$ and $\mathbf{P}(l^0, l^1, l^2)$ we see that the solutions of $(15.7)_2$ and (15.8) for rays are of the form

$$\mathbf{X} = \mathbf{P}(l^0, l^1, l^2). \tag{15.12}$$

In kinematic wave theory the tangent vector

$$\hat{\mathbf{c}} := \nabla_{\mathbf{k}} \Omega = -\frac{\nabla_{\mathbf{k}} F}{\partial_\omega F} \tag{15.13}$$

---

* In (15.9) the quantities $\mathbf{k}_0, \omega_0, \mathbf{X}(0)$ and G_0 are in general functions of (l^1, l^2).

to the ray is called the **group velocity**; the term **ray velocity** is also used, especially in the context of acceleration wave propagation. The vector $\hat{\mathbf{c}}$ is a measure of the rate of displacement of the surface point $(l^1, l^2) = \text{const}$ (we recall from Section 1.1 that the notation $\mathbf{c}$ is used for the velocity of displacement of a moving surface; $\hat{\mathbf{c}}$ is the counterpart of this in the reference configuration). In view of $(15.7)_2$ and (15.12) we have

$$\frac{\partial \mathbf{X}}{\partial l^0} = \frac{\partial \mathbf{P}}{\partial l^0} = \frac{\partial \mathbf{P}}{\partial t} = \hat{\mathbf{c}}, \tag{15.14}$$

and this should be compared with (1.13).

The tangent vectors

$$\mathbf{P}_{,\Delta} := \frac{\partial \mathbf{P}}{\partial l^\Delta}, \quad \Delta = 1, 2, \tag{15.15}$$

can be used in the determination of the Jacobian j of the transformation (15.12) since

$$j := \det(\nabla_1 \mathbf{P}) = (\mathbf{P}_{,1} \times \mathbf{P}_{,2}) \cdot \hat{\mathbf{c}}. \tag{15.16}$$

In the kinematic theory of wave propagation the quantity j is needed in the calculation of the wave amplitude (or intensity) through the transport equations mentioned above. From (15.16) we see that the Jacobian may be computed from the solution for the rays. However, since j represents a volume element convected along the rays it cannot be calculated by (15.16) if singularities in the geometry of the moving surface occur. Moreover, it turns out that j appears in the intensity part of the solutions to the transport equations (see, for example, Lewis, 1965) and carries all of the singular surface geometry required in the amplitude equation for acceleration waves. Later, an ordinary differential equation governing j along a ray will be derived.

To end the present analysis we note that by using the definition of $\hat{\mathbf{c}}$ equation $(15.7)_5$ for the variation of the phase G along a ray may be written

$$\frac{\mathrm{d}G}{\mathrm{d}l^0} = \hat{\mathbf{c}} \cdot \mathbf{k} - \omega. \tag{15.17}$$

The eikonal equation $(15.6)_1$ has the form of a Hamilton–Jacobi equation in which the phase G plays the role of an action,* while Ω may be identified with the Hamiltonian. Equations $(15.7)_{2,3}$ form a system of Hamiltonian equations for the generalized coordinates $\mathbf{X}$ and generalized momentum $\mathbf{k}$.

---

* According to Synge (1960, pp. 117–123), G should be called the characteristic or principal function.

The variation of the Hamiltonian Ω is given by the total derivative with respect to "time" l^0, that is

$$\frac{\mathrm{d}\Omega}{\mathrm{d}l^0} = \frac{\partial\Omega}{\partial t} + \nabla_{\mathbf{k}}\Omega\cdot\frac{\mathrm{d}\mathbf{k}}{\mathrm{d}t} + \nabla_{\mathbf{X}}\Omega\cdot\frac{\mathrm{d}\mathbf{X}}{\mathrm{d}l^0} = \frac{\partial\Omega}{\partial t}.$$

If Ω does not depend explicitly on time then $\mathrm{d}\Omega/dl^0 = 0$. This corresponds to conservation of energy in theoretical mechanics since the Hamiltonian represents the energy of the system. Thus, if $\partial\Omega/\partial t = 0$ we can say that the energy Ω is transported (or conserved) along a ray.

The Lagrangean L may then be defined by

$$L\left(\frac{\mathrm{d}\mathbf{X}}{\mathrm{d}l^0}, t, \mathbf{X}\right) = \mathbf{k}\cdot\frac{\mathrm{d}\mathbf{X}}{\mathrm{d}l^0} - \Omega(\mathbf{k}, t, \mathbf{X}),$$

and hence

$$\nabla_{\mathrm{d}\mathbf{X}/\mathrm{d}l^0}L = \mathbf{k}, \quad \nabla_{\mathbf{X}}L = -\nabla_{\mathbf{X}}\Omega, \quad \partial_t L = -\partial_t\Omega.$$

By use of L the function G is expressible in terms of the action integral (cf. (15.17))

$$G = \int L\,\mathrm{d}l^0.$$

The principle of stationary action is simply Fermat's principle* $\delta G = 0$ which implies that L satisfies the Euler–Lagrange equations

$$\frac{\mathrm{d}}{\mathrm{d}l^0}(\nabla_{\mathrm{d}\mathbf{X}/\mathrm{d}l^0}L) - \nabla_{\mathbf{X}}L = \mathbf{0},$$

We now return to the general form of the dispersion relation and consider particular physical situations which this describes. To this end we introduce the following:

Definition 15.1

We say that the medium characterized by the dispersion relation (15.6) is **steady** if the function Ω (or, equivalently, F) has no explicit dependence on time; the medium is termed **homogeneous** if $\nabla_{\mathbf{X}}\Omega = \nabla_{\mathbf{X}}F = \mathbf{0}$, and **isotropic** if $\nabla_{\mathbf{k}}\Omega\times\mathbf{k} = \mathbf{0}$, that is $\hat{\mathbf{c}}$ is parallel to $\mathbf{k}$.

Note that in a steady medium the frequency is constant along a ray, in a homogeneous medium $\mathbf{k}$ is constant along a ray, while in a steady homogeneous medium the rays are straight lines (by $(15.7)_2$), and integration of equations (15.7)–(15.9) leads to a closed-form solution. In a steady homogene-

* The principle of optics, formulated by Fermat in 1662 and called the principle of least time or the principle of least optical path, is physically incorrect; the correct statement is the principle of stationary action, called Fermat's principle.

ous but non-isotropic medium the ray velocity $\hat{\mathbf{c}}$ is constant along straight rays, which do not coincide with the normal trajectories. In the next section isotropic media are characterized by an isotropic dispersion relation (see equation (15.24)).

The remainder of this section is devoted to a discussion of the derivative with respect to the ray parameter l^0 as some kind of "time" derivative. In view of (15.11) and (15.14) one can define the derivative

$$\frac{\mathrm{d}h}{\mathrm{d}l^0} := \frac{\partial h}{\partial t} + \hat{\mathbf{c}} \cdot \operatorname{Grad} h \tag{15.18}$$

for an arbitrary function $h\colon \hat{\mathscr{N}} \to \boldsymbol{R}$. This derivative measures the rate of change of h along a ray. Recalling the definition of the displacement derivative

$$\frac{\delta h}{\delta t} := \frac{\partial h}{\partial t} + U_N \mathbf{N} \cdot \operatorname{Grad} h, \tag{15.19}$$

here referred to the reference placement, we see that in the case in which the ray velocity $\hat{\mathbf{c}}$ is equal to the normal velocity $U_N\mathbf{N}$ of the surface we have

$$\frac{\mathrm{d}}{\mathrm{d}l^0} = \frac{\delta}{\delta t}.$$

Of course, this is true if the rays coincide with the normal trajectories.

It may be more convenient to express (15.19) in terms of the gradient. This can be done by use of a coordinate transformation

$$(t, \mathbf{X}) \mapsto (\varphi, \mathbf{X})$$

which is invertible provided the normal velocity $U_N\mathbf{N}$ of the surface does not vanish (see (15.1)). In the new variables we define the function H according to

$$H(\varphi, \mathbf{X}) := h(\hat{t}(\varphi, \mathbf{X}), \mathbf{X}), \tag{15.20}$$

and for $\varphi = 0$ we set

$$\bar{h}(\mathbf{X}) := H(0, \mathbf{X}).$$

Note that

$$\bar{h}(\mathbf{X}) \equiv h(l^0 + t_0, \mathbf{X})|_{\mathbf{X}=\mathbf{P}(l^0, l^1, l^2)}.$$

In terms of $\bar{h}$ equation (15.19) becomes

$$\frac{\mathrm{d}\bar{h}}{\mathrm{d}l^0} = \hat{\mathbf{c}} \cdot \operatorname{Grad} \bar{h}. \tag{15.21}$$

For comparison with the displacement derivative, see equation (2.8).

If we set

$$\hat{c}^{\Delta} := \hat{g}^{\Gamma\Delta}\hat{\mathbf{c}}\cdot\frac{\partial\mathbf{P}}{\partial l^{\Gamma}},$$

where $\hat{g}^{\Gamma\Delta}$ are the contravariant components of the metric tensor of the surface $\Sigma(t)$, then

$$\frac{\mathrm{d}h}{\mathrm{d}l^{0}} = \frac{\delta h}{\delta t} + \hat{c}^{\Delta}\mathbf{P}_{,\Delta}\cdot\mathrm{Grad}\,h = \frac{\delta h}{\delta t} + \hat{c}^{\Delta}\tilde{h}_{,\Delta}, \tag{15.22}$$

where

$$\tilde{h}(l^{0}, l^{\Delta}) := h(l^{0}+t_{0}, \mathbf{P}(l^{0}, l^{\Delta})). \tag{15.23}$$

Finally we note that the gradient appearing on the right-hand side of (15.21) is not a surface derivative since it is performed at $G(t, \mathbf{X}) = 0$, with variable t. The surface derivative is the operator $\mathrm{grad}_{\mathscr{S}}$, defined by

$$\mathrm{grad}_{\mathscr{S}}\tilde{h}(l^{0},\cdot) := (\mathbf{1}-\mathbf{n}\otimes\mathbf{n})\,\mathrm{Grad}\,h(\mathbf{X})|_{\mathbf{X}=\mathbf{P}(l^{0},\cdot)},$$

which was used in equation (2.16) but with the differentiation with respect to the actual placement.

15.2 Particular Dispersion Relations

Different dynamical problems in physics described by partial differential equations lead to different dispersion relations. The dependence of the functions Ω and F on the wave vector $\mathbf{k}$ is crucial in the particular problem under consideration.

We write $\mathbf{k} = k\mathbf{N}$, recalling from (15.2) that $\mathbf{k}$ is the gradient of the phase G. The dispersion relation (15.5) is said to be isotropic if F is independent of the direction of $\mathbf{k}$, that is there exists a function F_i such that

$$F(\mathbf{k}, \omega, t, \mathbf{X}) = F_i(k, \omega, t, \mathbf{X}). \tag{15.24}$$

Then

$$\hat{\mathbf{c}} = -\nabla_{\mathbf{k}}F/\partial_{\omega}F = -(\partial_k F_i/\partial_{\omega}F_i)\mathbf{N}. \tag{15.25}$$

In explicit form an isotropic dispersion relation is characterized by

$$\omega = \Omega_i(k, t, \mathbf{X}), \qquad \hat{\mathbf{c}} = \partial_k\Omega_i\mathbf{N},$$

the index i referring to isotropy.

A partially isotropic dispersion relation may be identified according to

Definition 15.2

The dispersion relation (15.5) is **homogeneous in** k **with dispersion exponent** β if the function F has the property

$$F(k\mathbf{N}, \omega, t, \mathbf{X}) = F\left(\mathbf{N}, \frac{\omega}{k^{\beta}}, t, \mathbf{X}\right). \tag{15.26}$$

If $\beta = 0$ the dispersion relation is said to be **conical**. The following result explains this terminology.

Proposition 15.1

If a dispersion relation has the property (15.26) with $\beta = 0$ then

$$\hat{\mathbf{c}} \cdot \mathbf{k} = 0. \tag{15.27}$$

Proof

Differentiating both sides of (15.26) with respect to k we obtain

$$\partial_k F = \mathbf{N} \cdot \nabla_{\mathbf{k}} F = 0.$$

Comparing this equation with the definition (15.3) we see that (15.27) holds. □

The geometrical interpretation of (15.27) is as follows: the velocity $\hat{\mathbf{c}}$ is tangential to a ray, while $\mathbf{k}$ is normal to the surface of constant phase. Since the generator of a cone is orthogonal to the normal each ray is therefore a generator.

In the classical one-dimensional analysis of wave propagation the dispersion relation is usually expressed in the form

$$\omega = W(k). \tag{15.28}$$

The phase velocity c is then given as a function of the wave number,

$$c(k) = \frac{\omega}{k} = k^{-1} W(k). \tag{15.29}$$

The special case in which $c(k)$ is constant and the phase velocity is independent of the wave number is referred to as non-dispersive. Then $W(k) = kc$.

In kinematic wave theory the term non-dispersive refers to the property that $\hat{\mathbf{c}}$ does not change with a change in the magnitude of $\mathbf{k}$. This is reflected in the property that the tensor $\nabla_{\mathbf{k}} \nabla_{\mathbf{k}} \Omega$ is singular. Of course, this tensor may be singular in other ways and this makes the waves non-dispersive in another sense. After Hayes (1970) we therefore introduce

Definition 15.3

If, for the dispersion relation (15.4), the rank of the determinant of the tensor $\nabla_{\mathbf{k}} \nabla_{\mathbf{k}} \Omega$ is $3-l$, then we say that the multiplicity of non-dispersiveness is l.

Note that a dispersion relation homogeneous in $\mathbf{k}$ with dispersion exponent $\beta = 1$ is non-dispersive with multiplicity $\geqslant 1$. In the case of Alfvén waves in magnetohydrodynamics (see the example below) the tensor $\nabla_{\mathbf{k}} \nabla_{\mathbf{k}} \Omega$ is null and the waves are triply non-dispersive.

In what follows we summarize the main properties of homogeneous dispersion relations with dispersion exponent $\beta = 1$.

Proposition 15.2

If a dispersion relation has the property (15.26) with $\beta = 1$ then the following relations hold:

$$\begin{aligned}
&\hat{\mathbf{c}} \cdot \mathbf{N} = U_N, \qquad \text{or equivalently} \quad \hat{\mathbf{c}} \cdot \mathbf{k} - \omega = 0,\\
&\hat{\mathbf{c}} = U_N \mathbf{N} + (\mathbf{N} \otimes \mathbf{N} - \mathbf{1}) \nabla_{\mathbf{N}} F_1 / \partial_{U_N} F_1,\\
&\frac{\mathrm{d}U_N}{\mathrm{d}l^0} = \frac{-\partial_t F_1}{\partial_{U_N} F_1} - U_N \mathbf{N} \cdot \frac{\nabla_{\mathbf{X}} F_1}{\partial_{U_N} F_1},\\
&\frac{\mathrm{d}\mathbf{N}}{\mathrm{d}l^0} = (\mathbf{1} - \mathbf{N} \otimes \mathbf{N}) \frac{\nabla_{\mathbf{X}} F_1}{\partial_{U_N} F_1},
\end{aligned} \tag{15.30}$$

where F_1, as a function of $\mathbf{N}$, ω/k, t and $\mathbf{X}$, is defined by

$$F_1\left(\mathbf{N}, \frac{\omega}{k}, t, \mathbf{X}\right) \equiv F(k\mathbf{N}, \omega, t, \mathbf{X}). \tag{15.31}$$

Proof

Differentiation of (15.31) yields

$$\begin{aligned}
&\nabla_{\mathbf{N}} F_1 = k \nabla_{\mathbf{k}} F, \qquad \partial_{U_N} F_1 = k \partial_\omega F,\\
&\partial_t F_1 = \partial_t F, \qquad \nabla_{\mathbf{X}} F_1 = \nabla_{\mathbf{X}} F,\\
&\mathbf{N} \cdot \nabla_{\mathbf{k}} F = -\frac{\omega}{k^2} \partial_{U_N} F_1,
\end{aligned} \tag{15.32}$$

where

$$U_N = \frac{\omega}{k}. \tag{15.33}$$

In geometrical optics the right-hand side of (15.33) is called the phase velocity. In kinematic wave theory this scalar quantity must be distinguished from the vector $U_N \mathbf{N} = \dfrac{\omega}{k} \mathbf{N}$ and therefore the former is called the **phase speed** while the term phase velocity is retained for the latter. In view of (15.13) and (15.32) the group velocity $\hat{\mathbf{c}}$ may be expressed as

$$\hat{\mathbf{c}} = -\frac{\nabla_{\mathbf{N}} F_1}{\partial_{U_N} F_1} = \nabla_{\mathbf{N}} \Omega(\mathbf{N}, t, \mathbf{X}) \tag{15.34}$$

if the dispersion relation can be written in the explicit form

$$F_1(\mathbf{N}, U_N, t, \mathbf{X}) \equiv U_N - \Omega(\mathbf{N}, t, \mathbf{X}). \tag{15.35}$$

Use of $(15.32)_{1,5}$ and (15.33) leads to $(15.30)_1$. To obtain $(15.30)_2$ we apply the identity

$$\hat{\mathbf{c}} - U_N \mathbf{N} = \hat{\mathbf{c}} - (\hat{\mathbf{c}} \cdot \mathbf{N}) \mathbf{N} = \hat{\mathbf{c}} (\mathbf{1} - \mathbf{N} \otimes \mathbf{N})$$

together with (15.34). In order to obtain $(15.30)_{2,4}$ we note that

$$\frac{\mathrm{d}}{\mathrm{d}l^0}(\mathbf{N}\cdot\mathbf{k}) = \mathbf{N}\cdot\frac{\mathrm{d}\mathbf{k}}{\mathrm{d}l^0}$$

since **N** is a unit vector. Equation $(15.7)_3$ together with $(15.32)_{2,4}$ then yields

$$\frac{\mathrm{d}k}{\mathrm{d}l^0} = k\mathbf{N}\cdot\frac{\nabla_{\mathbf{X}}F_1}{\partial_{U_N}F_1}.$$

Finally, since

$$\frac{\mathrm{d}U_N}{\mathrm{d}l^0} = \frac{1}{k}\frac{\mathrm{d}\omega}{\mathrm{d}l^0} - \frac{\omega}{k^2}\frac{\mathrm{d}k}{\mathrm{d}l^0}, \qquad \frac{\mathrm{d}\mathbf{N}}{\mathrm{d}l^0} = \frac{1}{k}\frac{\mathrm{d}\mathbf{k}}{\mathrm{d}l^0} - \frac{\mathbf{N}}{k}\frac{\mathrm{d}k}{\mathrm{d}l^0},$$

we obtain $(15.30)_{3,4}$.

15.3 Bibliographical Notes

The mathematical theory of geometrical optics was formulated in the work of W. R. Hamilton during the years 1824–1844 (see Synge and Conway, 1931). Hamilton's main contribution was the idea of a characteristic function whose partial derivatives give the direction of a light ray at the point in question. Incidentally, two of the charactristic functions introduced by Hamilton were rediscovered by Bruns (1895) who coined the name eikonal.

Hamilton's theory had been preceded by two physical theories of light created in the seventeenth century by Ch. Huygens and I. Newton. In Huygen's geometrical "wave" theory the light is a longitudinal motion of ether spreading out from a point source at finite speed; the positions reached by the light at a given time form a surface which he called the front of the wave. In our model this is what we refer to as a singular surface. In a homogeneous isotropic medium the wave front is a sphere.

Newton, on the other hand, suggested that a source of light emits a stream of particles (distinct from the ether) in all directions in which the light propagates. In a homogeneous space (such as the ether) these particles travel in straight lines unless deflected by foreign bodies such as reflecting and refracting bodies.

A theory of light, called geometrical optics, was created on the basis of four laws which follow from one all-embracing principle of least time originally formulated by Fermat.

In his formulation of a deductive mathematical science of optics W. R. Hamilton was indifferent to the physical interpretations of Huygens and Newton. He showed, however, that from a knowledge of the characteristic functions all problems in optics involving, for example, lenses, mirrors, crystals, and light propagation in the atmosphere can be solved.*

Of course, our interest is in neither the theory of light nor the solution of the wave equation (or Maxwell's equations). For our purposes it is the discontinuities in the solution functions themselves which are of special interest.

The physical significance of discontinuities in the solution of partial differential equations for gas dynamics was pointed out by Hadamard (1903). However, it was Friedlander (1958) who showed that the discontinuity $[\![\partial u/\partial t]\!]$, where u is the solution of the wave equation $\Delta u = \partial^2 u/\partial t^2$, has the same formal properties as the intensity of a geometrical optics field.

For general systems of quasi-linear hyperbolic equations arising in mathematical physics the surfaces $G(t, \mathbf{X}) = 0$ on which discontinuities in the derivatives of the solutions appear are characteristic surfaces (cones), which we call wave fronts, and the bi-characteristics are curves along which the discontinuities propagate. The role of bi-characteristics (and rays) in the analysis of surface singularities is therefore clearer if one refers to the theory of partial differential equations of hyperbolic type.

Here we note that the derivation of the differential equations governing the rays from the eikonal equation is a special case of the general method in the theory of first-order partial differential equations; the elements of this theory can be found in the books by Courant (1962) and John (1978). The ordinary differential equations are called characteristic differential equations of the first-order partial differential equations. Since the eikonal equation (15.6) is itself a characteristic differential equation of the field equations, the system of bi-characteristic equations of the field equations, and its solutions, our rays, are called the bi-characteristics. (Examples of field equations will be given in Chapter 7.) It is better to distinguish between the terms bi-characteristics and rays (see Courant, 1962, p. 558) and to use the term bi-characteristics to refer to curves in space-time** lying on $G(t, \mathbf{X}) = 0$. The term ray should refer to curves in the space $\mathscr{E}^3$ (containing points $\mathbf{x}$) and transversal to $t = \tau(\mathbf{x})$. Geometrically, rays are projections of the bi-characteristics onto $\mathscr{E}^3$.

Returning to our historical discussion we note that a formal theory of geometrical optics and its connections with electromagnetic theory were first

* From Hamilton's work the equivalence of Fermat's principle and Huygens' principle is clear.

** That is, the solutions of the ordinary differential equations corresponding to the characteristic condition regarded as a partial differential equation in the four variables $(\mathbf{X}, t)$.

presented by Sommerfeld and Runge (1911). A modern treatment of this subject can be found in Kline and Kay (1965) and Born and Wolf (1959).

More recent publications concerning linear and non-linear geometrical acoustics are Friedlander (1958), Malecki (1969) and Musgrave (1970). The term geometrical acoustics (see Keller, 1962) refers to the study of discontinuities in pressure and deformation since in the linearized theory of acoustics the excess pressure satisfies the wave equations while in the exact theory the pressure is governed by a quasi-linear hyperbolic equation.

A unified presentation of the theory of bi-characteristics is given in Courant (1962), but mainly for linear systems; see also Jeffrey and Taniuti (1964). A simple variant of the technique described there can be used to determine the variation of discontinuities in all the first derivatives of the solution of a system of quasi-linear first-order hyperbolic differential equations. For such a system a characteristic surface is what we call a wave front and the concepts of wave (phase) velocity and ray (group) velocity can be defined in a natural way; moreover, the derivation of Bernoulli type transport equations governing variations in the discontinuities at the wave front is possible. The first derivation for a quite general system of quasi-linear hyperbolic equations was carried out by Boillat (1965) and independently by Varley and Cumberbatch (1965); see also Karpman (1973).

Some particular applications of ray theory to problems of continuum mechanics can be found in, for example, Varley (1965), Nariboli (1966), Bejda (1971b, 1977), Hayes and Rivlin (1972), Wright (1973, 1976), Wesołowski (1974, 1976), Scott (1975a, b), Ting (1976), Braun (1977, 1985), Borejko (1979), Hanyga (1984), Hayes (1984), Sobczyk (1984), and many others.

In addition to the paper concerning kinematic wave theory already mentioned we refer to Whitham (1960, 1961, 1974, 1979), Weinberg (1962), Lewis (1965), Brillouin (1960), Jeffrey and Taniuti (1964), Jeffrey (1976) and Boillat (1965). In Anile (1984) ray theory is applied to the analysis of weak shock propagation.

16 HOMOGENEOUS DISPERSION RELATIONS

For a number of problems in mathematical physics for which the methods of geometrical optics are in use, the dispersion relation is homogeneous of degree one in k and may be written explicitly as

$$\omega = k\Omega(\mathbf{N}, t, \mathbf{X}) \tag{16.1}$$

or, equivalently,

$$U_N = \Omega(\mathbf{N}, t, \mathbf{X}) \tag{16.2}$$

on use of (15.33). We observe that the dispersion relation (15.5) can be written in the explicit form (15.4) whenever $\partial_\omega F$ has constant value unity.

As a direct consequence of (15.30), (15.35) and (16.2) we have

Corollary 16.1

If the dispersion relation (15.4) has the property (16.1) then

$$\hat{\mathbf{c}} \cdot \mathbf{N} = U_N$$

and $(15.30)_{2,3,4}$ are replaced by

$$\begin{aligned} \hat{\mathbf{c}} &= U_N\mathbf{N} + (\mathbf{1} - \mathbf{N}\otimes\mathbf{N})\nabla_\mathbf{N}\Omega, \\ \frac{\mathrm{d}U_N}{\mathrm{d}l^0} &= \partial_t\Omega + U_N\mathbf{N}\cdot\nabla_\mathbf{X}\Omega, \\ \frac{\mathrm{d}\mathbf{N}}{\mathrm{d}l^0} &= (\mathbf{N}\otimes\mathbf{N} - \mathbf{1})\nabla_\mathbf{X}\Omega. \end{aligned} \tag{16.3}$$

Moreover, the derivatives of the propagation function Ω satisfy

$$\begin{aligned} &\mathbf{N}\cdot\nabla_\mathbf{N}\Omega = \Omega, \qquad (\nabla_\mathbf{X}\nabla_\mathbf{N}\Omega)\mathbf{N} = \nabla_\mathbf{X}\Omega, \\ &(\partial_t\nabla_\mathbf{N}\Omega)\cdot\mathbf{N} = \partial_t\Omega, \qquad (\nabla_\mathbf{N}\nabla_\mathbf{N}\Omega)\mathbf{N} = \mathbf{0}. \end{aligned} \tag{16.4}$$

In obtaining (16.4) from $(15.30)_1$ and (15.35) we have made use of the expression

$$\hat{\mathbf{c}} = \nabla_\mathbf{N}\Omega. \ \square \tag{16.5}$$

Equation $(16.4)_4$ shows that the tensor $\nabla_\mathbf{N}\nabla_\mathbf{N}\Omega$ is singular and the normal $\mathbf{N}$ is its right eigenvector, corresponding to zero eigenvalue. Thus, a dispersion relation, homogeneous with exponent one, is non-dispersive with multiplicity at least one.

16.1 Transport Equations for the Geometry

The aim of this section is to derive ordinary differential equations which govern the evolution of geometrical and kinematic quantities along a ray.

First, we derive a formula for the evolution of the Jacobian determinant j of the coordinate transformation (15.12); see (15.16). By the classical Euler formula for the derivative of a determinant, we have

$$\frac{\mathrm{d}j}{\mathrm{d}l^0} = j\,\mathrm{tr}\left\{(\nabla_1\mathbf{P})^{-1}\frac{\mathrm{d}}{\mathrm{d}l^0}(\nabla_1\mathbf{P})\right\} = j\,\mathrm{tr}\left\{(\nabla_\mathbf{X}\mathbf{l})\left(\nabla_1\frac{\mathrm{d}\mathbf{P}}{\mathrm{d}l_0}\right)\right\},$$

and hence

$$\frac{\mathrm{d}\ln j}{\mathrm{d}l^0} = \mathrm{tr}\left\{\nabla_\mathbf{X}\left(\frac{\mathrm{d}\mathbf{X}}{\mathrm{d}l^0}\right)\right\} = \mathrm{tr}(\nabla_\mathbf{X}\hat{\mathbf{c}}). \tag{16.6}$$

The ray velocity $\hat{\mathbf{c}}$ is defined only for surface points and is therefore a function of l^0, l^1, l^2. Specifically

$$\hat{\mathbf{c}} = \nabla_{\mathbf{N}}\Omega\left(\mathbf{N}(l^0, l^1, l^2), \hat{t}(l^0), \mathbf{P}(l^0, l^1, l^2)\right). \tag{16.7}$$

The differentiation with respect to $\mathbf{X}$ in (16.6) means that $\hat{\mathbf{c}}$ should be regarded there as a function of the variable $\mathbf{X}$ describing the position of the ray in space, that is it is a function of $\mathbf{X}$ expressed in terms of the inverse transformation of (15.2), namely

$$\mathbf{l} = \mathbf{P}^{-1}(\mathbf{X}).$$

To see this more clearly we note that if (15.12), being a solution to $(15.7)_2$ and representing a sequence of surfaces for increasing values of l^0, is univalent in some region of $\mathscr{E}^3$, then a fixed point $\mathbf{X}$ may be identified with the coordinates l^1, l^2 of a unique point on the initial surface $\Sigma(t_0)$, and the time of passage $l^0 = t - t_0$ of the surface (15.11). Since the gradient $\nabla_{\mathbf{l}}\mathbf{P}$ of the transformation (15.12) is given by

$$\mathbf{P}_{,\Delta} := \frac{\partial \mathbf{P}}{\partial l^\Delta}, \qquad \frac{\partial \mathbf{P}}{\partial l^0} = \hat{\mathbf{c}},$$

its inverse $\nabla_{\mathbf{X}}\mathbf{l}$ is given by

$$\begin{aligned} \nabla_{\mathbf{X}} l^0 &:= \operatorname{Grad} l^0 = U_N^{-1}\mathbf{N}, \\ \nabla_{\mathbf{X}} l^\Delta &:= \operatorname{Grad} l \;= (\mathbf{1} - U_N^{-1}\mathbf{N}\otimes\hat{\mathbf{c}})\mathbf{P}^\Delta, \end{aligned} \tag{16.8}$$

where $\mathbf{P}^\Delta$ are the dual basis vectors of $\mathbf{P}_{,\Delta}$, $\Delta = 1, 2$. In geometrical optics $U_N^{-1}\mathbf{N}$ is called the slowness vector.

The dispersion relation (16.2) may be written in the form

$$1 = \Omega(U_N^{-1}\mathbf{N}, t, \mathbf{X}). \tag{16.9}$$

Regarded as an equation in the space of slowness vectors this describes the so-called **slowness surface** for fixed t and $\mathbf{X}$. The geometry of the slowness surface and the curvature of the initial surface $\Sigma(t_0)$ together play an important role in connection with the focusing of waves.

The inverse of (15.12) is well defined if the Jacobian j is non-vanishing. Since $(\mathbf{P}_{,1} \times \mathbf{P}_{,2}) \cdot \mathbf{N} = (\hat{g})^{1/2}$ and $\hat{\mathbf{c}} \cdot \mathbf{N} = U_N$ we have

$$j = (\mathbf{P}_{,1} \times \mathbf{P}_{,2}) \cdot \hat{\mathbf{c}} = (\hat{g})^{1/2} U_N, \tag{16.10}$$

where $\hat{g}$ denotes the determinant of $\hat{g}_{\Delta\Gamma}$. From (16.10) we see that j is proportional to the cross-sectional area of a ray tube (where the section lies on the wave front $\Sigma\{\hat{t}(l^0)\}$) and to the normal speed.

For the surface element $\mathrm{d}S$ of a ray tube we have

$$\mathrm{d}S = (\mathbf{P}_{,1} \times \mathbf{P}_{,2}) \cdot \mathbf{N}\,\mathrm{d}l^1\mathrm{d}l^2 = (\hat{g})^{1/2}\mathrm{d}l^1\mathrm{d}l^2.$$

In the case of an acceleration wave with $U_N \neq 0$, j vanishes only when the cross-sectional area of the ray tube vanishes.

Let $\hat{\mathbf{b}}$ denote the second fundamental form of the moving surface $\{\Sigma\{\hat{t}(l^0)\}: l^0 > 0\}$ and its mixed components by $\hat{b}_{\Gamma\Delta}$. It follows from (1.7) that

$$\hat{\mathbf{b}} = \hat{b}^{\Delta}_{\Gamma}\mathbf{P}_{,\Delta}\otimes\mathbf{P}^{\Gamma} = -\mathbf{N}_{,\Delta}\otimes\mathbf{P}^{\Delta}. \tag{16.11}$$

Next, differentiating the basis vectors and their duals with respect to l^0 and using (16.7), we obtain

$$\begin{aligned}\frac{\mathrm{d}}{\mathrm{d}l^0}\mathbf{P}_{,\Delta} &= \frac{\partial\hat{\mathbf{c}}}{\partial l^0} = -(\nabla_{\mathbf{N}}\nabla_{\mathbf{N}}\Omega)\hat{b}^{\Gamma}_{\Delta}\mathbf{P}_{,\Gamma}+(\nabla_{\mathbf{X}}\nabla_{\mathbf{N}}\Omega)\mathbf{P}_{,\Delta},\\ \frac{\mathrm{d}}{\mathrm{d}l^0}\mathbf{P}^{\Gamma} &= g^{\Gamma\Delta}\frac{\partial\hat{\mathbf{c}}}{\partial l^{\Delta}} = -(\nabla_{\mathbf{N}}\nabla_{\mathbf{N}}\Omega)\hat{b}^{\Gamma\Delta}\mathbf{P}_{,\Delta}+(\nabla_{\mathbf{X}}\nabla_{\mathbf{N}}\Omega)\mathbf{P}^{\Gamma}.\end{aligned} \tag{16.12}$$

On use of (16.4), (16.11) and the identity $\mathbf{P}_{,\Delta}\otimes\mathbf{P}^{\Delta}+\mathbf{N}\otimes\mathbf{N} = \mathbf{1}$ we obtain

$$\begin{aligned}\frac{1}{2}\frac{\mathrm{d}}{\mathrm{d}l^0}\ln\hat{g} = \left(\frac{\mathrm{d}}{\mathrm{d}l^0}\mathbf{P}_{,\Delta}\right)\cdot\mathbf{P}^{\Delta} &= -\mathrm{tr}\,\{(\nabla_{\mathbf{N}}\nabla_{\mathbf{N}}\Omega)\hat{\boldsymbol{b}}\}\\ &+\mathrm{tr}(\nabla_{\mathbf{X}}\nabla_{\mathbf{N}}\Omega)-\mathbf{N}\cdot\nabla_{\mathbf{X}}\Omega.\end{aligned} \tag{16.13}$$

In view of (16.10), (16.13) and (16.3), equation (16.6) may be rewritten

$$\frac{\mathrm{d}}{\mathrm{d}l^0}\ln j = \mathrm{tr}(\nabla_{\mathbf{X}}\nabla_{\mathbf{N}}\Omega)-\mathrm{tr}\,\{(\nabla_{\mathbf{N}}\nabla_{\mathbf{N}}\Omega)\hat{\mathbf{b}}\}+U_N^{-1}\partial_t\Omega. \tag{16.14}$$

Use of (16.7) and (16.11) shows that

$$\mathrm{tr}\,\{\nabla_{\mathbf{X}}\nabla_{\mathbf{N}}\Omega-(\nabla_{\mathbf{N}}\nabla_{\mathbf{N}}\Omega)\hat{\mathbf{b}}\} = \mathrm{tr}\left\{\frac{\partial}{\partial l}(\nabla_{\mathbf{N}}\Omega)\otimes\mathbf{P}^{\Delta}\right\} = \mathrm{tr}(\hat{\mathbf{c}}_{,\Delta}\otimes\mathbf{P}^{\Delta}),$$

and (16.14) therefore simplifies to

$$\frac{\mathrm{d}}{\mathrm{d}l^0}\ln j = \mathrm{tr}(\hat{\mathbf{c}}_{,\Delta}\otimes\mathbf{P}^{\Delta})+U_N^{-1}\frac{\mathrm{d}\Omega}{\mathrm{d}l^0}. \tag{16.15}$$

The symmetric tensor $\nabla_{\mathbf{N}}\nabla_{\mathbf{N}}\Omega \equiv \nabla_{\mathbf{N}}\hat{\mathbf{c}}$, which appears in (16.14), together with the curvature tensor $\hat{\mathbf{b}}$ plays an important part in the analysis of caustics and their formation; see Wright (1973, Appendix).

In order to derive the transport equation for $\hat{\mathbf{b}}$ we note first that

$$\frac{\mathrm{d}}{\mathrm{d}l^0}(-\hat{b}^{\Gamma}_{\Delta}\mathbf{P}_{,\Gamma}) = \frac{\mathrm{d}}{\mathrm{d}l^0}\mathbf{N}_{,\Delta} = \frac{\partial}{\partial l^{\Delta}}\left(\frac{\mathrm{d}\mathbf{N}}{\mathrm{d}l^0}\right) = \frac{\partial}{\partial l^{\Delta}}\{(\mathbf{N}\otimes\mathbf{N}-\mathbf{1})\nabla_{\mathbf{X}}\Omega\}. \tag{16.16}$$

After some manipulations we obtain

$$\frac{\mathrm{d}\hat{b}^{\Gamma}_{\Delta}}{\mathrm{d}l^{0}} = \hat{b}^{\Lambda}_{\Delta}\hat{b}^{\Omega}_{\Lambda}\operatorname{tr}\{(\nabla_{\mathbf{N}}\nabla_{\mathbf{N}}\Omega)(\mathbf{P}_{,\Omega}\otimes\mathbf{P}^{\Gamma})\}$$
$$-\hat{b}^{\Lambda}_{\Delta}\operatorname{tr}\{(\nabla_{\mathbf{X}}\nabla_{\mathbf{N}}\Omega)(\mathbf{P}_{,\Lambda}\otimes\mathbf{P}^{\Gamma}+\mathbf{P}^{\Gamma}\otimes\mathbf{P}_{,\Lambda})\}$$
$$+\hat{b}^{\Gamma}_{\Delta}\mathbf{N}\cdot\nabla_{\mathbf{X}}\Omega+\operatorname{tr}\{(\nabla_{\mathbf{X}}\nabla_{\mathbf{X}}\Omega)(\mathbf{P}_{,\Delta}\otimes\mathbf{P}^{\Gamma})\}. \tag{16.17}$$

The invariants of $\hat{\mathbf{b}}$ are of more importance than $\hat{\mathbf{b}}$ itself, however, and the transport equations for the mean and Gaussian curvatures

$$K_M := \tfrac{1}{2}\hat{b}^{\Gamma}_{\Gamma}, \quad K_G := \det[\hat{b}^{\Gamma}_{\Delta}]$$

may be obtained directly from (16.17) in the forms

$$\frac{2\mathrm{d}K_M}{\mathrm{d}l^0} = \frac{\mathrm{d}\hat{b}^{\Delta}_{\Delta}}{\mathrm{d}l^0} = \operatorname{tr}\{(\nabla_{\mathbf{N}}\nabla_{\mathbf{N}}\Omega)\hat{\mathbf{b}}^2\} - 2\operatorname{tr}\{(\nabla_{\mathbf{X}}\nabla_{\mathbf{N}}\Omega)\hat{\mathbf{b}}\}$$
$$+(\operatorname{tr}\hat{\mathbf{b}})\mathbf{N}\cdot\nabla_{\mathbf{X}}\Omega+\operatorname{tr}\{(\nabla_{\mathbf{X}}\nabla_{\mathbf{X}}\Omega)(\mathbf{P}_{,\Delta}\otimes\mathbf{P}^{\Delta})\}, \tag{16.18}$$

$$\frac{\mathrm{d}K_G}{\mathrm{d}l^0} = \frac{\mathrm{d}}{\mathrm{d}l^0}(\det\hat{\mathbf{b}}) = (\det\hat{\mathbf{b}})\big\{\operatorname{tr}\{(\nabla_{\mathbf{N}}\nabla_{\mathbf{N}}\Omega)\hat{\mathbf{b}}\} - 2\operatorname{tr}(\nabla_{\mathbf{X}}\nabla_{\mathbf{N}}\Omega)$$
$$+4\mathbf{N}\cdot\nabla_{\mathbf{X}}\Omega+\operatorname{tr}\{(\nabla_{\mathbf{X}}\nabla_{\mathbf{X}}\Omega)\hat{\mathbf{b}}^{-1}\}\big\}. \tag{16.19}$$

It should be pointed out that in (16.19) the tensor $\hat{\mathbf{b}}^{-1}$ is defined by

$$\hat{\mathbf{b}}^{-1} \equiv (\hat{b}^{\Gamma}_{\Delta})^{-1}\mathbf{P}_{,\Gamma}\otimes\mathbf{P}^{\Delta},$$

where

$$(\hat{b}^{\Gamma}_{\Delta})^{-1}\hat{b}^{\Delta}_{\Gamma} = \hat{g}^{\Delta}_{\Delta}.$$

Since $\hat{\mathbf{b}}$ is planar in the three-dimensional space surrounding the surface $\Sigma\{\hat{t}(l^0)\}$ it does not have an inverse in the usual sense because, according to (16.11), $\hat{\mathbf{b}}\mathbf{N} = \mathbf{0}$, that is $\mathbf{N}$ is an eigenvector of $\hat{\mathbf{b}}$ corresponding to zero eigenvalue. The remaining two eigenvectors of $\hat{\mathbf{b}}$ are the directions of the principal curvatures and the corresponding eigenvalues are the principal curvatures. If $\hat{\mathbf{b}} = \mathbf{0}$ then the acceleration wave is plane.

Equations (16.14), (16.15) and (16.17)–(16.19) describe the variations along the rays of the important kinematical invariants associated with the propagation of a singular surface. Some results may be read off directly from the equations. To do this we require the following definition.

Definition 16.1

A point $(t, \mathbf{X})$ at which the Jacobian determinant j vanishes is called a **caustic point** and the locus of such points is called a **caustic** of the family of rays (15.12). If a caustic has just one point it is called a **focus**.

From (16.10) and the assumption $U_N > 0$ we see that a caustic forms where the rays converge and $\hat{g}$ vanishes. At such a point the transformation is no longer single-valued.

In addition to taking $U_N > 0$ it is usual to assume that Ω and its derivatives are bounded continuous functions of l^0. Under these conditions $\ln j > -\infty$ and $\ln \hat{g} > -\infty$ and hence $j > 0$ and $\hat{g} > 0$ along a ray for $l^0 < \infty$ if the curvatures remain finite. However, if j tends to zero along a ray at a finite value of l^0 then at least one of the curvatures becomes unbounded at this value, and $\ln j$ tends to $-\infty$, as can be seen from the second term in (16.14).

Proposition 16.1

(i) A plane wave remains plane if and only if

$$\mathrm{tr}\,\{\nabla_{\mathbf{X}}\nabla_{\mathbf{X}}\Omega)(\mathbf{P}_{,} \otimes \mathbf{P}^{\Gamma})\} = 0 \tag{16.20}$$

everywhere on the wave, or, equivalently,

$$\Omega_{;\Delta\Gamma} \equiv 0. \tag{16.21}$$

(ii) If the function Ω has the properties

$$\nabla_{\mathbf{X}}\Omega = e\mathbf{N}, \qquad e = e(\mathbf{N}, t, \mathbf{X}), \qquad \nabla_{\mathbf{N}}\Omega = U_N\mathbf{N} \tag{16.22}$$

everywhere on the wave then the moving surface $\{\Sigma\{\hat{t}(l^0)\}: l^0 > 0\}$ forms a family of parallel surfaces.

(iii) If Ω is finite and its derivatives are continuous bounded functions of l^0 then at least one of the principal curvatures of the surface becomes unbounded at a caustic.

Proof

Part (i) follows from (16.17) with $\hat{\mathbf{b}} = \mathbf{0}$ for a plane wave. The condition (16.22) for $\Sigma\{\hat{t}(l^0)\}$ to form a family of parallel surfaces follows from $(16.3)_{2,3}$. Under this condition the normal vector $\mathbf{N}$ is constant along a ray and the rays coincide with the normal trajectories. Part (iii) has already been proved. □

It is interesting to observe that any, not necessarily homogeneous, dispersion relation written in the explicit form $U_N = \Omega(\mathbf{N}, t, \mathbf{X})$ may be reduced to homogeneous form by the simple substitution

$$\Omega^*(\mathbf{N}, t, \mathbf{X}) = (\mathbf{N}\cdot\mathbf{N})^{1/2}\Omega\left\{\frac{\mathbf{N}}{(\mathbf{N}\cdot\mathbf{N})^{1/2}}, t, \mathbf{X}\right\}.$$

Of the better known dispersion relations for three-dimensional problems we mention that for Alfvén waves, $\omega^2 - (\mathbf{k}\cdot\hat{\mathbf{c}})^2 = 0$, and the relation $\omega^2 - k^2\hat{c}^2 = 0$ from magnetohydrodynamics (see Weinberg, 1962). Several examples of one-dimensional dispersion relations can be found in Whitham (1974).

16.2 Surfaces Moving into a Homogeneous Stationary State

If a singular surface propagates into a medium which is in a uniform (homogeneous) stationary state (see Definition 15.1) the propagation function Ω is independent of **X** and t and depends only on **N**. From $(16.3)_3$ it follows that U_N, **N** and $\hat{\mathbf{c}}$ are constant along a ray. Since the rays are then straight lines the solution (15.12) to equations $(15.7)_1$ and (15.8) may be expressed in the form

$$\mathbf{X} = \mathbf{P}(l^0, l^1, l^2) \equiv \mathbf{P}(t_0, l^1, l^2) + l^0\hat{\mathbf{c}}, \tag{16.23}$$

where $\hat{\mathbf{c}}$ is a function of l^1 and l^2 only. The function $\mathbf{P}(t_0, l^1, l^2)$ gives the parametric representation of the initial surface $\Sigma(t_0)$.

It is easier to work out the geometry and kinematics of the moving surface $\{\Sigma\{\hat{t}(l^0)\}: l^0 > 0\}$ directly from (16.23) rather than from the general equations. Let $\mathbf{P}^0_{,\Delta}$, $\Delta = 1, 2$, denote the tangent vectors to $\Sigma(t_0)$, that is $\mathbf{P}^0_{,\Delta} := \partial\mathbf{P}(t_0, l^1, l^2)/\partial l^\Delta$. Then, for the tangent vectors $\mathbf{P}_{,\Delta}$ to the surface $\Sigma(t)$ at any instant $t = \hat{t}(l^0) = l^0 + t_0$, we have

$$\mathbf{P}_{,\Delta} = \mathbf{P}^0_{,\Delta} + l^0\hat{\mathbf{c}}_{,\Delta}. \tag{16.24}$$

In this expression $\hat{\mathbf{c}}(l^\Delta) := \nabla_{\mathbf{N}}\Omega(\mathbf{N}(l^\Delta))$ and in order to calculate $\hat{\mathbf{c}}_{,\Delta}$ one can use the invariants and geometry of $\Sigma(t_0)$. Since the normal vector is independent of l^0 we have

$$\mathbf{N}_{,\Delta} = -\hat{\mathbf{b}}\mathbf{P}_{,\Delta} = -\hat{\mathbf{b}}^0\mathbf{P}^0_{,\Delta}, \tag{16.25}$$

where $\hat{\mathbf{b}}^0$ is the second fundamental form of $\Sigma(t_0)$. It follows that

$$\hat{\mathbf{c}}_{,\Delta} = -(\nabla_{\mathbf{N}}\nabla_{\mathbf{N}}\Omega)\hat{\mathbf{b}}\mathbf{P}_{,\Delta} = -(\nabla_{\mathbf{N}}\nabla_{\mathbf{N}}\Omega)\hat{\mathbf{b}}^0\mathbf{P}^0_{,\Delta}. \tag{16.26}$$

We define the surface tensor $\hat{\mathbf{d}}$ by

$$\hat{\mathbf{d}} = \hat{d}^\Gamma_\Delta\mathbf{P}^0_{,\Gamma}\otimes\mathbf{P}^{0\Delta} \equiv -(\nabla_{\mathbf{N}}\nabla_{\mathbf{N}}\Omega)\hat{\mathbf{b}}^0. \tag{16.27}$$

This is a constant along a ray. As we shall see, $\hat{\mathbf{d}}$ controls the kinematics of the propagation of $\{\Sigma\{\hat{t}(l^0)\}: l^0 > 0\}$. With the help of $\hat{\mathbf{d}}$ the derivatives $\hat{\mathbf{c}}_{,\Delta}$ and $\mathbf{P}_{,\Delta}$ may be written

$$\begin{aligned} \hat{\mathbf{c}}_{,\Delta} &= \hat{\mathbf{d}}\mathbf{P}^0_{,\Delta} = \hat{d}^\Gamma_\Delta\mathbf{P}^0_{,\Gamma}, \\ \mathbf{P}_{,\Delta} &= \mathbf{P}^0_{,\Delta} + l^0\hat{\mathbf{d}}\mathbf{P}^0_{,\Delta} = (\mathbf{1}_{\mathscr{S}} + l^0\hat{\mathbf{d}})\mathbf{P}^0_{,\Delta}. \end{aligned} \tag{16.28}$$

Since $\mathbf{N}\hat{\mathbf{d}} = \mathbf{0}$, $\hat{\mathbf{d}}$ is planar with null vector **N** and we have used the symbol $\mathbf{1}_{\mathscr{S}}$ to denote the surface identity tensor. From (16.25) and $(16.28)_2$ we obtain

$$\hat{\mathbf{b}}(\mathbf{1}_{\mathscr{S}} + l^0\hat{\mathbf{d}})\mathbf{P}^0_{,\Delta} = \hat{\mathbf{b}}^0\mathbf{P}^0_{,\Delta}, \quad \Delta = 1, 2, \tag{16.29}$$

and hence

$$\hat{\mathbf{b}}(\mathbf{1}_{\mathscr{S}} + l^0\hat{\mathbf{d}}) = \hat{\mathbf{b}}^0, \quad \hat{\mathbf{b}} = \hat{\mathbf{b}}^0(\mathbf{1}_{\mathscr{S}} + l^0\hat{\mathbf{d}})^{-1}, \tag{16.30}$$

where

$$(\mathbf{1}_{\mathscr{S}}+l^0\hat{\mathbf{d}})(\mathbf{1}_{\mathscr{S}}+l^0\hat{\mathbf{d}})^{-1} = \mathbf{1}_{\mathscr{S}}.$$

In some problems it is convenient to have $\hat{\mathbf{b}}$ expressed in terms of its initial value $\hat{\mathbf{b}}^0$ and the symmetric surface tensor $\hat{\mathbf{M}} := -\nabla_N\nabla_N\Omega$. Since

$$\hat{\mathbf{d}} = \hat{\mathbf{M}}\hat{\mathbf{b}}^0 \tag{16.31}$$

we have

$$\hat{\mathbf{b}} = \hat{\mathbf{b}}^0(\mathbf{1}_{\mathscr{S}}+l^0\hat{\mathbf{M}}\hat{\mathbf{b}}^0)^{-1}. \tag{16.32}$$

From the Cayley–Hamilton theorem we obtain

$$(\mathbf{1}_{\mathscr{S}}+l^0\hat{\mathbf{M}}\hat{\mathbf{b}}^0)^{-1} = \frac{\mathbf{1}_{\mathscr{S}}+l^0(I_d\mathbf{1}_{\mathscr{S}}-\hat{\mathbf{d}})}{\det_{\mathscr{S}}(\mathbf{1}_{\mathscr{S}}+l^0\hat{\mathbf{M}}\hat{\mathbf{b}}^0)} = \frac{\mathbf{1}_{\mathscr{S}}+l^0\,II_d\hat{\mathbf{d}}^{-1}}{\det_{\mathscr{S}}(\mathbf{1}_{\mathscr{S}}+l^0\hat{\mathbf{M}}\hat{\mathbf{b}}^0)},$$

where

$$I_d \equiv \operatorname{tr}_{\mathscr{S}}\hat{\mathbf{d}} = \operatorname{tr}_{\mathscr{S}}(\hat{\mathbf{M}}\hat{\mathbf{b}}^0), \qquad II_d \equiv \det_{\mathscr{S}}\hat{\mathbf{d}} = \det_{\mathscr{S}}(\hat{\mathbf{M}}\hat{\mathbf{b}}^0).$$

Hence

$$\hat{\mathbf{b}} = \frac{(1+I_d l^0)\hat{\mathbf{b}}^0 - l^0\hat{\mathbf{b}}^0\hat{\mathbf{M}}\hat{\mathbf{b}}^0}{1+I_d l^0+II_d(l^0)^2}, \tag{16.33}$$

where

$$1+I_d l^0+II_d(l^0)^2 = \det_{\mathscr{S}}(\mathbf{1}_{\mathscr{S}}+l^0\hat{\mathbf{d}}).$$

Later, the expression

$$\hat{\mathbf{M}}\hat{\mathbf{b}} = \frac{\hat{\mathbf{M}}\hat{\mathbf{b}}^0+II_d l^0\,\mathbf{1}_{\mathscr{S}}}{1+I_d l^0+II_d(l^0)^2} \tag{16.34}$$

will prove useful.

Under the present assumptions the transport equation (16.14) takes the form

$$\frac{\mathrm{d}}{\mathrm{d}l^0}\ln j = -\operatorname{tr}\{(\nabla_N\nabla_N\Omega)\hat{\mathbf{b}}\}.$$

In view of $(16.27)_2$ and (16.31) and the fact that $\hat{\mathbf{M}}$ and $\hat{\mathbf{d}}$ are regarded as two-dimensional surface tensors on $\Sigma\{\hat{t}(l^0)\}$, we obtain

$$\frac{\mathrm{d}}{\mathrm{d}l^0}\ln j = \operatorname{tr}_{\mathscr{S}}(\hat{\mathbf{M}}\hat{\mathbf{b}}) = \operatorname{tr}_{\mathscr{S}}\{\hat{\mathbf{d}}(\mathbf{1}_{\mathscr{S}}+l^0\hat{\mathbf{d}})^{-1}\}. \tag{16.35}$$

Finally, equation (16.35) may be written

$$\frac{\mathrm{d}}{\mathrm{d}l^0}\ln j = \frac{\mathrm{tr}_{\mathscr{S}}\hat{\mathbf{d}}+2l^0\det_{\mathscr{S}}\hat{\mathbf{d}}}{\det_{\mathscr{S}}(\mathbf{1}_{\mathscr{S}}+l^0\hat{\mathbf{d}})}. \tag{16.36}$$

Its solution has the form

$$\frac{j}{j_0} = \det_{\mathscr{S}}(\mathbf{1}_{\mathscr{S}}+l^0\hat{\mathbf{d}}), \tag{16.37}$$

where j_0 is the initial value of j (see (15.10)).

The kinematic and geometrical invariants needed for determining solutions to transport equations can be found directly from the results obtained so far. In particular, from $(16.30)_2$ we obtain

$$\begin{aligned} 2K_M &= \hat{b}^{\Gamma}_{\Gamma} \equiv \mathrm{tr}_{\mathscr{S}}\hat{\mathbf{b}} = \frac{\mathrm{tr}_{\mathscr{S}}\mathbf{b}^0+l^0(\det_{\mathscr{S}}\hat{\mathbf{d}})\,\mathrm{tr}_{\mathscr{S}}(\hat{\mathbf{b}}^0\hat{\mathbf{d}}^{-1})}{\det_{\mathscr{S}}(\mathbf{1}_{\mathscr{S}}+l^0\hat{\mathbf{d}})} \\ K_G &= \det_{\mathscr{S}}\hat{\mathbf{b}} = \frac{\det_{\mathscr{S}}\hat{\mathbf{b}}^0}{\det_{\mathscr{S}}(\mathbf{1}_{\mathscr{S}}+l^0\hat{\mathbf{d}})}. \end{aligned} \tag{16.38}$$

We now formulate the conditions required for the formation of a caustic. From the form of j we deduce the following proposition.

Proposition 16.2

If the surface tensor $\hat{\mathbf{d}}$ has a negative real eigenvalue $d_1 < 0$ then a caustic forms at $l^0 = |d_1|^{-1}$, and if $d := d_1 = d_2$ are two equal eigenvalues of $\hat{\mathbf{d}}$ then a focus forms at $l^0 = |d|^{-1}$. □

It would be interesting to examine the characteristic equation for $\hat{\mathbf{d}}$ and to determine the nature of the eigenvalues for various kinds of slowness surface (defined by (16.9)) and the singular surface Σ.

16.3 Bibliographical Notes

The derivation of equations (16.3) and (16.4) can be found in Hayes (1970) and Wright (1973). Definition 16.1 is similar to that in Lewis (1965) and Hayes (1970, 1974). The transport equation (16.14) has been derived in, for example, Hayes (1970) and Wright (1973). The remaining transport equations are contained in Wright (1973); see also Hayes (1970).

The analysis appropriate to a uniform state was carried out by Wright (1973), who also presented conditions for the formation of caustics; see also Nariboli (1966) and Hayes (1970). However, the term $b := \det\hat{\mathbf{b}}$, rather than j, is considered in Nariboli (1966). Since $\det\hat{\mathbf{b}}^0$ may vanish without j_0 vanishing it seems clear that j is preferable as the basic kinematic variable, as Hayes (1970) pointed out.

Equations (16.38) provide generalizations of the well-known formulae for the mean and Gaussian curvatures of parallel surfaces (recall (2.58)) which arise in propagation problems for isotropic media (Thomas, 1965). Note, in particular, that (16.32) holds along rays, not along orthogonal trajectories.

The derivation of equations (16.33) and (16.34) can be found in Scott (1975b). The integrated transport equations (16.37) and (16.38) were given by Wright (1973); see also Hayes (1970).

The characteristic equation for $\hat{\mathbf{d}}$ for various kinds of slowness surface has been examined by Wright (1973). The extensive discussion of geometrical optics methods, Fermat principle, reflection and refraction of waves can be found in Hanyga (1984), together with the analysis of caustic formation. However most of the results presented there deal with linear seismic waves.

Chapter 7

Examples

In this chapter we discuss some particular examples in order to illustrate the theory described in the previous chapters. We shall begin with acoustic waves in a barotropic medium without heat condition. First, using the displacement derivative, an equation for the wave amplitude is derived; kinematic wave theory is then applied to show that the transport equations along a ray provide a more useful description than the equations along the normal trajectories derived previously.

In the second example shock waves in a barotropic medium are analysed. The time derivative along a curve which differs from the normal trajectory of the shock wave (a surface carrying discontinuities in the particle velocity and the entropy) is more useful than the displacement derivative in this example. Separate transport equations for the derivatives along the curve and tangential to the surface are derived. Unfortunately, integration of these equations is not possible, even when the region ahead of the wave is undisturbed. This is due to a coupling between the transport equations along a ray and the field values behind the shock. The resulting equations are of functional-differential type.

The reason for such complications in the description of shock wave propagation is that, in non-linear problems, the surface of discontinuity of the velocity field does not coincide with that of the acceleration. The latter is the characteristic surface of the field equations, that is the initial system of partial differential equations describing the problem under consideration. It is not possible to find the motion (and the wave velocity) first and then the amplitude (or intensity) of the shock wave because these quantities are strongly coupled.

17 ACOUSTIC WAVES IN A BAROTROPIC MEDIUM

The results of Chapter 5 are universal in that they hold for any deformable continuum, independently of the material. In Section 9 (Chapter 4) the basic equations for a deformable body were formulated without reference to the material constituting the body. Roughly speaking a deformable body is a material body if there is a relation between the dynamic, thermodynamic and kinematic quantities used in its description. Among the dynamic quantities are the stress tensor **T** and the body force **b**, while the specific entropy and the temperature are thermodynamic quantities. The main kinematic quantities were introduced in Section 9.

We are not concerned here with a general or precise characterization of a material, and we shall confine attention to some representative examples of materials.

17.1 Barotropic Media

We consider a deformable body $\mathscr{B}$ endowed with an additional structure defined by the constitutive relation

$$\mathbf{T} = -p(\varrho, \eta)\mathbf{1}, \tag{17.1}$$

where $p(\varrho, \eta)$, called the hydrostatic pressure, is a scalar-valued function of the mass density ϱ and specific entropy η. Equation (17.1) is called a **stress constitutive relation** for a barotropic medium and it defines the material of the body $\mathscr{B}$ if it holds for any motion χ_t, $t \in I$, and time-dependent fields **T**, ϱ and η defined on $\mathscr{B}$.

In what follows we assume that the whole body $\mathscr{B}$ is composed of material defined by the constitutive relation (17.1) so that we do not admit surface constitutive relations which could describe surface concentration effects.

Let $\mathscr{S}$ be a hypersurface in the region $\mathscr{N}$ (recall (14.1)–(14.3)).

Definition 17.1

We say that a motion χ_t, $t \in I$, is accompanied by an **acoustic wave** if the hypersurface $\mathscr{S}$ is a carrier of first-order singularities in the fields **v**, ϱ and η.

Clearly, an acoustic wave is a singular surface of second order for the motion χ_t. Hence we are concerned with the kind of singularity referred to earlier as an acceleration wave.

We now list the balance equations for the medium under consideration. From (10.8) the mass conservation equation is

$$\frac{\partial \varrho}{\partial t} + \operatorname{div}(\varrho \mathbf{v}) = 0 \tag{17.2}$$

while, in the absence of body forces, the momentum conservation law (10.31) becomes

$$\varrho \frac{\partial \mathbf{v}}{\partial t} + \varrho(\mathbf{v} \cdot \mathbf{grad})\mathbf{v} = \operatorname{div}\mathbf{T}. \tag{17.3}$$

If there is no heat conduction or volume heat supply the energy conservation law (10.46) takes the form

$$\varrho\left(\frac{\partial e}{\partial t}+\mathbf{v}\cdot\mathbf{grad}\,e\right) = \mathrm{tr}\,\{(\mathbf{grad\,v})\mathbf{T}^{\mathrm{T}}\}. \tag{17.4}$$

The entropy balance law (10.54) becomes

$$\varrho\left(\frac{\partial \eta}{\partial t}+\mathbf{v}\cdot\mathbf{grad}\,\eta\right) = 0 \tag{17.5}$$

when there is no entropy flux, and, finally, the entropy production inequality (11.4) reduces to

$$\varrho\dot{\eta} \geqslant 0. \tag{17.6}$$

Note that (17.5) has to be derived from (17.6) when (17.2)–(17.4) are supplied with the constitutive equation below. The assumptions mean that effects caused by the existence of production, supply or efflux of thermodynamic quantities are excluded.

The material structure of the medium is enriched by the constitutive equation for the internal energy

$$e = e(\varrho, \eta) \tag{17.7}$$

with the property (to be derived from (17.6))

$$\varrho^2\frac{\partial e}{\partial \varrho} = p(\varrho, \eta). \tag{17.8}$$

Then it follows that the energy equation (17.4) is satisfied identically iff (17.5) holds. We make the additional assumption that the functions p and e are twice continuously differentiable with respect to ϱ and η.

17.2 Relations Satisfied on an Acoustic Wave

According to our assumptions the balance laws for points not on the surface $\mathscr{S}$ are

$$\begin{aligned}
&\frac{\partial \varrho}{\partial t}+\mathbf{v}\cdot\mathbf{grad}\,\varrho+\varrho\,\mathrm{div}\,\mathbf{v} = 0,\\
&\varrho\frac{\partial \mathbf{v}}{\partial t}+\mathbf{grad}\,p+\varrho(\mathbf{v}\cdot\mathbf{grad})\mathbf{v} = \mathbf{0},\\
&\frac{\partial \eta}{\partial t}+\mathbf{v}\cdot\mathbf{grad}\,\eta = 0,
\end{aligned} \tag{17.9}$$

and on $\mathscr{S}$

$$\left[\!\left[\frac{\partial \varrho}{\partial t}\right]\!\right]+[\![\mathbf{grad}\varrho]\!]\cdot\mathbf{v}+\varrho[\![\mathrm{div}\mathbf{v}]\!] = 0,$$
$$\varrho\left[\!\left[\frac{\partial \mathbf{v}}{\partial t}\right]\!\right]+[\![\mathbf{grad}p]\!]+\varrho[\![\mathbf{grad}\mathbf{v}]\!]\mathbf{v} = \mathbf{0}, \tag{17.10}$$
$$\left[\!\left[\frac{\partial \eta}{\partial t}\right]\!\right]+[\![\mathbf{grad}\eta]\!]\cdot\mathbf{v} = 0.$$

Because of the compatibility conditions (6.10) we may write

$$\left[\!\left[\frac{\partial \varrho}{\partial t}\right]\!\right] = -u_n\left[\!\left[\frac{\partial \varrho}{\partial n}\right]\!\right], \qquad [\![\mathbf{grad}\varrho]\!] = \left[\!\left[\frac{\partial \varrho}{\partial n}\right]\!\right]\mathbf{n},$$
$$[\![\mathbf{grad}\mathbf{v}]\!] = \left[\!\left[\frac{\partial \mathbf{v}}{\partial n}\right]\!\right]\otimes\mathbf{n}, \qquad \left[\!\left[\frac{\partial \mathbf{v}}{\partial t}\right]\!\right] = -u_n\left[\!\left[\frac{\partial \mathbf{v}}{\partial n}\right]\!\right],$$
$$[\![\mathbf{grad}p]\!] = \left[\!\left[\frac{\partial p}{\partial n}\right]\!\right]\mathbf{n}, \qquad \left[\!\left[\frac{\partial \eta}{\partial t}\right]\!\right] = -u_n\left[\!\left[\frac{\partial \eta}{\partial n}\right]\!\right], \tag{17.11}$$
$$[\![\mathbf{grad}\eta]\!] = \left[\!\left[\frac{\partial \eta}{\partial n}\right]\!\right]\mathbf{n}.$$

Note that (17.11) implies that

$$[\![\mathbf{grad}p]\!] = \frac{\partial p}{\partial \varrho}[\![\mathbf{grad}\varrho]\!]+\frac{\partial p}{\partial \eta}[\![\mathbf{grad}\eta]\!]. \tag{17.12}$$

Substituting (17.11) into (17.10) we obtain

$$(\mathbf{v}\cdot\mathbf{n}-u_n)\left[\!\left[\frac{\partial \varrho}{\partial n}\right]\!\right]+\varrho\left[\!\left[\frac{\partial \mathbf{v}}{\partial n}\right]\!\right]\cdot\mathbf{n} = 0,$$
$$\varrho(\mathbf{v}\cdot\mathbf{n}-u_n)\left[\!\left[\frac{\partial \mathbf{v}}{\partial n}\right]\!\right]+\left[\!\left[\frac{\partial p}{\partial n}\right]\!\right]\mathbf{n} = \mathbf{0}, \tag{17.13}$$
$$(\mathbf{v}\cdot\mathbf{n}-u_n)\left[\!\left[\frac{\partial \eta}{\partial n}\right]\!\right] = 0.$$

It follows from the last of these equations that

$$U \equiv u_n-v_n = 0 \quad \text{or} \quad \left[\!\left[\frac{\partial \eta}{\partial n}\right]\!\right] = 0. \tag{17.14}$$

Note that $(17.14)_1$ leads to $[\![\partial\mathbf{v}/\partial n]\!] = \mathbf{0}$ and $[\![\partial p/\partial n]\!] = 0$. Since $\mathscr{S}(t)$ is a material surface in this case we reject this possibility and assume that $(17.14)_2$ holds.

We now formulate an interesting result which holds for any medium in which Cauchy's first law of motion has the form $(17.9)_2$.

Lemma 17.1

In a barotropic medium any acceleration wave is longitudinal.

Proof

The equation of motion $(17.9)_2$ may be written in the form

$$\varrho[\![\ddot{\mathbf{x}}]\!] = -[\![\operatorname{grad} p]\!].$$

From $(17.11)_5$ the result follows (recall the definition (14.19)). □

Using this result together with (14.24) we obtain

$$\left[\!\!\left[\frac{\partial \mathbf{v}}{\partial n}\right]\!\!\right] = a\mathbf{n}, \tag{17.15}$$

where $a \equiv [\![\partial \mathbf{v}/\partial n]\!] \cdot \mathbf{n}$; the quantity a is called the **intensity** of the acceleration wave. Using this we may write $(17.13)_{1,2}$ in the form

$$U\left[\!\!\left[\frac{\partial \varrho}{\partial n}\right]\!\!\right] = \varrho a, \qquad \left(\varrho U a - \left[\!\!\left[\frac{\partial \varrho}{\partial n}\right]\!\!\right]\right)\mathbf{n} = \mathbf{0}. \tag{17.16}$$

Under the condition $(17.14)_2$ equation (17.12) leads to

$$\left[\!\!\left[\frac{\partial p}{\partial n}\right]\!\!\right] = \frac{\partial p}{\partial \varrho}\left[\!\!\left[\frac{\partial \varrho}{\partial n}\right]\!\!\right].$$

Finally, we note that equations (17.16) have non-zero solutions for a if

$$U^2 = c_d^2, \tag{17.17}$$

where c_d denotes the square root of $\partial p/\partial \varrho$. The quantity c_d is called the local speed of sound in the medium. Choosing the value c_d for U we obtain

Proposition 17.1

The normal component of velocity of an acoustic wave propagating in the direction of the vector **n** is given by

$$u_n = \mathbf{v}\cdot\mathbf{n} + c_d, \qquad c_d = \sqrt{\frac{\partial p}{\partial \varrho}}. \quad \square \tag{17.18}$$

In what follows we shall be concerned with the derivation of an amplitude equation for the acoustic wave, and this will involve the displacement derivative. First, however, we note that equation (17.18) is a homogeneous dispersion relation of order one. In fact, writing $u_n = \omega/k$, we can transform $(17.18)_1$ into

$$\omega = \Omega(\mathbf{k}, t, \mathbf{x}) \equiv \mathbf{v}\cdot\mathbf{k} + k c_d \tag{17.19}$$

(recall (16.2)).

17.3 The Amplitude Equation

Under the assumption that fields defined on the sets $\mathscr{N}^+\cup\mathscr{S}$ and $\mathscr{N}^-\cup\mathscr{S}$ are continuously differentiable we differentiate the continuity equation $(17.9)_1$ and the equation of motion $(17.9)_2$ with respect to **x**. Taking differences of the results on $\mathscr{S}$, we obtain

$$\begin{aligned}
&\left[\!\left[\mathbf{grad}\left(\frac{\partial\varrho}{\partial t}\right)\right]\!\right]+\mathbf{v}\cdot[\![\mathbf{grad}\,\mathbf{grad}\,\varrho]\!]+[\![(\mathbf{grad}\,\varrho)(\mathbf{grad}\,\mathbf{v})]\!]\\
&+\varrho[\![\mathbf{grad}\cdot\mathbf{grad}\,\mathbf{v}]\!]+[\![(\mathrm{div}\,\mathbf{v})\,\mathbf{grad}\,\varrho]\!]=\mathbf{0},\\
&\left[\!\left[\mathbf{grad}\,\varrho\otimes\frac{\partial\mathbf{v}}{\partial t}\right]\!\right]+\varrho\left[\!\left[\mathbf{grad}\left(\frac{\partial\mathbf{v}}{\partial t}\right)\right]\!\right]+[\![\mathbf{grad}\,\mathbf{grad}\,p]\!]\\
&+[\![\mathbf{grad}\,\varrho\otimes(\mathbf{grad}\,\mathbf{v})\mathbf{v}]\!]+\varrho[\![(\mathbf{grad}\,\mathbf{v})(\mathbf{grad}\,\mathbf{v})^{\mathrm{T}}]\!]\\
&+\varrho[\![\mathrm{tr}_{13}(\mathbf{v}\otimes\mathbf{grad}\,\mathbf{grad}\,\mathbf{v})]\!]=\mathbf{0}.
\end{aligned}\tag{17.20}$$

The symbol tr_{13} denotes the trace with respect to the first and third indices of the tensor in the brackets.

In our calculations we shall use the compatibility conditions (16.2) for the second derivatives, and also for the jump of the product of two functions, f and h, the identity

$$[\![fh]\!]=[\![f]\!][\![h]\!]+(f)^+[\![h]\!]+[\![f]\!](h)^+.\tag{17.21}$$

Equations (17.20) may then be written

$$\begin{aligned}
&\frac{\delta b}{\delta t}+b_{,\alpha}(\mathbf{v}\cdot\boldsymbol{\varphi}^\alpha)-U\left[\!\left[\frac{\partial^2\varrho}{\partial n^2}\right]\!\right]+2ab+2a\left\{\left(\frac{\partial\varrho}{\partial n}\right)^+-\varrho K_m\right\}\\
&+\varrho\left[\!\left[\frac{\partial^2\mathbf{v}}{\partial n^2}\right]\!\right]\cdot\mathbf{n}+b\left\{\left(\frac{\partial\mathbf{v}}{\partial n}\right)^+\cdot\mathbf{n}+(\mathrm{div}\,\mathbf{v})^+\right\}=0,\\
&\varrho\left\{\frac{\delta a}{\delta t}+a_{,\alpha}(\mathbf{v}\cdot\boldsymbol{\varphi}^\alpha)\right\}-\varrho U\left[\!\left[\frac{\partial^2\mathbf{v}}{\partial n^2}\right]\!\right]\cdot\mathbf{n}-Ua\left(\frac{\partial\varrho}{\partial n}\right)^+\\
&+\left\{\left(\frac{\partial\mathbf{v}}{\partial n}\right)^++(\mathbf{grad}\,\mathbf{v})\mathbf{v}\right\}\cdot\mathbf{n}b+\left[\!\left[\frac{\partial^2 p}{\partial n^2}\right]\!\right]+2\varrho\left(\frac{\partial\mathbf{v}}{\partial n}\right)^+\cdot\mathbf{n}a=0,
\end{aligned}\tag{17.22}$$

where $b:=[\![\partial p/\partial n]\!]$. The constitutive relation (17.1) enables us to write

$$\begin{aligned}
\mathbf{grad}\,\mathbf{grad}\,p=&\frac{\partial^2 p}{\partial\varrho^2}\,\mathbf{grad}\,\varrho\otimes\mathbf{grad}\,\varrho+\frac{\partial p}{\partial\varrho}\,\mathbf{grad}\,\mathbf{grad}\,\varrho\\
&+\frac{\partial^2 p}{\partial\varrho\,\partial\eta}\,(\mathbf{grad}\,\varrho\otimes\mathbf{grad}\,\eta+\mathbf{grad}\,\eta\otimes\mathbf{grad}\,\varrho)\\
&+\frac{\partial p}{\partial\eta}\,\mathbf{grad}\,\mathbf{grad}\,\eta+\frac{\partial^2 p}{\partial\eta^2}\,\mathbf{grad}\,\eta\otimes\mathbf{grad}\,\eta.
\end{aligned}\tag{17.23}$$

On use of $(17.14)_2$ and (17.11) we obtain

$$\left[\!\left[\frac{\partial^2 p}{\partial n^2}\right]\!\right] = \frac{\partial^2 p}{\partial \varrho^2}\left\{\left[\!\left[\frac{\partial \varrho}{\partial n}\right]\!\right]^2 + 2\left(\frac{\partial \varrho}{\partial n}\right)^+\left[\!\left[\frac{\partial \varrho}{\partial n}\right]\!\right]\right\} + U^2\left[\!\left[\frac{\partial^2 \varrho}{\partial n^2}\right]\!\right]$$
$$+\frac{\partial p}{\partial \eta}\left[\!\left[\frac{\partial^2 \eta}{\partial n^2}\right]\!\right] + 2\frac{\partial^2 p}{\partial \varrho\,\partial \eta}\frac{\partial \eta}{\partial n}\left[\!\left[\frac{\partial \varrho}{\partial n}\right]\!\right]. \tag{17.24}$$

Differentiating $(17.9)_3$ with respect to t and making use of the compatibility condition (6.15) we obtain

$$\left[\!\left[\frac{\partial^2 \eta}{\partial n^2}\right]\!\right] = \frac{a}{U}\frac{\partial \eta}{\partial n}. \tag{17.25}$$

Substitution of (17.25) into (17.24) leads to

$$\left[\!\left[\frac{\partial^2 p}{\partial n^2}\right]\!\right] - U^2\left[\!\left[\frac{\partial^2 \varrho}{\partial n^2}\right]\!\right] = \frac{\partial^2 p}{\partial \varrho^2}\left\{\left[\!\left[\frac{\partial \varrho}{\partial n}\right]\!\right]^2 + 2\left(\frac{\partial \varrho}{\partial n}\right)^+\left[\!\left[\frac{\partial \varrho}{\partial n}\right]\!\right]\right\}$$
$$+2\frac{\partial^2 p}{\partial \varrho\,\partial \eta}\frac{\partial \eta}{\partial n}\left[\!\left[\frac{\partial \varrho}{\partial n}\right]\!\right] + \frac{\partial p}{\partial \eta}\frac{\partial \eta}{\partial n}\frac{a}{U}.$$

Finally, be eliminating the terms $[\![\partial^2\rho/\partial n^2]\!]$, $[\![\partial^2 p/\partial n^2]\!]$ and $[\![\partial^2\mathbf{v}/\partial n^2]\!]\cdot\mathbf{n}$ from (17.22), we obtain the amplitude equation

$$\varrho\left\{\frac{\delta a}{\delta t} + a_{,\alpha}(\mathbf{v}\cdot\boldsymbol{\varphi}^\alpha)\right\} + U\left\{\frac{\delta b}{\delta t} + b_{,\alpha}(\mathbf{v}\cdot\boldsymbol{\varphi}^\alpha)\right\}$$
$$+b\left\{2\frac{\partial^2 p}{\partial \eta\,\partial \varrho}\frac{\partial \eta}{\partial n} + \frac{\partial p}{\partial n}\frac{\partial \eta}{\partial n}\frac{1}{\varrho} + 2\frac{\partial^2 p}{\partial \varrho^2}\left(\frac{\partial \varrho}{\partial n}\right)^+ + \frac{\partial p}{\partial \varrho}\left(\frac{\partial \varrho}{\partial n}\right)^+\frac{1}{\varrho}\right.$$
$$\left.-2U^2K_m + 3U\left(\frac{\partial \mathbf{v}}{\partial n}\right)^+\cdot\mathbf{n} + \left(\frac{\partial \mathbf{v}}{\partial t}\right)^+\cdot\mathbf{n} + \{(\mathbf{grad}\,\mathbf{v})\mathbf{v}\}^+\cdot\mathbf{n} + U(\operatorname{div}\mathbf{v})^+\right\}$$
$$+b^2\frac{1}{\varrho^2}\frac{\partial}{\partial \varrho}\left(\varrho^2\frac{\partial p}{\partial \varrho}\right) = 0, \tag{17.26}$$

which holds along a normal trajectory of the wave.

Only in a few special cases is it possible to solve this equation; for example, when a and b are independent of the surface parameters l^α. This difficulty arises because of the appearance of the tangential derivatives $a_{,\alpha}$ and $b_{,\alpha}$ in (17.26). If $\mathbf{v} = \mathbf{0}$ on $\mathscr{S}$ then the terms containing these derivatives vanish; this corresponds to the situation in which the acoustic wave propagates into a region in a stationary uniform state since, by (17.9), if $\mathbf{v} = \mathbf{0}$ in the region $\mathscr{N}^+\cup\mathscr{S}$ then also

$$\frac{\partial \varrho}{\partial t} = \frac{\partial \eta}{\partial t} = 0, \qquad \mathbf{grad}\,p = \mathbf{0}, \qquad \frac{\partial p}{\partial t} = 0,$$

$$\frac{\partial p}{\partial \varrho} \operatorname{grad} \varrho + \frac{\partial p}{\partial \eta} \operatorname{grad} \eta = \mathbf{0}$$

in that region.

Another situation in which the tangential derivatives disappear is when $\mathbf{v}$ is normal to the surface, so that $\mathbf{v} \cdot \boldsymbol{\varphi}_{,\alpha} = 0$.

The above discussion provides conditions under which the main disturbances of an acoustic wave in a barotropic medium travel along the normal trajectories of the singular surface $\mathscr{S}$.

17.4 Rays and Normal Trajectories

We now return to the dispersion relation for an acoustic wave, namely

$$u_n = \mathbf{v} \cdot \mathbf{n} + c_d.$$

If the region $\mathscr{N}^+$ is in a stationary state then $\mathbf{v} = \mathbf{0}$ and this reduces to

$$u_n = c_d.$$

Clearly, the right-hand side is independent of the normal $\mathbf{n}$.

Lemma 17.2

If an acoustic wave propagates into a region which is in a stationary state then the rays are straight lines and they coincide with the normal trajectories of the wave front $\mathscr{S}(t)$; the latter travels with constant speed u_n equal to the local sound speed in the medium. □

In the lemma we have not assumed that the state is uniform; we have only used the homogeneity of the stress (pressure) distribution, $\operatorname{grad} p = \mathbf{0}$, with $\operatorname{grad} \varrho$ and $\operatorname{grad} \eta$ not necessarily vanishing.

Corollary 17.1

If, in the region $\mathscr{N}^+ \cup \mathscr{S}$, $\mathbf{v}$ is parallel to the normal vector $\mathbf{n}$ to the wave front then the rays coincide with the normal trajectories.

Proof

If $\mathbf{v}$ is parallel to $\mathbf{n}$ then differentiation of the propagation function $\Omega(\mathbf{n}, t, \mathbf{x}) := \mathbf{v} \cdot \mathbf{n} + c_d$ with respect to $\mathbf{n}$ leads to the expression

$$\mathbf{c} = u_n \mathbf{n} + (\mathbf{1} - \mathbf{n} \otimes \mathbf{n}) \nabla_{\mathbf{n}} \Omega = u_n \mathbf{n}$$

for the ray velocity $\mathbf{c}$ (see $(16.3)_1$). This means that the rays are orthogonal to the wave front. □

This result does not, however, imply that the wave speed u_n and the normal vector are constant along a ray. In other words, the condition $\mathbf{v} = |\mathbf{v}|\mathbf{n}$ does not imply that the normal trajectories are straight lines, in distinction from the case covered by Lemma 17.2.

17.5 The Transport Equations

We now pass to the general case in which the velocity $\mathbf{v}$ in front of the wave is unconstrained and derive the ray equation* for the dispersion relation (17.19). From $(15.7)_2$ and (16.3) we obtain

$$\frac{\mathrm{d}\mathbf{x}}{\mathrm{d}l^0} = \mathbf{v}+U\mathbf{n}, \qquad \mathbf{c} = u_n\mathbf{n}+\mathbf{v}-(\mathbf{v}\cdot\mathbf{n})\mathbf{n} = \mathbf{v}+U\mathbf{n}. \tag{17.27}$$

The tangential components c^α of the ray velocity $\mathbf{c}$ are expressed as

$$c^\alpha \equiv \mathbf{c}\cdot\boldsymbol{\varphi}^\alpha = \mathbf{v}\cdot\boldsymbol{\varphi}^\alpha. \tag{17.28}$$

Recalling from (15.22) the connection between the displacement derivative and the derivative along a ray of an arbitrary function h, namely

$$\frac{\mathrm{d}h}{\mathrm{d}l^0} = \frac{\delta h}{\delta t}+c^\alpha \tilde{h}_{,\alpha},$$

enables us to represent the expressions in the first two brackets of equation (17.26) in the form

$$\frac{\delta a}{\delta t}+a_{,\alpha}\mathbf{v}\cdot\boldsymbol{\varphi}^\alpha = \frac{\mathrm{d}a}{\mathrm{d}l^0},$$

$$\frac{\delta b}{\delta t}+b_{,\alpha}\mathbf{v}\cdot\boldsymbol{\varphi}^\alpha = \frac{\mathrm{d}b}{\mathrm{d}l^0},$$

using (17.28). The right-hand sides represent the derivative along a ray.

Using the equality $Ub = \varrho a$ we may reduce (17.26) to an equation for b:

$$\frac{\mathrm{d}}{\mathrm{d}l^0}\left(\frac{b^2U}{\varrho}\right)+\frac{\mu(l^0)}{\varrho}b^2+\frac{1}{\varrho^3}\frac{\partial}{\partial\varrho}\left(\varrho^2\frac{\partial p}{\partial\varrho}\right)b^3 = 0, \tag{17.29}$$

where

$$\begin{aligned}\mu(l^0) = {} & 2\frac{\partial^2 p}{\partial\varrho^2}\left(\frac{\partial\varrho}{\partial n}\right)^+ +2\frac{\partial^2 p}{\partial\varrho\,\partial\eta}\frac{\partial\eta}{\partial n}+\frac{1}{\varrho}\frac{\partial p}{\partial\eta}\frac{\partial\eta}{\partial n}+\frac{1}{\varrho}\frac{\partial p}{\partial\varrho}\left(\frac{\partial\varrho}{\partial n}\right)^+ \\ & -2U^2K_m+3U\left(\frac{\partial\mathbf{v}}{\partial n}\right)^+\cdot\mathbf{n}+\left(\frac{\partial\mathbf{v}}{\partial t}\right)^+\cdot\mathbf{n} \\ & +\{(\mathrm{grad}\,\mathbf{v})\mathbf{v}\}^+\cdot\mathbf{n}+U(\mathrm{div}\,\mathbf{v})^+. \end{aligned} \tag{17.30}$$

Equation (17.29) is of Bernoulli type with variable coefficients. Under initial conditions ϱ_0, U_0, b_0 at $l^0 = 0$ its solution is of the form

---

* Here the spatial description is used and there is therefore a difference in the independent variables, although the ray parameter is still denoted by l^0.

$$b(l^0) = b_0 \left(\frac{\varrho U_0}{\varrho_0 U}\right)^{1/2} \exp\left(-\int_0^{l^0} \frac{\mu(l)}{2U(l)} \, \mathrm{d}l\right)$$

$$\times \left\{1 + \frac{b_0}{2}\left(\frac{U_0}{\varrho_0}\right)^{1/2} \int_0^{l^0} (\varrho U)^{-3/2} \frac{\partial}{\partial \varrho}\left(\varrho^2 \frac{\partial p}{\partial \varrho}\right)\right.$$

$$\left. \times \left(\exp\left(-\int_0^{l} \frac{\mu(\tau)}{2U(\tau)} \, \mathrm{d}\tau\right)\right) \mathrm{d}l\right\}^{-1} . \tag{17.31}$$

It is obvious that the solution b is different from zero if $b_0 \neq 0$. On the other hand, vanishing of the term in braces leads to unbounded growth of b. In what follows we assume that $U \neq 0$ and that the numerator in (17.31) is not zero and seek conditions which lead to this singularity.

We assume that the constitutive function p satisfies the condition

$$\frac{\partial}{\partial \varrho}\left(\varrho^2 \frac{\partial p}{\partial \varrho}\right) > 0. \tag{17.32}$$

It is easy to check that this inequality is equivalent to the condition

$$\frac{\partial}{\partial \varrho}(\varrho c_d) > 0$$

often used in physical problems dealing with (ideal) gases and (non-viscous) fluids.

Lemma 17.3

If the initial value b_0 is negative then at the finite value l^* of the ray parameter l^0, given by

$$\frac{2}{b_0}\sqrt{\frac{\varrho_0}{U_0}} = -\int_0^{l^*} (\varrho U)^{-3/2} \frac{\partial}{\partial \varrho}\left(\varrho^2 \frac{\partial p}{\partial \varrho}\right) \exp\left(-\int_0^{l} \frac{\mu(s)}{2U(s)} \, \mathrm{d}s\right) \mathrm{d}l, \tag{17.33}$$

$-b$ becomes unbounded, that is

$$\lim_{l^0 \to l^*} b(l^0) = -\infty . \quad \square$$

We now examine the consequence of b_0 being negative. If the acoustic wave results in compression of the medium then

$$\left(\frac{\partial \varrho}{\partial t}\right)^- > \left(\frac{\partial \varrho}{\partial t}\right)^+ ,$$

that is

$$\left[\!\left[\frac{\partial \varrho}{\partial t} \right]\!\right] > 0. \tag{17.34}$$

From the definition of b we conclude that this corresponds to negative b since $[\![\partial\varrho/\partial t]\!] = -u_n[\![\partial\varrho/\partial n]\!] = -u_n b$ will then be positive. Thus the condition $b_0 < 0$ means that the acoustic wave is compressive.

Corollary 17.2

If the material of the medium satisfies the inequality (17.32) then the compressive wave will "break" in the finite time $l^0 = l^*$, that is a singularity appears in the wave. □

This singularity could mean that a caustic forms if additionally K_m becomes unbounded or a shock wave is generated.

Next we consider an expansive wave, for which $b_0 > 0$. Unbounded growth of b at a finite value l^0 is achieved if

$$\frac{\partial}{\partial \varrho}\left(\varrho^2 \frac{\partial p}{\partial \varrho}\right) < 0,$$

Theorem 17.1

The inequality

$$b_0 \frac{\partial}{\partial \varrho}\left(\varrho^2 \frac{\partial p}{\partial \varrho}\right) < 0 \tag{17.35}$$

is sufficient for the existence of a finite value l^*, given by (17.33), with the property

$$\lim_{l^0 \to l^*} \frac{b(l^0)}{b_0} = \infty. \;\square \tag{17.36}$$

17.6 Acoustic Waves at Rest

We consider an acoustic wave propagating in a medium in a homogeneous stationary state. Then the coefficient $\mu(l^0)$ in (17.30) reduces to the simple form

$$\mu(l^0) = -2U^2 K_m$$

with

$$U = u_n = c_d = \text{const}.$$

Further simplifications are then implied by Lemma 17.2. Since the normal trajectories are straight lines and u_n is constant the moving surface $\{\mathscr{S}(t)\}_{t\in I}$ forms a family of parallel surfaces. The transport equations for the wave amplitudes then take the forms

$$\frac{\delta a}{\delta t} - a u_n K_m + \frac{a^2}{2U_n^2} \frac{\partial}{\partial \varrho}\left(\varrho^2 \frac{\partial p}{\partial \varrho}\right) = 0,$$
$$\frac{\delta b}{\delta t} - b u_n K_m + \frac{b^2}{2\varrho u_n} \frac{\partial}{\partial \varrho}\left(\varrho^2 \frac{\partial p}{\partial \varrho}\right) = 0. \tag{17.37}$$

From the results of Section 2 for the mean curvature K_m we obtain

$$K_m(t) = \frac{K_m^0 - K_g^0 u_n(t-t_0)}{1 - 2K_m^0 u_n(t-t_0) + K_g^0 (u_n(t-t_0))^2}, \tag{17.38}$$

where $K_m^0 := K_m(t_0)$.

Let n be the distance measured from $\mathscr{S}(t_0)$ along the normal trajectories to the family of surfaces $\mathscr{S}(t)$ in the direction of propagation. Then

$$n = u_n(t-t_0) \tag{17.39}$$

and the scalars a and b can be regarded as functions of n along each normal trajectory. The amplitude equations can therefore be rewritten in the form

$$\frac{da}{dn} = aK_m - \frac{a^2}{2u_n^3} \frac{\partial}{\partial \varrho}\left(\varrho^2 \frac{\partial p}{\partial \varrho}\right),$$
$$\frac{db}{dn} = bK_m - \frac{b^2}{2\varrho u_n^2} \frac{\partial}{\partial \varrho}\left(\varrho^2 \frac{\partial p}{\partial \varrho}\right). \tag{17.40}$$

For plane waves $K_m = 0$ and the solutions of (17.40) are

$$a = a_0 \left\{1 + \frac{a_0 n}{2u_n^3} \bar{\beta}\right\}^{-1}, \quad \bar{\beta} = \frac{\partial}{\partial \varrho}\left(\varrho^2 \frac{\partial p}{\partial \varrho}\right),$$
$$b = b_0 \left\{1 + \frac{b_0 n}{2\varrho u_n^2} \bar{\beta}\right\}^{-1}, \quad n = u_n(t-t_0). \tag{17.41}$$

Similarly to the general case described in Theorem 17.1 a singularity in the amplitude will appear if one of the inequalities*

$$a_0 \frac{\partial}{\partial \varrho}\left(\varrho^2 \frac{\partial p}{\partial \varrho}\right) < 0 \quad \text{or} \quad b_0 \frac{\partial}{\partial \varrho}\left(\varrho^2 \frac{\partial p}{\partial \varrho}\right) < 0 \tag{17.42}$$

holds. The critical value n^* at which the singularity appears is given by

$$n^* = -\frac{2\varrho u_n^2}{b_0 \bar{\beta}} = -\frac{2u_n^3}{a_0 \bar{\beta}}.$$

If neither of the inequalities (17.42) holds the amplitudes will approach zero monotonically as $n \to \infty$.

If $\{\mathscr{S}(t)\}_{t \in I}$ forms a family of concentric spheres then the mean cur-

---

* From the conditions $Ub = \varrho a$ and $U > 0$ it follows that $ab \geqslant 0$.

vature* K_m is $-1/r$, where r denotes the radius of a sphere, assuming that r increases with l^0. The amplitude equations may then be written

$$\frac{\mathrm{d}a}{\mathrm{d}r} = -\frac{a}{r} - \frac{a^2}{2u_n^3}\bar{\beta}, \quad \frac{\mathrm{d}b}{\mathrm{d}r} = -\frac{b}{r} - \frac{b^2}{2\varrho u_n^2}\bar{\beta}. \tag{17.43}$$

Integrating (17.43) we find

$$\frac{1}{ar} = \frac{1}{a_0 r_0} + \frac{\bar{\beta}}{2u_n^3}\ln\left(\frac{r}{r_0}\right), \quad \frac{1}{br} = \frac{1}{b_0 r_0} + \frac{\bar{\beta}}{2\varrho u_n^2}\ln\left(\frac{r}{r_0}\right), \tag{17.44}$$

where $1/r_0 = -K_m^0 = -K_m(t_0)$. If $a_0\bar{\beta} < 0$ or $b_0\bar{\beta} < 0$ then, for the value r^* of the radius r given by

$$\ln\left(\frac{r^*}{r_0}\right) = \frac{-2u_n^3}{a_0 r_0 \bar{\beta}} = \frac{2\varrho u_n^2}{b_0 r_0 \bar{\beta}}, \tag{17.45}$$

the spherical acoustic wave becomes unbounded and at this value the acoustic wave turns into a spherical shock wave.

Finally, we consider the general case in which K_m and K_g are non-zero. The solutions of the amplitude equations are then

$$a = a_0\left\{\varphi(n)\left(1 + \frac{\bar{\beta}}{2u_n^3}\int_0^n \frac{\mathrm{d}s}{\varphi(s)}\right)\right\}^{-1},$$
$$b = b_0\left\{\varphi(n)\left(1 + \frac{\bar{\beta}}{2\varrho u_n^2}\int_0^n \frac{\mathrm{d}s}{\varphi(s)}\right)\right\}^{-1}, \tag{17.46}$$

with

$$\bar{\beta} = \frac{\partial}{\partial\varrho}\left(\varrho^2\frac{\partial p}{\partial\varrho}\right), \quad \varphi(n) = \sqrt{1 - 2K_m^0 n + K_g^0 n^2}.$$

Introducing the notation $1/r_1$, $1/r_2$ for the principal curvatures of the initial surface $\mathscr{S}(t_0)$, with r_1 and r_2 as the corresponding radii of curvature, the mean and Gaussian curvatures become $K_m^0 = \frac{1}{2}(1/r_1 + 1/r_2)$ and $K_g^0 = 1/(r_1 r_2)$, and hence

$$\varphi^2(n) = 1 - 2K_m^0 n + K_g^0 n^2 = \frac{(r_1 - n)(r_2 - n)}{r_1 r_2}. \tag{17.47}$$

We now consider conditions for the formation of singularities. They depend on the form of the integral occurring in (17.46) and this in turn depends on the sign of the Gaussian curvature.

---

* Since **n** denotes the outward normal to the sphere the mean curvature has the negative value $-1/r$.

Case 1

If $K_g^0 > 0$ then

$$\int_0^n \frac{ds}{\varphi(s)} = \sqrt{r_1 r_2}\ \ln\left\{\frac{2n-(r_1+r_2)+2\sqrt{(r_1-n)(r_2-n)}}{2\sqrt{r_1 r_2}-(r_1+r_2)}\right\}. \tag{17.48}$$

Case 2

If $K_g^0 < 0$ and the surface $\mathscr{S}(t_0)$ is not minimal, then

$$\int_0^n \frac{ds}{\varphi(s)} = \sqrt{-r_1 r_2}\left\{\arcsin\frac{2n+r_1+r_2}{\sqrt{(r_1-r_2)^2}} - \arcsin\frac{r_1+r_2}{\sqrt{(r_1-r_2)^2}}\right\}. \tag{17.49}$$

Case 3

If $K_g^0 = 0$ then $\mathscr{S}(t_0)$ is a developable surface and

$$\int_0^n \frac{ds}{\varphi(s)} = \frac{1}{2K_m}\left\{1-\sqrt{1-2K_m^0 n}\right\}. \tag{17.50}$$

In each case the critical value n^* satisfies the condition $\varphi(n^*) = 0$. Some elementary but lengthy analysis leads to the proposition that n^* is a singular point of the solution of the amplitude equation.

17.7 Bibliographical Notes

The example of an acoustic wave, called a sonic wave by Thomas (1957b) propagating into a barotropic medium (gas or liquid) has been considered in order to illustrate the methods of Thomas and ray theory.

The derivation of the amplitude equations (17.26) in the general case is contained in Elcrat (1977) along with the transport equations. In our derivation of (17.25) and (17.30) we have used a different definition of the jump discontinuity. Consequently the forms of some of our coefficients differ from those in Elcrat (1977); moreover, in Elcrat's derivation there are some terms missing and some minor errors.

The case of a sonic wave propagating into a uniform gas at rest has been considered in detail by Thomas (1957b), who analysed the formation of singularities due to the curvature of the initial wave front $\mathscr{S}(t_0)$. This analysis is represented by Cases 1–3 above.

18 SHOCK WAVES IN A BAROTROPIC MEDIUM

The constitutive equation (17.1) is used again here and the balance equations for points not on the singular surface are as used in Section 17. The moving surface $\{\mathscr{S}(t)\}_{t\in I}$ now represents a shock wave propagating in a barotropic medium; the surface is therefore the locus of points at which thermodynamic quantities and the first derivatives of the motion $\boldsymbol{\chi}_t$ suffer jump discontinuities. The form of the balance law on $\mathscr{S}(t)$ is of great importance in this case.

18.1 Conservation Laws on a Wave

Since surface effects have been excluded the required balance laws result from equations derived in Section 12, namely

$$\begin{gathered}[\![\varrho U]\!] = 0, \quad [\![\varrho U\mathbf{v}+\mathbf{Tn}]\!] = \mathbf{0}, \\ [\![\varrho U(e+\tfrac{1}{2}\mathbf{v}\cdot\mathbf{v})+\mathbf{v}\cdot\mathbf{Tn}]\!] = 0, \quad [\![\varrho U\eta]\!] \geqslant 0.\end{gathered} \tag{18.1}$$

The balance law $(18.1)_1$ is the Stokes–Christoffel condition (10.9). The local normal speed of the wave $U = u_n - \mathbf{v}\cdot\mathbf{n}$ is a discontinuous quantity on the shock wave, that is

$$[\![U]\!] \neq 0,$$

since it measures the normal shock speed relative to the medium on either side of the shock (recall (13.24) and Definition 13.2).

Lemma 18.1

In a barotropic medium the mass and momentum conservation laws $(18.1)_{1,2}$ imply that the jump of the particle velocity is parallel to the wave front normal $\mathbf{n}$.

Proof

In a barotropic medium $(18.1)_2$ takes the form

$$[\![p]\!]\mathbf{n} = [\![\varrho U\mathbf{v}]\!].$$

The right-hand side may be split up into products,

$$[\![\varrho U\mathbf{v}]\!] = [\![\varrho U]\!][\![\mathbf{v}]\!]+(\varrho U)^+[\![\mathbf{v}]\!]+[\![\varrho U]\!](\mathbf{v})^+,$$

and using $(18.1)_1$ we obtain

$$(\varrho U)^- = (\varrho U)^+, \quad [\![p]\!]\mathbf{n} = (\varrho U)^+[\![\mathbf{v}]\!]. \ \square \tag{18.2}$$

Note that this may be formulated in terms of the characteristic vector $\mathbf{s}$ of the shock wave (see (13.22), Corollaries 13.1, 13.2 and Lemma 13.1).

18.2 Consequences of the Stokes–Christoffel Condition for a General Medium

Assuming only that the mass balance equation on the surface is $(18.1)_1$ we obtain the following result which is stronger than Proposition 13.2.

Corollary 18.1

In a material medium in which the mass balance equation has the form of the Stokes–Christoffel condition $(18.1)_1$ a shock wave cannot be a one-sided material surface.

Proof

If a wave is a one-sided material surface then $(U)^+(U)^- = 0$. Since $(\varrho)^+(\varrho)^- \neq 0$ equation $(18.1)_1$ implies that $(U)^+ = (U)^- = 0$. This contradicts the Definition 13.2 of a shock wave and means that the surface $\mathscr{S}(t_0)$ is an absolutely material surface (see (13.11) and Definition 13.1). □

From the definition of the local speed U we obtain

$$[\![U]\!] = -[\![\mathbf{v}]\!]\cdot\mathbf{n} \equiv -[\![v_n]\!] \tag{18.3}$$

and hence

$$\begin{aligned}[\![\varrho]\!] &= \frac{(\varrho)^+}{(U)^-}[\![\mathbf{v}]\!]\cdot\mathbf{n} = \frac{(\varrho)^-}{(U)^+}[\![\mathbf{v}]\!]\cdot\mathbf{n} = -\frac{(\varrho)^-}{(U)^+}[\![U]\!],\\ (\varrho U)^+[\![v_n]\!] &= (\varrho U)^-[\![v_n]\!] = (U^2)^-\frac{(\varrho)^-}{(\varrho)^+}[\![\varrho]\!] = (U^2)^+\frac{(\varrho)^+}{(\varrho)^-}[\![\varrho]\!].\end{aligned} \tag{18.4}$$

For the specific volume $\nu := 1/\varrho$ we obtain

$$[\![\nu]\!] = -\frac{[\![\varrho]\!]}{(\varrho)^-(\varrho)^+}\,. \tag{18.5}$$

On use of (10.13) in $(18.2)_1$ we find that for the Jacobian J

$$\frac{(J)^-}{(J)^+} = \frac{(U)^-}{(U)^+}, \tag{18.6}$$

where we have used the fact that the initial values ϱ_0 and J_0, as quantities defined in the reference placement, are continuous on the wave. Since $J > 0$ the local speed $U = u_n - v_n$ has the same sign on both sides of the wave. Defining the intensity δ of the shock wave by the relation

$$\delta \equiv \frac{[\![\varrho]\!]}{(\varrho)^-} = 1 - \frac{(\varrho)^+}{(\varrho)^-} = \frac{[\![v_n]\!]}{(U)^+}, \tag{18.7}$$

we see that the wave results in compression of the medium behind the wave if $\delta > 0$. The wave is then said to be **compressive**. The case $\delta < 0$ corresponds to expansion of the medium and the wave is then called a **rarefaction** wave. Since the mass density is positive a compressive shock wave is characterized by $[\![\varrho]\!] > 0$ and a rarefaction wave by $[\![\varrho]\!] < 0$.

18.3 The Normal Speed and the Hugoniot Relation

For a barotropic medium we now derive some further consequences of the balance laws on a shock wave. In terms of the intensity δ the velocity jump may be written

$$[\![\mathbf{v}]\!] = \delta(U)^+\mathbf{n}, \qquad (\mathbf{v})^- = (\mathbf{v})^+ + \delta(U)^+\mathbf{n}. \tag{18.8}$$

For the mass density, specific volume and normal speed we have the relations

$$(\varrho)^- = \frac{(\varrho)^+}{1-\delta}, \qquad (v)^- = (v)^+(1+\delta), \qquad (U)^- = (U)^+(1-\delta). \tag{18.9}$$

Using the pressure p and the intensity δ we write the momentum and energy balance laws in the form

$$\begin{aligned} (\varrho)^+(U)^{+2}\delta &= [\![p]\!], \\ (\varrho)^+[\![e]\!] &= \tfrac{1}{2}\{(p)^+ + (p)^-\}\delta. \end{aligned} \tag{18.10}$$

The first of these leads to the equation

$$(\varrho)^+(U)^{+2} = \frac{[\![p]\!]}{\delta} \tag{18.11}$$

for the wave velocity, while the second is the so-called **Hugoniot relation.** For known forms of the constitutive functions p and e with given values $(\varrho)^+$ and $(\eta)^+$ the Hugoniot relation restricts the set of all possible pairs $(\varrho)^-$, $(\eta)^-$ to those which are compatible with $(18.10)_2$.

We denote by $\bar{\eta}$ the magnitude of the entropy jump defined by

$$(\eta)^- = (\eta)^+(1+\bar{\eta}). \tag{18.12}$$

In view of the entropy production inequality $(18.1)_4$ and the positiveness of the local speed U we have $\bar{\eta} \geqslant 0$. From $(18.9)_1$ and the fact that the pressure p and the energy e were originally taken to be functions of ϱ and η it is clear that immediately behind the shock $(p)^-$ and $(e)^-$ may now be regarded as functions of δ and $\bar{\eta}$ with $(\varrho)^+$ and $(\eta)^+$ as known parameters. Equations (18.10) then provide relations among the variables δ, $\bar{\eta}$ and $(U)^+$.

The continuous differentiability of p and e may be used in (18.10) to eliminate two of these variables in terms of the third. For example, assuming that δ and $\bar{\eta}$ are differentiable functions of $(U)^+$ for fixed $(\varrho)^+$ and $(\eta)^+$, we obtain

$$\frac{\mathrm{d}\delta}{\mathrm{d}U^+} = \frac{p_\eta^- \dfrac{\mathrm{d}\eta^-}{\mathrm{d}U^+} - 2\delta\varrho^+U^+}{\varrho^+U^{+2} - p_\varrho^- \dfrac{\varrho^+}{(1-\delta)^2}}$$

after differentiating $(18.10)_1$, while differentiation of $(18.10)_2$ results in

$$(\varrho^+ e_\eta^- - \tfrac{1}{2} p_\eta^- \delta)\frac{d\eta^-}{dU^+} = \tfrac{1}{2}\delta\left\{\frac{\varrho^+ p_\varrho^-}{(1-\delta)^2} - \varrho^+ U^{+2}\right\}\frac{d\delta}{dU^+}.$$

Note that for the time being we omit the parenthesis in the notation for limit values from either side of the wave front. Also the partial derivatives are denoted by the appropriate subscript.

The above two equations yield

$$\frac{d\delta}{dU^+} = \frac{U^+\delta(\delta p_\eta^- - 2\varrho^+ e_\eta^-)}{e_\eta^-\left\{\varrho^+ U^{+2} - p_\varrho^- \dfrac{\varrho^+}{(1-\delta)^2}\right\}},$$
$$\frac{d\bar{\eta}}{dU^+} = \frac{U^+\delta^2}{\eta^+ e_\eta^-}. \tag{18.13}$$

It should be remembered that the solution of these differential equations will depend parametrically on ϱ^+ and η^+.

18.4 The Amplitude Equation

The aim of this section is to supplement the system (18.13) by a third independent equation among the three quantities δ, $\bar{\eta}$ and U^+ with time as a parameter. Then the evolution of these quantities during shock propagation can be completely determined, at least in principle. Clearly, the required equation must involve a coupling between the motion of the medium, its dynamics and the motion of the singular surface itself. It is obtained by using the conservation equations for each of the regions $\mathcal{N}^+$ and $\mathcal{N}^-$. Equation (17.3) leads to

$$\left[\!\!\left[\varrho\left\{\frac{\partial \mathbf{v}}{\partial t} + (\mathrm{grad}\,\mathbf{v})\mathbf{v}\right\}\right]\!\!\right] + [\![\mathrm{grad}\,p]\!] = \mathbf{0},$$

use having been made of the material time derivative

$$\dot{\mathbf{v}} \equiv \frac{\partial \mathbf{v}}{\partial t} + (\mathrm{grad}\,\mathbf{v})\mathbf{v}.$$

From (17.9) we obtain

$$[\![\dot{\varrho}]\!] + [\![\varrho\,\mathrm{div}\,\mathbf{v}]\!] = 0,$$
$$[\![\varrho\dot{\mathbf{v}}]\!] + [\![p_\varrho\,\mathrm{grad}\,\varrho]\!] + [\![p_\eta\,\mathrm{grad}\,\eta]\!] = \mathbf{0}, \tag{18.14}$$
$$[\![\dot{\eta}]\!] = 0.$$

Although $(18.14)_3$ implies that the material time derivative of the entropy is continuous on the shock wave, discontinuity of its spatial derivatives is not precluded.

If the surface $\mathscr{S}(t)$ has the representation (1.22) for any time t, that is

$$\mathscr{S}(t)\colon\ t = \tau(\mathbf{x}),$$

then the limit values $(f)^-$ and $(f)^+$ of any field on $\mathscr{N}$ may be regarded as functions of the variable $\mathbf{x}$. If $\mathscr{S} = \bigcup_{t\in I}\{t\}\times\mathscr{S}(t)$ is a hypersurface of discontinuity for the functions ϱ, $\mathbf{v}$ and η then we assume that the functions $\varrho\|_{\mathscr{N}^-\cup\mathscr{S}}$, $\varrho\|_{\mathscr{N}^+\cup\mathscr{S}}$, $\mathbf{v}\|_{\mathscr{N}^-\cup\mathscr{S}}$, $\mathbf{v}\|_{\mathscr{N}^+\cup\mathscr{S}}$, $\eta\|_{\mathscr{N}^-\cup\mathscr{S}}$ and $\eta\|_{\mathscr{N}^+\cup\mathscr{S}}$ are continuously differentiable in the sense of Definition 4.1. Specifically, for the function $\varrho^- \equiv (\varrho)^-$ defined on $\mathscr{S}$, we may write

$$\operatorname{grad}\varrho^- = \frac{\mathbf{n}}{u_n}\left(\frac{\partial\varrho}{\partial t}\right)^- + (\operatorname{grad}\varrho)^-$$

(recall $(2.2)_3$). Using the material time derivative we may rewrite this as

$$\operatorname{grad}\varrho^- = \left(\mathbf{1} - \frac{1}{u_n}\,\mathbf{n}\otimes\mathbf{v}^-\right)(\operatorname{grad}\varrho)^- + \frac{\mathbf{n}}{u_n}\,(\dot{\varrho})^-. \tag{18.15}$$

On replacing the superscript $-$ by $+$ in this we obtain a corresponding relation for ϱ^+. Using this together with (18.15) rearranged in the form

$$(\operatorname{grad}\varrho)^- = \left(\mathbf{1} + \frac{\mathbf{n}\otimes\mathbf{v}^-}{U^-}\right)\left(\operatorname{grad}\varrho^- - \frac{\mathbf{n}}{u_n}\,(\dot{\varrho})^-\right) \tag{18.16}$$

and equations (17.2), (18.8) and $(18.9)_{1,2}$ we obtain

$$\begin{aligned}(\operatorname{grad}\varrho)^- = &\left\{\mathbf{1} + \frac{\mathbf{n}\otimes(\mathbf{v}^+ + \delta U^+\mathbf{n})}{U^+(1-\delta)}\right\}\left\{\frac{\operatorname{grad}\varrho^+}{1-\delta} + \frac{\varrho^+\operatorname{grad}\delta}{(1-\delta)^2}\right.\\ &\left. + \frac{\mathbf{n}}{u_n}\,\frac{\varrho^+}{(1-\delta)}\,(\operatorname{div}\mathbf{v})^-\right\}.\end{aligned} \tag{18.17}$$

The last term in (18.17) may be expressed in the form

$$\begin{aligned}(\operatorname{div}\mathbf{v})^- = &\operatorname{div}(\mathbf{v}^+ + \delta U^+\mathbf{n}) - \frac{\mathbf{n}\cdot\dot{\mathbf{v}}^-}{U^+(1-\delta)}\\ &+ \frac{\mathbf{n}}{U^+(1-\delta)}\cdot\{(\mathbf{v}^+ + \delta U^+\mathbf{n})\cdot\operatorname{grad}(\mathbf{v}^+ + \delta U^+\mathbf{n})\},\end{aligned} \tag{18.18}$$

where we have used the definition of U together with $(18.8)_2$, $(18.9)_3$ and the material derivative. Care must be taken to distinguish the spatial derivative of a quantity defined on the surface and that defined in the region ahead of (or behind) the shock but evaluated on the shock.

Using (18.15) with the $+$ label instead of $-$ and a similar equation for $\mathbf{v}^+$ we obtain

$$(\operatorname{grad}\varrho)^- = \left\{\frac{\mathbf{1}}{(1-\delta)} + \frac{\delta}{(1-\delta)^2}\mathbf{n}\otimes\mathbf{n}\right\}(\operatorname{grad}\varrho)^+ - \frac{\varrho^+}{(U^+)^2(1-\delta)^3}$$

$$\times\ \mathbf{n}\otimes\mathbf{n}[\![\dot{\mathbf{v}}]\!] + \frac{\varrho^+}{(1-\delta)}\left\{\mathbf{1} + \frac{1+\delta}{1-\delta}\mathbf{n}\otimes\mathbf{n} + \frac{2\mathbf{n}\otimes\mathbf{v}}{U^+(1-\delta)}\right\}\operatorname{grad}\delta$$

$$+ \frac{\delta\varrho^+}{U^+(1-\delta)^2}\,\mathbf{n}\left\{\frac{U^+\mathbf{n}+\mathbf{v}^+}{U^+(1-\delta)}\right\}\cdot\operatorname{grad}U^+ + \frac{\delta\varrho^+}{(1-\delta)^2}\,\mathbf{n}\operatorname{div}\mathbf{n}$$

$$- \frac{\delta\varrho^+\mathbf{n}\otimes\mathbf{n}}{U^+(1-\delta)^3}\{(\operatorname{grad}\mathbf{v})^+\mathbf{n}\} \tag{18.19}$$

in place of (18.17), where $(\dot{\varrho})^{\pm} = -(\varrho\operatorname{div}\mathbf{v})^{\pm}$ has been used. It is obvious that $\operatorname{div}\mathbf{n}$ can be replaced by $-2K_m$ since, in view of (2.4) and $(1.6)_2$, we have

$$\operatorname{grad}\mathbf{n} = \mathbf{n}_{,\alpha}\otimes\boldsymbol{\varphi}^{\alpha} + (\operatorname{grad}\mathbf{n})\mathbf{n}\otimes\mathbf{n} = \mathbf{n}_{,\alpha}\otimes\boldsymbol{\varphi}^{\alpha},$$

$$\operatorname{div}\mathbf{n} = \mathbf{n}_{,\alpha}\cdot\boldsymbol{\varphi}^{\alpha} = -b^{\alpha}_{\alpha}.$$

In a similar way to the derivation of (18.19) we can derive an expression for $(\operatorname{grad}\eta)^-$. The result is simpler, however, since $\dot{\eta} = 0$. The final form of this expression is

$$(\operatorname{grad}\eta)^- = (1+\overline{\eta})\left(\mathbf{1} + \frac{\delta}{1-\delta}\mathbf{n}\otimes\mathbf{n}\right)(\operatorname{grad}\eta)^+$$

$$+\eta^+\left\{\mathbf{1} + \frac{\delta}{1-\delta}\mathbf{n}\otimes\mathbf{v}^+\left(1+\frac{1}{\delta U^+}\right)\right\}\operatorname{grad}\overline{\eta}. \tag{18.20}$$

Substituting (18.19) and (18.20) into $(18.14)_2$ we obtain the required relationship, namely

$$\frac{\varrho^+}{1-\delta}[\![\dot{\mathbf{v}}]\!] - \{p^+_{\varrho}(\operatorname{grad}\varrho)^+ + p^+_{\eta}(\operatorname{grad}\eta)^+\}\frac{1}{1-\delta}$$

$$+p^-_{\varrho}(\operatorname{grad}\varrho)^- + p^-_{\eta}(\operatorname{grad}\eta)^- = \mathbf{0}. \tag{18.21}$$

This is too complicated to be useful. However, inspection of (18.19) and (18.21) shows that the gradients of the surface quantities δ, $\overline{\eta}$ and U^+ are accompanied by the factor $U^+\mathbf{n}+\mathbf{v}^+$. Since no choice has yet been made for the parametrization of the surface and the velocity in (1.13) we now set

$$\mathbf{c} \equiv \frac{\partial\boldsymbol{\varphi}}{\partial t} = U^+\mathbf{n}+\mathbf{v}^+ = U^-\mathbf{n}+\mathbf{v}^-, \tag{18.22}$$

the last equality following from the reduction

$$\mathbf{v}^- + U^-\mathbf{n} = \mathbf{v}^+ + \delta U^+\mathbf{n} + U^+(1-\delta)\mathbf{n} = \mathbf{v}^+ + U^+\mathbf{n}$$

which makes use of $(18.8)_2$ and $(18.9)_3$.

We note that the velocity of displacement of the singular surface (the shock front) **c** defined above coincides with the ray velocity of the acoustic wave considered previously. Hence, for any quantity f given on $\mathscr{N}$ we may define its rate of change along a curve $(l^1, l^2) = \text{const}$ by the formula

$$\frac{\mathrm{d}_c f}{\mathrm{d}t} = \frac{\partial \tilde{f}}{\partial t} = \mathbf{c}\cdot\operatorname{grad}\bar{f} = \frac{\partial f}{\partial t} + \mathbf{c}\cdot\operatorname{grad} f, \tag{18.23}$$

obtained by use of (2.7), (2.8) and (15.21), where the notation from Section 2 has been used to distinguish f as a function of t and $\mathbf{x}$ from its restriction to $\mathscr{S}$, namely $\tilde{f}$, given in terms of t and l^α, and $\bar{f}$ given in terms of $\mathbf{x}$ (because of (1.22)). Note that $\mathrm{d}_c/\mathrm{d}t$ is the counterpart of $\mathrm{d}/\mathrm{d}l^0$ given in Section 15.1. In terms of (18.23) equation (18.19) may be used to obtain

$$\begin{aligned}
&\frac{2(\varrho^-)^2}{\varrho^+}p_\varrho^-\frac{\mathrm{d}_c\delta}{\mathrm{d}t} + \eta^+ p_\eta^- \frac{\mathrm{d}_c\bar{\eta}}{\mathrm{d}t} + \frac{\delta(\varrho^-)^2}{\varrho^+ U^+}p_\varrho^-\frac{\mathrm{d}_c U^+}{\mathrm{d}t} \\
&+\delta(1-\delta)\frac{(\varrho^-)^2}{\varrho^+}p_\varrho^- U^+\operatorname{div}\mathbf{n} + \left(U^+\varrho^+ - \frac{(\varrho^-)^2}{\varrho^+ U^+}p_\varrho^-\right)[\![\dot{\mathbf{v}}]\!]\cdot\mathbf{n} \\
&-U^+\{(\varrho^+)^2 p_\varrho^+ - (\varrho^-)^2(1-\delta)p_\varrho^-\}\frac{\partial\varrho^+}{\partial n} + U^+\{p_\eta^-(1+\bar{\eta}) - p_\eta^+\}\frac{\partial\eta^+}{\partial n} \\
&+\frac{\delta}{\varrho^+}(\varrho^-)^2 p_\varrho^- \frac{\partial\mathbf{v}^+}{\partial n}\cdot\mathbf{n} = 0.
\end{aligned} \tag{18.24}$$

Two further equations for the derivatives $\mathrm{d}_c\delta/\mathrm{d}t$, $\mathrm{d}_c\bar{\eta}/\mathrm{d}t$ and $\mathrm{d}_c U^+/\mathrm{d}t$ may be found, assuming, as in (18.13), that δ and $\bar{\eta}$ are functions of U^+. Using the identities

$$\frac{\mathrm{d}_c\varrho}{\mathrm{d}t} = \dot{\varrho} + U\frac{\partial\varrho}{\partial n}, \qquad \frac{\mathrm{d}_c\eta}{\mathrm{d}t} = U\frac{\partial\eta}{\partial n}, \tag{18.25}$$

and remembering that δ and $\bar{\eta}$ are functions of time for $(l^1, l^2) = \text{const}$, we obtain

$$\begin{aligned}
\frac{\mathrm{d}_c\delta}{\mathrm{d}t} &= \frac{\delta\gamma^2}{1-\gamma^2}\left(2 - \frac{\delta p_\eta^-}{\varrho^+ e_\eta^-}\right)\frac{1}{U^+}\frac{\mathrm{d}_c U^+}{\mathrm{d}t} + \frac{\varrho^+ N_1}{(\varrho^-)^2 p_\varrho^-(1-\gamma^2)} \\
&\quad + \frac{p_\eta^-(N_2 - \frac{1}{2}\delta N_1)}{(\varrho^-)^2 e_\eta^- p_\varrho^-(1-\gamma^2)},
\end{aligned} \tag{18.26}$$

$$\frac{\mathrm{d}_c\bar{\eta}}{\mathrm{d}t} = \frac{\delta^2 U^+}{\eta^+ e_\eta^-}\frac{\mathrm{d}_c U^+}{\mathrm{d}t} + \frac{(\frac{1}{2}\delta N_1 - N_2)}{\eta^+ e_\eta^-}.$$

Here, N_1 and N_2 are given by

$$\begin{aligned}
N_1 = &-\{(\varrho^-)^2(\delta-1)p_\varrho^- + (\varrho^+)^2 p_\varrho^+ + \delta(\varrho^+ U^+)^2\}\left(\frac{U^+}{(\varrho^+)^2}\frac{\partial\varrho^+}{\partial n}\right. \\
&\left.-\frac{1}{\varrho^+}\operatorname{div}\mathbf{v}^+\right) + \{(1+\bar{\eta})p_\eta^- + p_\eta^+\}U^+\frac{\partial\eta^+}{\partial n},
\end{aligned} \tag{18.27$_1$}$$

$$N_2 = [\tfrac{1}{2}\delta\{(1-\delta)(\varrho^-)^2 p_\varrho^- + (\varrho^+)^2 p_\varrho^+\}$$

$$+\delta(1-\tfrac{1}{2}\delta)(\varrho^+ U^+)^2]\left(\frac{U^+}{(\varrho^+)^2}\frac{\partial\varrho^+}{\partial n} - \frac{1}{\varrho^+}\operatorname{div}\mathbf{v}^+\right)$$

$$-[\tfrac{1}{2}\delta\{(1+\bar{\eta})p_\eta^- + p_\eta^+\} - \varrho^+(1+\bar{\eta})e_\eta^- - \varrho^+ e_\eta^+]\,U^+\frac{\partial\eta^+}{\partial n}, \quad (18.27)_2$$

and they vanish if the shock wave propagates in a homogeneous state. The quantity γ appearing in (18.26) and defined by

$$\gamma^2 = \left(\frac{U^-}{c^-}\right)^2 = \frac{(\varrho^+ U^+)^2}{(\varrho^-)^2 p_\varrho^-} \quad (18.28)$$

is called the **Mach number.** Here c^- is the local sound speed (see $(17.18)_2$) relative to the medium behind the shock wave. The final form of the equation for the time variation of the local shock speed U^+ is

$$\left(\frac{1+3\gamma^2}{\gamma^2(1-\gamma^2)}\,\delta - \frac{1+\gamma^2}{1-\gamma^2}\,\frac{p_\eta^-}{\varrho^+ e_\eta^-}\,\delta^2\right)\frac{\mathrm{d}_c U^+}{\mathrm{d}t}$$

$$+\delta(1-\delta)\left(\frac{U^+}{\gamma}\right)^2 \operatorname{div}\mathbf{n} = \frac{1-\gamma^2}{\gamma^2}-[\![\dot{\mathbf{v}}]\!]\cdot\mathbf{n}+N_0, \quad (18.29)$$

where

$$N_0 = \frac{\varrho^+ U^+}{-\gamma^2}\,\frac{1+\gamma^2}{1-\gamma^2}\left[\left\{1-\left(\frac{\varrho^+}{\varrho^-}\right)^2\frac{p_\varrho^+}{p_\varrho^-} - \frac{1+3\gamma^2}{1+\gamma^2}\,\delta\right\}\right.$$

$$\left.-\delta\gamma^2\frac{p_\eta^-}{\varrho^+ e_\eta^-}\left\{1-\delta-\frac{1}{(U^+)^2}\,p_\varrho^+\right\}\right]\left(\frac{U^+}{(\varrho^+)^2}\frac{\partial\varrho^+}{\partial n} - \frac{1}{\varrho^+}\operatorname{div}\mathbf{v}^+\right)$$

$$+\frac{p_\eta^+}{\varrho^+ U^+}\,\frac{1+\gamma^2}{1-\gamma^2}\left(\frac{p_\eta^- e_\eta^+}{p_\eta^+ e_\eta^-} - 1 + \delta\,\frac{p_\eta^-}{\varrho^+ e_\eta^-}\right)U^+\frac{\partial\eta^+}{\partial n}$$

$$-\frac{U^+}{\gamma^2}\left\{\delta\mathbf{n}\otimes\mathbf{n} + \left(1-\delta-\left(\frac{\varrho^+}{\varrho^-}\right)^2\frac{p_\varrho^+}{p_\eta^-}\right)\mathbf{1}\right\}\cdot\operatorname{grad}\mathbf{v}^+.$$

Equation (18.29) may be called the transport equation for U^+ since it governs the evolution of U^+ along a curve $(l^1, l^2) = \text{const}$. As can be seen from (18.22) these curves do not coincide with the orthogonal trajectories of the shock front.

Since ϱ^+, η^+ and $\mathbf{v}^+$ are known ahead of the shock the solution of (18.29) together with (18.13) determine the evolution of all the thermodynamic quantities appearing in the description of the problem under consideration.

Of course, equation (18.22), in the form

$$\frac{\partial \boldsymbol{\varphi}}{\partial t} = U^{+}\mathbf{n}+\mathbf{v}^{+},$$

together with (18.29) should be solved simultaneously.

However, equation (18.13) and a knowledge of the state ahead of the shock are not sufficient to solve equations (18.22) and (18.29) since one has to determine the scalar quantity $[\![\dot{\mathbf{v}}]\!] \cdot \mathbf{n}$ appearing on the right-hand side of (18.29). The quantity provides the only coupling of the shock wave propagation with the motion of the medium behind the shock front. There are some *ad hoc* methods which give estimates of this coupling term, and it can also be obtained under special conditions such as when the shock wave is (assumed) steady.

The acceleration $\dot{\mathbf{v}}$ may not be prescribed arbitrarily behind the shock since it is possible to obtain equations connecting the transverse derivatives $\delta_{,\alpha}$, $\overline{\eta}_{,\alpha}$ and $U^{+}_{,\alpha}$ with the transverse derivatives of ϱ^{+}, $\overline{\eta}^{+}$ and the components $\boldsymbol{\varphi}_{,\alpha} \cdot [\![\dot{\mathbf{v}}]\!]$. Thus, in either case a coupling term containing the jump $[\![\dot{\mathbf{v}}]\!]$ occurs. Moreover, the derivative $\dot{\mathbf{v}}$ appearing under the jump sign is not known in terms of quantities intrinsic to the shock wave, but depends on initial and boundary data. This derivative depends on the history of the motion of the whole body, and equation (18.29) should therefore be regarded as a functional-differential equation rather than a differential equation.

Note that in the special case in which the wave propagates into a region in a stationary state with $\mathbf{v}^{+} = \mathbf{0}$ equation (18.29) describes the variation of U^{+} along a normal trajectory; moreover $\mathrm{d}_c/\mathrm{d}t$ then turns out to be identical to the Thomas displacement derivative.

18.5 Bibliographical Notes

Literature dealing with three-dimensional shock wave propagation is rather scarce as far as original solutions are concerned. The example presented here comes from Wright (1976) along with his ideas dealing with the intrinsic description of unsteady shock waves in three dimensions. He also showed that the transport equation (18.29) together with a kinematical relation between the local shock speed and the normal to the shock front form a hyperbolic system of equations if it is assumed that the value of the coupling term is known exactly. The intrinsic approach presented here would appear to have considerable potential and should be particularly useful for various special circumstances (such as steady shocks) since it is exact, no assumptions beyond the balance equations and jump conditions are required, and all the terms have been retained in exact form.

In the paper by Chen and Wright (1975) a growth equation for the shock amplitude in an inviscid simple elastic fluid was derived in the context of the mechanical theory under the assumption that the fluid ahead of the shock was in a uniform stationary state. Extension of that problem to account for thermodynamic effects but with vanishing heat flux vector has been discussed by Bowen, Chen and McCarthy (1976). The results of Chen and Wright (1975) have been reproduced in Chen's book (1976); see also Ting (1976) and Turhan and Mengi (1977).

There are several papers dealing with interactions of waves. A comprehensive theory of such interactions is contained in Brun (1975). Some new results on the interaction of waves and shocks are given by Boillat and Ruggeri (1980) and Morro (1981).

19 LOVE WAVES IN A MEDIUM WITH SURFACE TENSION

The next example deals with the propagation of waves in a half-space $\{\mathbf{x}: x_3 \geqslant 0\}$ consisting of linearly elastic material. It is assumed that the medium is isotropic and homogeneous in some reference placement and that its boundary (the surface $x_3 = 0$) supports a surface (interfacial) stress $\mathbf{S}$ and has surface mass concentration $\varrho_{\mathscr{S}}$.

19.1 Constitutive Equations

Constitutive assumptions concerning the material of the medium are given in terms of a relation between the stress $\mathbf{T}$ acting in the body and the gradient of the displacement vector $\hat{\mathbf{u}}$. Here we take the material to be linearly elastic with constitutive relation

$$\mathbf{T} = \lambda(\operatorname{tr}\mathbf{E})\mathbf{1}+2\mu\mathbf{E}, \tag{19.1}$$

where

$$\mathbf{E} = \tfrac{1}{2}\{\operatorname{grad}\hat{\mathbf{u}}+(\operatorname{grad}\hat{\mathbf{u}})^{\mathrm{T}}\} \tag{19.2}$$

is the infinitesimal strain tensor. The (positive) scalars λ and μ are the Lamé moduli.

The material of the boundary $x_3 = 0$ is defined by a constitutive equation for the surface stress tensor $\mathbf{S}$, written in component form as

$$\begin{aligned} S_{\alpha\beta} &= \{\sigma+(\lambda_{\mathscr{S}}+\sigma)\hat{u}_{\gamma,\gamma}\}\delta_{\alpha\beta}+(\mu_{\mathscr{S}}-\sigma)(\hat{u}_{\alpha,\beta}+\hat{u}_{\beta,\alpha})+\sigma\hat{u}_{\alpha,\beta}, \\ S_{3\beta} &= \sigma\hat{u}_{3,\beta}, \qquad \alpha, \beta = 1, 2, \end{aligned} \tag{19.3}$$

where σ is the residual surface tension (the stress prevailing in the absence of strain), and $\lambda_{\mathscr{S}}$ and $\mu_{\mathscr{S}}$ are the (positive) Lamé moduli of the surface $\mathscr{S}$

$:= \{\mathbf{x}:\ x_3 = 0\}$. Orthogonal Cartesian coordinates are used and the surface parameters (l^1, l^2) are taken to be the Cartesian coordinates (x_1, x_2).

The outward unit normal vector $\mathbf{n}$ to the boundary is then $(0, 0, -1)$. In the absence of body forces $\mathbf{b}$ and $\mathbf{b}_{\mathscr{S}}$ the equations of motion for the bulk medium and its boundary take the forms

$$\begin{aligned} \operatorname{div}\mathbf{T} &= \varrho\ddot{\mathbf{u}} \quad \text{in} \quad x_3 > 0, \\ \operatorname{div}_{\mathscr{S}}\mathbf{S} - \mathbf{Tn} &= \varrho_{\mathscr{S}}\ddot{\mathbf{u}} \quad \text{on} \quad x_3 = 0, \end{aligned} \tag{19.4}$$

from $(10.31)_2$ and (10.36), where ϱ and $\varrho_{\mathscr{S}}$ are the volume and surface mass densities respectively.

In index notation these equations can be written as

$$\begin{aligned} T_{ij,j} &= \varrho\ddot{u}_i, \\ S_{i\alpha,\alpha} + T_{i3} &= \varrho_{\mathscr{S}}\ddot{u}_i, \quad i = 1, 2, 3, \ \alpha = 1, 2. \end{aligned} \tag{19.5}$$

19.2 SH-Waves

We denote the square of the shear wave speed in the bulk medium and in the surface by $c_T^2 = \mu/\varrho$ and $c_{\mathscr{S}}^2 = \mu_{\mathscr{S}}/\varrho_{\mathscr{S}}$ respectively. Consider a disturbance of the medium characterized by the displacement vector $\hat{\mathbf{u}} = (u_i)$ with

$$u_i = \delta_{i2} A \exp(-\alpha x_3)\sin\hat{\theta}, \tag{19.6}$$

where α is a constant and $\hat{\theta}$ denotes the phase (see Section 15) subject to

$$\begin{aligned} a > 0, &\quad \hat{\theta} = \mathbf{k}\cdot\mathbf{x} - \omega t, \\ \mathbf{k} = k\mathbf{N}, &\quad \mathbf{N} = (1, 0, 0). \end{aligned} \tag{19.7}$$

We seek conditions under which (19.6) provides a solution of (19.5). Clearly, the disturbances characterized by (19.6) decay exponentially with distance from the surface and involve displacements perpendicular to the x_1-axis (the direction of propagation) but parallel to the surface. Classically, such disturbances are called Love waves in an elastic medium. In a general medium they are called SH-waves. It is well known that in a half-space of homogeneous linear elastic material SH-waves cannot exist. Such waves are observed on the surface of the earth, however.

19.3 The Dispersion Relation for Love Waves

Here we prove that Love waves exist in a medium whose boundary supports surface stresses and carries surface mass density.

To this end we note that, in view of (19.1)–(19.3) and (19.6), the only non-zero stress components are

$$\begin{aligned} T_{12} &= T_{21} = \mu k A \exp(-\alpha x_3) \cos\hat{\theta}, \\ T_{23} &= T_{32} = -\mu\alpha A \exp(-\alpha x_3) \sin\hat{\theta}, \\ S_{12} &= (\mu_{\mathscr{S}} - \sigma) k A \cos\hat{\theta}, \qquad S_{21} = \mu_{\mathscr{S}} k A \cos\hat{\theta}, \\ S_{11} &= S_{22} = \sigma. \end{aligned} \tag{19.8}$$

Because of $(19.8)_{1,2}$ equations $(19.5)_1$ reduce to

$$T_{21,1} + T_{23,3} = \varrho \ddot{u}_2. \tag{19.9}$$

This equation is satisfied if

$$\alpha^2 = k^2 - \frac{\omega^2}{(c_T)^2}. \tag{19.10}$$

Because of $(19.8)_{2,3}$ equations $(19.5)_2$ simplify to

$$S_{21,1} + T_{23} = \varrho \ddot{u}_2 \tag{19.11}$$

and this is satisfied if

$$\alpha = \frac{\mu_{\mathscr{S}}}{\mu}\left(\frac{\omega^2}{c_{\mathscr{S}}^2} - k^2\right). \tag{19.12}$$

Equations (19.10), (19.12) and $(19.7)_1$ imply the inequalities

$$c_{\mathscr{S}} < U_N < c_T. \tag{19.13}$$

Corollary 19.1

In the medium considered SH-waves (Love waves) can propagate only if

$$c_{\mathscr{S}} < c_T. \; \square \tag{19.14}$$

The phase speed U_N satisfies (19.13). Hence, eliminating α from (19.10) and (19.12), we obtain the dispersion relation in the implicit form

$$F(\mathbf{k}, \omega) \equiv l^2 \left(\frac{\omega^2}{c_{\mathscr{S}}^2} - k^2\right)^2 - \left(k^2 - \frac{\omega^2}{c_T^2}\right) = 0, \qquad l = \frac{\mu_{\mathscr{S}}}{\mu}, \tag{19.15}$$

where the constant l has the dimension of length.

The fact that F is independent of t and $\mathbf{x}$ is obvious since disturbances propagate into a region in a homogeneous stationary state. Moreover, since the medium is isotropic the relation (19.15) is homogeneous in $\mathbf{k}$ (Definition 15.1).

From the general ray equations (15.7) applied to the present problem it follows that the frequency ω and the wave vector $\mathbf{k}$ are constant. The group (ray) velocity $\hat{\mathbf{c}}$ is given by

$$\hat{\mathbf{c}} = -\frac{\nabla_{\mathbf{k}} F}{\partial_\omega F} = \frac{c_T^2}{\omega}\mathbf{k}\left\{\frac{1+2l^2\left(\dfrac{\omega^2}{c_{\mathscr{S}}^2}-k^2\right)}{1+2l^2\dfrac{c_T^2}{c_{\mathscr{S}}^2}\left(\dfrac{\omega^2}{c_{\mathscr{S}}^2}-k^2\right)}\right\}. \tag{19.16}$$

The isotropy of the medium leads to the fact that $\hat{\mathbf{c}}$ is parallel to $\mathbf{k}$, as equation (19.16) shows.

The dispersion tensor $\nabla_{\mathbf{k}}\hat{\mathbf{c}}$ has the form

$$\nabla_{\mathbf{k}}\hat{\mathbf{c}} = \frac{c_T^2}{\omega}\left\{\frac{1+2l^2\left(\dfrac{\omega^2}{c_{\mathscr{S}}^2}-k^2\right)}{1+2l^2\dfrac{c_T^2}{c_{\mathscr{S}}^2}\left(\dfrac{\omega^2}{c_{\mathscr{S}}^2}-k^2\right)}\right\}\mathbf{1}$$
$$+\frac{c_T^2}{\omega}\left\{\frac{4l^2\left(\dfrac{c_T^2}{c_{\mathscr{S}}^2}-k^2\right)}{\left[1+2l^2\dfrac{c_T^2}{c_{\mathscr{S}}^2}\left(\dfrac{\omega^2}{c_{\mathscr{S}}^2}-k^2\right)\right]^2}\right\}\mathbf{k}\otimes\mathbf{k}. \tag{19.17}$$

The group velocity vanishes for $k^2 = k_1^2 := \omega^2/c_{\mathscr{S}}^2+1/2l^2$ and possesses a singularity at $k^2 = k_2^2 := \omega^2/c_{\mathscr{S}}^2+c_{\mathscr{S}}^2/2l^2c_T^2$. The dispersion tensor has singularities at k_1 and k_2.

After suitable rearrangement (19.15) may be written to express the explicit dependence of the wave number k on the phase velocity U_N, that is

$$k^2 = \left(\frac{\varrho}{\varrho_{\mathscr{S}}}\right)^2\frac{1-\left(\dfrac{U_N}{c_T}\right)^2}{\left[\left(\dfrac{U_N}{c_T}\right)^2-\left(\dfrac{c_{\mathscr{S}}}{c_T}\right)^2\right]^2}. \tag{19.18}$$

The wave number vanishes when $U_N = c_T$ and becomes unbounded when $U_N = c_{\mathscr{S}}$. Moreover, the derivative $\mathrm{d}k/\mathrm{d}U_N$ is negative, becoming

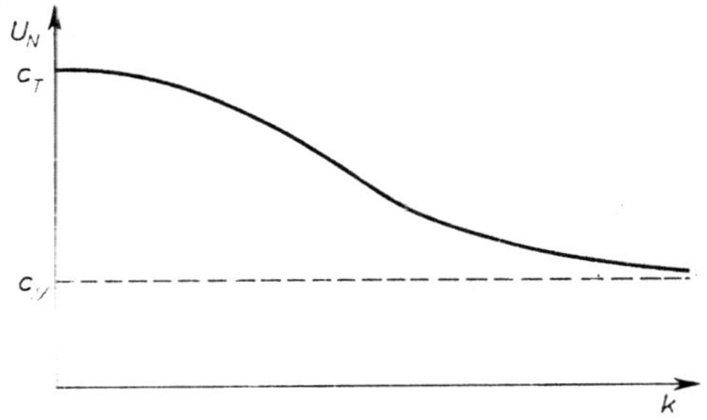

Fig. 19.1

unbounded when U_N approaches either $c_{\mathscr{S}}$ or c_T. The second derivative vanishes at just one vlaue of U_N. On the basis of this analysis one can sketch a graph of the relationship between U_N and k, as shown in Figure 19.1 (after Murdoch, 1976).

19.4 Bibliographical Notes

The example of Love waves in an unbounded linear elastic medium with surface tension is similar to that described by Murdoch (1976), who also considered Rayleigh waves. Murdoch (1977) then examined the effect of surface stress on the propagation of Stoneley waves. Surface waves centred on the interface between two homogeneous linearly elastic half-space which remain in contact without slipping have been investigated extensively. As in the present section the common boundary between the media is assumed to be an elastic material interface occupying the plane $\mathscr{S}$. It has been shown that disturbances having the character of Love and Rayleigh waves can propagate for appropriate relative values of the interfacial and bulk parameters. The waves are dispersive and for Love waves the residual surface stress represented by σ is of no consequence, as can be observed in the problem considered in Section 19.3.

Another approach was presented by Blinowski (1970) who studied Rayleigh waves in a half-space with surface tension. He also treated the modified shear wave in an elastic slab bounded by two material surfaces with non-vanishing stresses.

It should be remarked that equations (19.3) may be related to equations occurring in the work of Mindlin (1963, 1965), who, in considering waves in thin plates, devised an approximation in which the three-dimensional equations were replaced by equations for a two-dimensional continuum. This approximation was used by Tiersten (1969) for studying surface waves in a thin stratum adhering to a semi-infinite body, boundary conditions for the latter being assumed. Mention should be made of the relevance of problems of this kind to ultrasonic signal processing devices (see Thurston, 1974, Tiersten *et al.*, 1981).

20 COUPLED THERMOMECHANICAL WAVES

The aim of this section is to illustrate the application of the theory introduced in the preceding chapters to wave problems in thermoviscoplasticity. The material model that we have chosen provides some examples of the analysis of acceleration waves and of recent investigations concerning heat conduction in solids.

Most mechanical theories of the nineteenth and twentieth centuries that took account of heat transport were based on classical heat conduction theory, in which the heat flux is proportional to the temperature gradient. This theory, known as Fourier's law of heat conduction, was introduced by J. B. J. Fourier in 1882. The result of the theory is that the temperature distribution in the body is governed by a linear parabolic partial differential equation. Consequently, thermal disturbances have an infinite velocity of propagation. This fact is in disagreement with physical observations at very low temperatures and also with the relativistic theory. Maxwell, in 1867, was first to introduce a modified (hyperbolic) equation for heat conduction. Unfortunately, he then cancelled the term in his equation related to the velocity of propagation of thermal disturbances and recovered Fourier's law.

In a number of publications an evolutionary type equation for the heat flux can be found. It is cllaed the **Maxwell-Cattaneo equation.**

Here, we shall introduce a hyperbolic system of equations for a thermo-viscoplastic material described by internal state variables. For a three-dimensional medium necessary and sufficient conditions for the existence of real wave speeds will be derived.

20.1 Materials with Internal State Variables

First we define what is meant by an inelastic material. It is a material (medium) which exhibits rheological effects such as viscosity, relaxation and susceptibility to permanent (irreversible) deformation, as well as elastic behaviour. There are many mathematical models of inelastic materials. The following model is worthy of particular note because of its simplicity and universality. It is a model with internal state variables, and it has certain advantages:

— it is easily specialized to an elastic or hyperelastic model;

— it provides the possibility of a simple description of viscoelastic materials;

— under certain assumptions it is equivalent to the model of a simple material with (fading) memory;

— the possibility of describing elastic-viscoplastic behaviour;

— under certain assumptions it is equivalent to the model of a rate-type material;

— the possibility of describing fluids with chemical reactions and gases with vibratory relaxation.

The model with internal variables is obtained by enriching the description of the state of a material by additional quantities called **internal state variables.** Besides the constitutive relations, certain kinetic equations for the internal variables are postulated. The additional equations are evolutionary differential equations of order one.

In the theory of thermoviscoplasticity the reaction of a material is described by means of a set of functions $\{\psi, \mathsf{T}, \eta, \mathbf{q}\}$, where ψ is the free energy defined by the formula

$$\psi = e - \vartheta\eta, \tag{20.1}$$

T is the first Piola–Kirchhoff stress tensor, η is the entropy, $\mathbf{q}$ the heat flux and e the internal energy. Each of the functions depends on the displacement gradient $\mathbf{E} = \mathrm{Grad}\,\hat{\mathbf{u}}$, the temperature $\vartheta > 0$, the temperature gradient $\mathbf{g} = \mathrm{Grad}\,\vartheta$ and the internal parameters (variables) $\boldsymbol{\alpha}$, $\boldsymbol{\beta}$, γ and $\varkappa$:

$$\begin{aligned} \psi &= \Psi(\mathbf{E}, \vartheta, \mathbf{g}, \boldsymbol{\alpha}, \boldsymbol{\beta}, \gamma, \varkappa), \\ \mathsf{T} &= \mathsf{T}(\mathbf{E}, \vartheta, \mathbf{g}, \boldsymbol{\alpha}, \boldsymbol{\beta}, \gamma, \varkappa), \\ \eta &= \mathsf{N}(\mathbf{E}, \vartheta, \mathbf{g}, \boldsymbol{\alpha}, \boldsymbol{\beta}, \gamma, \varkappa), \\ \mathbf{q} &= \mathsf{Q}(\mathbf{E}, \vartheta, \mathbf{g}, \boldsymbol{\alpha}, \boldsymbol{\beta}, \gamma, \varkappa). \end{aligned} \tag{20.2}$$

The selection of internal parameters and their physical interpretation depends on the inelastic properties of the considered material. The parameter $\boldsymbol{\alpha}$ describes the viscoplastic deformation, $\varkappa$ is the hardening parameter and γ the viscosity parameter. The evolution equations for $\boldsymbol{\alpha}$, $\varkappa$ and γ are particular cases of the evolution equations introduced by Perzyna (1978), namely

$$\begin{aligned} \dot{\boldsymbol{\alpha}} &= \mathsf{A}(\mathbf{E}, \vartheta, \boldsymbol{\alpha}, \boldsymbol{\beta}, \gamma, \varkappa) := \gamma\Phi\left(\frac{\sqrt{\Pi_T}}{\varkappa} - 1\right)\frac{\mathsf{T}}{\sqrt{\Pi_T}} \quad \text{for } \Pi_T > \varkappa^2, \\ \dot{\gamma} &= \mathrm{tr}(\boldsymbol{\Gamma}\mathsf{A}), \qquad \dot{\varkappa} = \mathrm{tr}(\mathbf{K}\mathsf{A}), \end{aligned} \tag{20.3}$$

where Φ is a function of $\sqrt{\Pi_T}/\varkappa - 1$ and Π_T is the second invariant of the Cauchy stress tensor $\mathbf{T}$ (cf. (13.37)), while $\boldsymbol{\Gamma}$ and $\mathbf{K}$ are functions of the state variables.

The parameter $\boldsymbol{\beta}$ will describe the thermal properties of the material, and we assume that

$$\dot{\boldsymbol{\beta}} = \mathsf{B}_1(\mathbf{E}, \vartheta, \boldsymbol{\alpha}, \boldsymbol{\beta}, \gamma, \varkappa)\mathrm{Grad}\,\vartheta + \mathsf{B}_2(\mathbf{E}, \vartheta, \boldsymbol{\alpha}, \boldsymbol{\beta}, \gamma, \varkappa). \tag{20.4}$$

From the dissipation inequality

$$-\dot{\psi} - \eta\dot{\vartheta} + \frac{1}{\varrho_0}\mathrm{tr}(\mathsf{T}\dot{\mathbf{E}}) - \frac{1}{\varrho_0\vartheta}\mathbf{q}\cdot\mathrm{Grad}\,\vartheta \geqslant 0, \tag{20.5}$$

where ϱ_0 is the mass density, we deduce the existence of the constitutive functions

$$\begin{aligned} \psi &= \Psi(\mathbf{E}, \vartheta, \boldsymbol{\alpha}, \boldsymbol{\beta}, \gamma, \varkappa), \\ \mathsf{T} &= \varrho_0\,\partial_{\mathbf{E}}\Psi(\mathbf{E}, \vartheta, \boldsymbol{\alpha}, \boldsymbol{\beta}, \gamma, \varkappa), \\ \eta &= -\partial_\vartheta\Psi(\mathbf{E}, \vartheta, \boldsymbol{\alpha}, \boldsymbol{\beta}, \gamma, \varkappa), \\ \mathbf{q} &= -\varrho_0\vartheta B_1\,\partial_{\boldsymbol{\beta}}\Psi(\mathbf{E}, \vartheta, \boldsymbol{\alpha}, \boldsymbol{\beta}, \gamma, \varkappa). \end{aligned} \tag{20.6}$$

We note that the Maxwell–Cattaneo relation for the heat flux $\mathbf{q}$ (see Wołoszyńska, 1981) results from linearization of (20.4) and (20.6).

20.2 Coupled Acceleration Waves; Symmetry and Hyperbolicity

The basic system of equations for a thermoviscoplastic body consists of the equation of motion,

$$\varrho_0 \dot{\mathbf{v}} - \mathrm{Div}\mathsf{T} - \varrho_0 \mathbf{b} = \mathbf{0} \tag{20.7}$$

with mass density ϱ_0 and velocity of displacement $\mathbf{v} = \partial\hat{\mathbf{u}}/\partial t$, the compatibility equation

$$\dot{\mathbf{E}} - \mathrm{Grad}\,\mathbf{v} = \mathbf{0}, \tag{20.8}$$

the energy equation (the first law of thermodynamics)

$$\vartheta \partial_\vartheta \mathsf{N} \dot{\vartheta} + \vartheta \,\mathrm{tr}\,\{(\partial_{\mathbf{E}} \mathsf{N})\dot{\mathbf{E}}\} + (\vartheta \partial_{\mathbf{h}} \mathsf{N} + \partial_{\mathbf{h}} \Psi) \cdot \dot{\mathbf{h}} + \frac{1}{\varrho_0} \mathrm{Div}\,\mathbf{q} = \varrho_0 r, \tag{20.9}$$

where r is the heat source, and the evolution equations

$$\begin{aligned} &\dot{\boldsymbol{\beta}} - \mathsf{B}_1\,\mathrm{Grad}\,\vartheta - \mathsf{B}_2 = \mathbf{0},\\ &\dot{\boldsymbol{\alpha}} - \mathsf{A} = \mathbf{0},\\ &\dot{\gamma} - \mathrm{tr}(\boldsymbol{\Gamma}\mathsf{A}) = 0,\\ &\dot{\varkappa} - \mathrm{tr}(\mathsf{K}\mathsf{A}) = 0, \end{aligned} \tag{20.10}$$

where the internal parameters are denoted collectively as $\mathbf{h} := (\boldsymbol{\alpha}, \boldsymbol{\beta}, \gamma, \varkappa)$.

We shall consider how an acoustic (acceleration) wave propagates in $\mathscr{B}$. Let $\{\Sigma(t)\}_{t\in I}$ be a discontinuity surface for the vector $\dot{\mathbf{v}}$, the temperature rate $\dot{\vartheta}$ and the temperature gradient $\mathrm{Grad}\,\vartheta$. From the notation of (14.17) and the compatibility conditions (14.18) we have

$$\begin{aligned} &[\![E^l_{L,K}]\!] = \hat{a}^l N_L N_K,\\ &[\![v^l_{,L}]\!] = -U_N \hat{a}^l N_L,\\ &[\![\dot{v}^l]\!] = U_N^2 \hat{a}^l,\\ &[\![\vartheta_{,L}]\!] = \hat{e} N_L,\\ &[\![\beta_{K,L}]\!] = -\frac{1}{U_N} \hat{e} B^M_{1K} N_M N_L, \end{aligned} \tag{20.11}$$

where $\hat{e} := [\![\partial\vartheta/\partial N]\!]$ is called the thermal amplitude.

Assuming that $\mathbf{b}$, r, A, B_2, $\boldsymbol{\Gamma}$ and K are continuous we can rewrite equations (20.7)–(20.10) as

$$
\begin{aligned}
&[\![\dot v^l]\!] - \frac{1}{\varrho_0} A_n^{lKM}[\![E^n_{M,K}]\!] - \frac{1}{\varrho_0} P^{lK}[\![\vartheta_{,K}]\!] - \frac{1}{\varrho_0}[\![\alpha^n_{M,K}]\!] B^{lK}{}_n{}^M \\
&\qquad - \frac{1}{\varrho_0} C^{lKM}[\![\beta_{M,K}]\!] - \frac{1}{\varrho_0} D^{lL}[\![\gamma_{,L}]\!] - \frac{1}{\varrho_0} G^{lL}[\![\varkappa_{,L}]\!] = 0, \\
&[\![\dot E^l_L]\!] - [\![v^l{}_{,L}]\!] = 0, \\
&[\![\dot\vartheta]\!] - c_W^{-1} P^K_l [\![v^l{}_{,K}]\!] + c_W^{-1}(W^K + I^K)[\![\vartheta_{,K}]\!] + c_W^{-1} R^L{}_n{}^M [\![E^n_{M,L}]\!] \\
&\qquad + c_W^{-1} N^{LM}[\![\beta_{M,L}]\!] + c_W^{-1} M^L{}_n{}^M [\![\alpha^n_{M,L}]\!] \qquad (20.12) \\
&\qquad + c_W^{-1} L^K [\![\gamma_{,K}]\!] + c_W^{-1} t^K [\![\varkappa_{,K}]\!] = 0, \\
&[\![\dot\beta_K]\!] - B^L_{1K}[\![\vartheta_{,L}]\!] = 0, \\
&[\![\dot\alpha^n_M]\!] = 0, \\
&[\![\dot\gamma]\!] = 0, \quad [\![\dot\varkappa]\!] = 0.
\end{aligned}
$$

The coefficients in (20.12) are defined by

$$
\begin{aligned}
&c_W := \varrho_0 \vartheta \partial_\vartheta \mathsf{N}, \quad A^{lL}{}_n{}^N := \frac{\partial T^{lL}}{\partial E^n_N}, \quad P^{lK} := \frac{\partial T^{lK}}{\partial \vartheta}, \\
&B^{lL}{}_n{}^N := \frac{\partial T^{lL}}{\partial \alpha^n_N}, \quad C^{lKM} := \frac{\partial T^{lK}}{\partial \beta_M}, \quad D^{lL} := \frac{\partial T^{lL}}{\partial \gamma}, \\
&G^{lL} := \frac{\partial T^{lL}}{\partial \varkappa}, \quad W^K := \varrho_0 B^K_{1M}\left(\vartheta \frac{\partial \mathsf{N}}{\partial \beta_M} + \frac{\partial \Psi}{\partial \beta_M}\right), \qquad (20.13) \\
&R^L{}_n{}^M := \frac{\partial Q^L}{\partial E^n_M}, \quad I^L := \frac{\partial Q^L}{\partial \vartheta}, \quad M^L{}_n{}^M := \frac{\partial Q^L}{\partial \alpha^n_M}, \\
&N^{LM} := \frac{\partial Q^L}{\partial \beta_M}, \quad L^K := \frac{\partial Q^K}{\partial \gamma}, \quad t^K := \frac{\partial Q^K}{\partial \varkappa}.
\end{aligned}
$$

Here $\mathbf{A}$ is the elasticity tensor, c_W is the specific heat of the material and $\mathbf{P}$ is the temperature coefficient of stress.

Using (20.11) equation (20.12) can be reduced to

$$
\begin{aligned}
&(\mathbf{Q} - \varrho_0 U_N^2 \mathbf{1})\hat{\mathbf{a}} + \left(\mathbf{p} - \frac{1}{U_N}\mathbf{\Lambda}\right)\hat e = \mathbf{0}, \qquad (20.14) \\
&(U_N^2 \vartheta \mathbf{p} + U_N \boldsymbol{\omega}) \cdot \hat{\mathbf{a}} + (U_N \pi - Z - c_W U_N^2)\hat e = 0,
\end{aligned}
$$

where

$$
\begin{aligned}
&Q^l_n := A^{lK}{}_n{}^M N_K N_M, \quad p_l := P^K_l N_K, \\
&\Lambda^l := C^{lKM} B^L_{1M} N_K N_L, \quad \omega_l := R^L{}_n{}^M N_L N_M \delta^n_l, \qquad (20.15) \\
&\pi := (W^K + I^K) N_K, \quad Z := N^{LM} B^K_{1M} N_L N_K.
\end{aligned}
$$

Non-trivial values of **a** and $\hat{e}$ exist when

$$\det\begin{bmatrix} Q_1^1-\varrho_0 U_N^2 & Q_2^1 & Q_3^1 & p_1-\frac{1}{U_N}\Lambda_1 \\ Q_1^2 & Q_2^2-\varrho_0 U_N^2 & Q_3^2 & p_2-\frac{1}{U_N}\Lambda_2 \\ Q_1^3 & Q_2^3 & Q_3^3-\varrho_0 U_N^2 & p_3-\frac{1}{U_N}\Lambda_3 \\ U_N^2\vartheta p_1+U_N\omega_1 & U_N^2\vartheta p_2+U_N\omega_2 & U_N^2\vartheta p_3+U_N\omega_3 & U_N\pi-Z-c_W U_N^2 \end{bmatrix}=0. \tag{20.16}$$

This is an algebraic equation of the eight degree. In this case acceleration waves are not symmetric since, in general, $+U_r$ and $-U_r$ are not both roots of (20.16). Non-symmetric waves have also been found in the papers by Gurtin and Pipkin (1968), Chen (1969), Bogy and Naghdi (1970) and Müller (1973). On the other hand Suliciu (1975) assumed that the waves were symmetric and obtained the resulting restrictions on the constitutive functions. In our case symmetry is obtained if a certain coupling between mechanical and thermal properties is neglected so that $\mathbf{\Lambda}$, $\boldsymbol{\omega}$ and π vanish (Kosiński and Szmit, 1977). We conclude that the asymmetry is caused mainly by thermal effects.

To show this we consider a rigid conductor for which only the arguments ϑ and $\boldsymbol{\beta}$ remain in the constitutive equations (20.2) and in B_1 and B_2. Then there is one amplitude, $\hat{e}$, and equations (20.14) reduce to

$$c_W U_N^2 - U_N(\mathbf{W}+\mathbf{I})\cdot\mathbf{N}+Z = 0. \tag{20.17}$$

There are two acceleration waves; we assume that waves propagate equally in the direction $\mathbf{N}$ and $-\mathbf{N}$ so that

$$Z := \left(\frac{\partial \mathbf{Q}}{\partial \boldsymbol{\beta}}\mathsf{B}_1\mathbf{N}\right)\cdot\mathbf{N} < 0. \tag{20.18}$$

Following Gurtin and Pipkin (1968) we can write the roots of (20.17) in the form

$$\begin{aligned} \vec{U}_{1N} &= U_0\left\{\sqrt{1+\left[\frac{(\mathbf{W}+\mathbf{I})\cdot\mathbf{N}}{2U_0 c_W}\right]^2}+\frac{(\mathbf{W}+\mathbf{I})\cdot\mathbf{N}}{2U_0 c_W}\right\} > 0, \\ \vec{U}_{2N} &= U_0\left\{-\sqrt{1+\left[\frac{(\mathbf{W}+\mathbf{I})\cdot\mathbf{N}}{2U_0 c_W}\right]^2}+\frac{(\mathbf{W}+\mathbf{I})\cdot\mathbf{N}}{2U_0 c_W}\right\} < 0, \end{aligned} \tag{20.19}$$

where $U_0 := \sqrt{Z/c_W}$. Here the arrow indicates that these speeds are in the direction $\mathbf{N}$. The speeds (20.19) are invariant under a change from $\mathbf{N}$ to $-\mathbf{N}$ in the sense that

$$\begin{aligned} \overleftarrow{U}_{1N} &= -\overrightarrow{U}_{2N} > 0, \\ \overleftarrow{U}_{2N} &= -\overrightarrow{U}_{1N} < 0. \end{aligned} \tag{20.20}$$

Thus

$$|\overrightarrow{U}_{1N}|-|\overleftarrow{U}_{2N}| = \overrightarrow{U}_{1N}-\overleftarrow{U}_{2N} = \frac{1}{c_W}(\mathbf{W}+\mathbf{I})\cdot\mathbf{N} = \frac{\pi}{c_W}. \tag{20.21}$$

It follows that if $(\mathbf{W}+\mathbf{I})\cdot\mathbf{N} = 0$ then acceleration waves in the directions $\mathbf{N}$ and $-\mathbf{N}$ are symmetric to each other. For $(\mathbf{W}+\mathbf{I})\cdot\mathbf{N} \neq 0$ we can see that either $|\overrightarrow{U}_{1N}| < |\overleftarrow{U}_{2N}|$ or $|\overrightarrow{U}_{1N}| > |\overleftarrow{U}_{2N}|$. If, in particular, $\partial\mathbf{N}/\partial\boldsymbol{\beta} = \mathbf{0}$ and B_1 is a spherical tensor then the vector $\mathbf{W}+\mathbf{I}$ is proportional to the heat flux $\mathbf{q}$, that is

$$\mathbf{W}+\mathbf{I} = \left(\frac{\partial \mathsf{B}_1}{\partial\vartheta}\right)\mathbf{q}\Big/ B_1, \quad \text{when} \quad \mathsf{B}_1 = B_1\mathbf{1}. \tag{20.22}$$

Let $(\partial\mathsf{B}_1/\partial\vartheta)/B_1 > 0$; then we conclude that the speed of a wave propagating in the direction $\mathbf{q}$ is greater than that of a wave propagating in the direction $-\mathbf{q}$, and

$$|\overrightarrow{U}_{1N}|-|\overleftarrow{U}_{2N}| = \frac{1}{c_W}\left(\frac{\partial \mathsf{B}_1}{\partial\vartheta}\right)(\mathbf{q}\cdot\mathbf{N})\Big/ \mathsf{B}_1 \tag{20.23}$$

(see Gurtin and Pipkin, 1968; and Müller, 1973).

Next, we assume (1) that the material ahead of the acceleration wave is in a homogeneous asymptotic equilibrium state $(\vartheta^*, \boldsymbol{\beta}^*)$ with $\boldsymbol{\beta}^* = \mathbf{0}$, in which $\partial_{\boldsymbol{\beta}}\Psi(\vartheta^*, \mathbf{0}) = \mathbf{0}$ (see Wołoszyńska, 1981), and (2) that the material has a centre of symmetry. This means that $-\mathbf{1} \in \mathscr{G}$, where $\mathscr{G}$ is the symmetry group of the body $\mathscr{B}$. Thus

$$\begin{aligned} \mathsf{N}(\vartheta, \boldsymbol{\beta}) &= \mathsf{N}(\vartheta, -\boldsymbol{\beta}), \\ \Psi(\vartheta, \boldsymbol{\beta}) &= \Psi(\vartheta, -\boldsymbol{\beta}), \\ \mathsf{Q}(\vartheta, \boldsymbol{\beta}) &= \mathsf{Q}(\vartheta, -\boldsymbol{\beta}), \end{aligned} \tag{20.24}$$

and

$$\partial_{\boldsymbol{\beta}}\Psi(\vartheta, \mathbf{0}) = \partial_{\boldsymbol{\beta}}\mathsf{N}(\vartheta, \mathbf{0}) = \partial_{\boldsymbol{\beta}}\mathsf{Q}(\vartheta, \mathbf{0}) = \mathbf{0}. \tag{20.25}$$

Using these assumptions we can calculate the squared wave speed as

$$(U_N)^2 = (\mathsf{B}_1^{\#}\,\partial^2_{\boldsymbol{\beta\beta}}\Psi^{\#}\mathsf{B}_1^{\#}\mathsf{N})\cdot\mathsf{N}/\partial_{\vartheta}\mathsf{N}^{\#} =: (U_0^{\#})^2. \tag{20.26}$$

It follows that for any adiabatic process in a thermoviscoplastic material acceleration waves are symmetric. Indeed, in such a process $\mathbf{q} = \mathbf{0}$ and hence $\mathsf{B}_1\,\partial_{\boldsymbol{\beta}}\Psi = \mathbf{0}$; if $\det\mathsf{B}_1 \neq 0$ we deduce that

$$\partial_{\boldsymbol{\beta}}\Psi = \mathbf{0}. \tag{20.27}$$

Moreover, by equations (20.13) and (20.15)

$$\boldsymbol{\omega} = \mathbf{0}, \quad Z = 0, \quad \boldsymbol{\Lambda} = \mathbf{0}, \quad \pi = 0 \tag{20.28}$$

and equation (20.16) takes the form

$$\varrho_0^3 c_W z^3 - \varrho_0 z^2 \{\vartheta \mathbf{p}\cdot\mathbf{p} + \varrho_0 c_W \operatorname{tr}\mathbf{Q}\} + \varrho_0 z \{\vartheta \mathbf{p}\cdot\mathbf{p}\operatorname{tr}\mathbf{Q} - \vartheta \mathbf{p}\cdot(\mathbf{Q}\mathbf{p}) - c_W \mathrm{II}_Q\} - \vartheta\{(\mathbf{Q}\mathbf{p})\cdot(\mathbf{Q}\mathbf{p}) - \mathbf{p}\cdot\mathbf{p}\,\mathrm{II}_Q - (\mathbf{p}\otimes\mathbf{p})\cdot\mathbf{Q}\operatorname{tr}\mathbf{Q}\} + c_W \det\mathbf{Q} = 0, \tag{20.29}$$

where

$$\mathrm{II}_Q = \tfrac{1}{2}\{\operatorname{tr}(\mathbf{Q}^2) - (\operatorname{tr}\mathbf{Q})^2\}, \qquad z := (U_N)^2.$$

Lemma 20.1

For positive definite **Q** and positive c_W, a necessary and sufficient condition for the existence of real roots of the polynomial (20.29) is

$$S := m^2 + n^3 < 0,$$

where

$$2m = -\frac{2(C_2)^3}{27(C_3)^3} + \frac{C_2 C_1}{2(C_3)^2} - \frac{C_0}{C_3}, \qquad 3n = \frac{3C_1 C_3 - (C_2)^2}{3(C_3)^2}$$

and C_0, C_1, C_2, C_3 are the coefficients in (20.29) given by

$$\begin{aligned} C_3 &:= \varrho_0^3 c_W,\\ C_2 &:= \varrho_0\{\vartheta\mathbf{p}\cdot\mathbf{p} + \varrho_0 c_W \operatorname{tr}\mathbf{Q}\},\\ C_1 &:= \varrho_0\{\vartheta\mathbf{p}\cdot\mathbf{p}\operatorname{tr}\mathbf{Q} - \vartheta\mathbf{p}\cdot(\mathbf{Q}\mathbf{p}) - c_W\mathrm{II}_Q\},\\ C_0 &:= \vartheta\{(\mathbf{Q}\mathbf{p})\cdot(\mathbf{Q}\mathbf{p}) - \mathbf{p}\cdot\mathbf{p}\,\mathrm{II}_Q - (\mathbf{p}\otimes\mathbf{p})\cdot\mathbf{Q}\operatorname{tr}\mathbf{Q}\}\\ &\quad + c_W \det\mathbf{Q}. \ \square \end{aligned} \tag{20.30}$$

The proof is based on the theory of algebraic equations.

From the assumptions of the lemma we have

$$C_0 > 0, \quad C_1 > 0, \quad C_2 > 0, \quad C_3 > 0. \tag{20.31}$$

The inequality $S < 0$ is a necessary and sufficient condition for the existence of real roots of equation (20.29). From (20.31) it follows that these roots are positive.

Similarly, for a thermoviscoplastic material in a state of equilibrium possessing a centre of symmetry, the faster wave is symmetric. Therefore, by evaluating the coefficients in equation (20.14) at an asymptotic equilibrium state $(\mathbf{E}^{\#}, \vartheta^{\#}, \boldsymbol{\alpha}^{\#}, \mathbf{0}, \gamma^{\#}, \varkappa^{\#})$ and using conditions similar to (20.24) and (20.25) we obtain

$$\boldsymbol{\omega} = \mathbf{0}, \quad \boldsymbol{\Lambda} = \mathbf{0}, \quad \pi = 0,$$

and the polynomial (20.16) reduces to

$$w(z) = d_4 z^4 - d_3 z^3 + d_2 z^2 - d_1 z + d_0, \tag{20.32}$$

where

$$\begin{aligned}
d_4 &:= \varrho_0^3 c_W^\#,\\
d_3 &:= \varrho_0(\vartheta^\# \mathbf{p}^\# \cdot \mathbf{p}^\# - Z^\# \varrho_0^2 + \varrho_0 c_W^\# \operatorname{tr} \mathbf{Q}^\# \},\\
d_2 &:= \varrho_0 \{\vartheta^\# \mathbf{p}^\# \cdot \mathbf{p}^\# \operatorname{tr} \mathbf{Q}^\# - \vartheta^\# \mathbf{p}^\# \cdot \mathbf{Q}^\# \mathbf{p}^\#\\
&\quad - Z^\# \varrho_0 \operatorname{tr} \mathbf{Q}^\# - c_W^\# \mathrm{II}_{\mathbf{Q}^\#} \},\\
d_1 &:= \vartheta^\# \{(\mathbf{Q}^\# \mathbf{p}^\#) \cdot (\mathbf{Q}^\# \mathbf{p}^\#) - \mathbf{p}^\# \cdot \mathbf{p}^\# \mathrm{II}_{\mathbf{Q}^\#} - (\mathbf{p}^\# \otimes \mathbf{p}^\#) \cdot \mathbf{Q}^\# \operatorname{tr} \mathbf{Q}^\#\\
&\quad + Z^\# \varrho_0 \mathrm{II}_{\mathbf{Q}^\#} + c_W^\# \det \mathbf{Q}^\#,\\
d_0 &:= Z^\# \det \mathbf{Q}^\#, \qquad Z^\# := \mathbf{N} \cdot (\partial_\beta \mathbf{Q}^\# \mathsf{B}_1^\# \mathbf{N}).
\end{aligned}$$

Lemma 20.2

Let $\mathbf{Q}$ be positive definite, $c_W^\# > 0$ and $Z^\# > 0$. Then, necessary and sufficient conditions for the existence of real roots of the polynomial (20.32) are

(a) $\bar{S} := \bar{m}^2 + \bar{n}^3 < 0$,

(b) $w(z_1) < 0, \quad w(z_2) > 0, \quad w(z_3) < 0$,

where z_1, z_2, z_3 are the roots of the polynomial

$$w'(z) = 4d_4 z^3 - 3d_3 z^2 + 2d_2 z - d_1$$

such that $z_1 < z_2 < z_3$, and

$$2\bar{m} = -\frac{(d_3)^2}{32(d_4)^2} + \frac{d_2 d_3}{8(d_4)^2} - \frac{d_1}{4d_4},$$

$$3\bar{n} = \frac{8d_2 d_4 - 3(d_3)^2}{16(d_4)^2}.$$

Proof

From the assumptions we have $d_p > 0$ for $p = 0, 1, \ldots, 4$. The roots z_1, z_2, z_3 are positive by Lemma 20.1. □

20.3 Bibliographical Notes

The idea of introducing internal state variables* for describing the internal friction (viscosity) of the material is due to Biot (1954). The concept of internal state variables has been widely used in the thermodynamics of irreversible processes, for example by Kestin (1966); see also Schapery (1964). The internal state variable approach has already been used in the description of such phenomena as viscoelasticity, viscoplasticity, generalized Navier–Stokes–Fourier flow, rate-independent plasticity, dielectric and magnetic relaxation,

* Some authors use the term "hidden variables" or "hidden coordinates". In the Polish literature the term "internal parameter" is often used.

and vibration viscosity in gases. Mathematical and thermodynamic versions of the internal state variable approach have been presented by, for example, Coleman and Gurtin (1967), Bowen (1968), Valanis (1968), Truesdell (1969), Gurtin (1973), Kosiński and Perzyna (1973), Perzyna and Kosiński (1973), Day (1976), Perzyna (1978), Morro (1980), Kosiński (1983a, b) and Frischmuth and Perzyna (1983). The reader interested in the physical interpretations of internal variables and examples of their applications to solids and fluids may consult the articles by Kratochvíl, Kröner, Mandel, Mróz, Perzyna, Teodosiu and Valanis in the volumes edited by Sawczuk (1973, 1974), the monograph by Truesdell (1969) and the literature cited there.

It seems that the first article on acceleration waves propagating in a fluid described by internal state variables was written by Bürger (1966). This was followed by the article of Coleman and Gurtin (1967a). The literature dealing with wave analysis in materials with internal state variables is fairly extensive; we cite, for example, Bürger (1968), Becker and Schmitt (1968), Doria and Bowen (1970), Chen and Gurtin (1971), Bowen and Wang (1971b), Schmitt (1972), Kosiński and Perzyna (1972), Nunziato (1973), Kosiński (1974, 1975, 1976, 1980), Kennedy and Nuziato (1976), Ram and Pandey (1979), Singh and Sharma (1979), Bampi and Morro (1980) and Morro (1980a, b).

There are several papers in which it is assumed that thermal disturbances can propagate with a finite speed. Of course, such a result cannot be obtained under the assumption that Fourier's law of heat conduction is obeyed. Several authors have proposed alternative equations to replace Fourier's law and to obtain finite wave speeds for thermal disturbances. First, Maxwell (1867), then Cattaneo (1948, 1958), Vernotte (1958) and Kaliski (1965) used a modified form of Fourier's law in order to obtain a non-vanishing thermal relaxation time and hence a finite speed of propagation.

An interesting contribution to the subject for its simplicity, both physically and mathematically, was provided by Müller (1967). This work was also the basis for further developments; Carrassi and Morro (1972); in the relativistic context see Israel (1976) and Kranyš (1972). Different approaches to the problem are given by Gurtin and Pipkin (1968), Fox (1969) and Bogy and Naghdi (1970).

Approaches employing internal state variables can be found in Kosiński and Perzyna (1972), Kosiński (1975), Suliciu (1975), Mihăilescu and Suliciu (1976), Morro (1980a, b) and Bampi and Morro (1980). In all these papers the temperature ϑ was governed by a hyperbolic equation. Müller (1971) also obtained a hyperbolic equation for ϑ by introducing the so-called coldness function.

Thermomechanical waves with the Maxwell–Cattaneo law for a linear elastic material were investigated by Achenbach (1968), for a non-linear elastic material by Beevers (1973), and for hyperelastic and plastic materials by Tokuoka (1973a, b, 1974).

Problems in which the thermal wave speed is finite have been studied for the last forty years, and references to this work can be found in Kosiński (1975), Wołoszyńska (1979) and Ignaczak (1979) in respect of elastic and inelastic materials. One of the simplest ideas for describing this phenomenon was presented by Kosiński and Szmit (1977) and the extension of this idea to viscoplastic materials presented here follows the paper by Wołoszyńska (1981).

Chapter 8

One-dimensional waves

Numerous practical problems for deformable media may be simplified, by means of appropriate justifiable assumptions, so that the medium can be regarded as a one-dimensional geometrical object. Also, there are many dynamic problems for which it is justifiable to assume that the field functions depend on only one space variable and time. These and other physical situations occurring in continuum mechanics, when considered in the context of the propagation of disturbances, lead to the theory of one-dimensional waves.

In order to analyse one-dimensional waves we assume the existence of a curve in the plane of the variables x and t as a counterpart of a moving surface in the three-dimensional theory. Across this curve the kinematic and thermodynamic quantities or their derivatives suffer jump discontinuities.

In this chapter we present the basic equations of the theory of one-dimensional waves, particularly acceleration and shock waves. We return here to the concept of a material with internal state variables, which represents a commonly used model of non-elastic materials with energy dissipation, to finish the chapter with an analysis of wave propagation in an elastic-viscoplastic material with temperature effects. In this example we stress that the modified equation of heat conduction is used and the finite speed of propagation of thermal disturbances is taken into account.

21 BASIC EQUATIONS

All of the mechanical and thermodynamic quantities appearing from now on are functions of only two independent variables, time t and particle X. Both X and t vary over certain intervals of the real axis $\boldsymbol{R}$. We assume that $X \in B \subset \boldsymbol{R}$ and $t \in I \equiv (a, b) \subset \boldsymbol{R}$, where a and b may be finite or infinite real numbers.

The quantities describing the kinematics of the medium are now scalars. In what follows we do not include point concentrations of thermodynamic or mechanical quantities.

21.1 Conservation Equations

Let the deformable body be identified with the interval B in some reference placement. We denote the mass density in this placement by $\varrho_0(X), X \in B$; note that $\varrho_0(X) = \varrho_0 = \text{const}$ for a homogeneous body. The motion of the body is described by a function χ whose value $x = \chi(X, t)$ defines the location of the particle at time t. We introduce the notation

$$F(X, t) = \frac{\partial}{\partial X}\chi(X, t), \quad v(X, t) = \frac{\partial}{\partial t}\chi(X, t),$$
$$\dot{v}(X, t) = \frac{\partial^2}{\partial t^2}\chi(X, t), \quad E(X, t) = F(X, t)-1. \tag{21.1}$$

If the body force $b(X, t)$ is given then the pair $\{F(X, t), T(X, t)\}$ for $(X, t) \in B\times I$ will be called a **dynamical process** for the body in the motion χ if both the mass conservation equation

$$\varrho_0(X) = \varrho(X, t)F(X, t) \tag{21.2}$$

and the equation of motion

$$\frac{\partial T}{\partial X}+\varrho_0 b = \varrho_0\dot{v} \tag{21.3}$$

are satisfied.

It should be noted that the local form of the equation of motion (21.3) is a consequence of the integral form of the momentum balance equation which, for every pair $X_1, X_2 \in B$ and every $t \in I$, demands that the equality

$$\frac{\mathrm{d}}{\mathrm{d}t}\int_{X_1}^{X_2} \varrho_0 v\,\mathrm{d}X = \int_{X_1}^{X_2} \varrho_0 b\,\mathrm{d}X+T(X_2, t)-T(X_1, t) \tag{21.4}$$

should hold.

The dynamic quantity $T(\cdot, \cdot)$ plays the same role in the one-dimensional theory as the first Piola–Kirchhoff stress tensor in the three-dimensional theory.

If the heat source $r(X, t)$ is given in addition to the body force then the functions $F(X, t)$, $T(X, t)$, $e(X, t)$ and $q(X, t)$ for $(X, t) \in B\times I$ are collectively called a **thermodynamic process** in the motion χ if the first pair form a dynamical process, and the energy balance equation

$$\dot{e} = T\dot{F}-\frac{\partial q}{\partial X}+\varrho_0 r \tag{21.5}$$

is satisfied.

The local conservation equation (21.5) is a consequence of the equation of motion (21.3) and the global energy balance equation

$$\frac{\mathrm{d}}{\mathrm{d}t}\int\limits_{X_1}^{X_2}(e+\tfrac{1}{2}\varrho_0 v^2)\,\mathrm{d}X = \int\limits_{X_1}^{X_2}\varrho_0(vb+r)\,\mathrm{d}X + T(X_2,t)v(X_2,t)$$

$$-T(X_1,t)v(X_1,t)+q(X_1,t)-q(X_2,t). \tag{21.6}$$

The quantity $q(\cdot\,,\,\cdot)$ is called the heat flux and $e(\cdot\,,\,\cdot)$ the internal energy. Here e is referred to the unit of volume.

The entropy balance equation will be derived. We assume, however, that the entropy production inequality takes the form of the Clausius–Duhem inequality, namely

$$\frac{\mathrm{d}}{\mathrm{d}t}\int\limits_{X_1}^{X_2}\varrho_0\eta\,\mathrm{d}X \geqslant \int\limits_{X_1}^{X_2}\frac{r}{\vartheta}\varrho_0\,\mathrm{d}X + \frac{q(X_1,t)}{\vartheta(X_1,t)} - \frac{q(X_2,t)}{\vartheta(X_2,t)}, \tag{21.7}$$

where $\eta(\cdot\,,\,\cdot)$ is the entropy and $\vartheta(\cdot\,,\,\cdot)$ is the absolute temperature.

21.2 Curves of Discontinuity and Singularity

In the one-dimensional theory a wave is represented by a curve $\mathscr{C}$ in the set $B\times I$ which carries discontinuities in the derivatives of χ and possibly also of certain thermodynamic quantities.

Let a curve of singularity be described by the function

$$X = Y(t), \qquad t\in I,$$

in the reference placement. Then

$$V(t) := \frac{\mathrm{d}Y(t)}{\mathrm{d}t} \tag{21.8}$$

is the **speed of propagation** of the discontinuity point $X = Y(t)$. Note that the counterpart here of the surface $\Sigma(t)$ in the three-dimensional theory is the point $Y(t)$. It is assumed that $V(t) > 0$ is associated with propagation in the direction of increasing values of X.

The image of the curve $\mathscr{C}$ in the actual placement is obtained from

$$x = y(t) = \chi\big(Y(t),t\big),$$

and

$$u(t) := \frac{\mathrm{d}y(t)}{\mathrm{d}t} \tag{21.9}$$

is the speed of displacement of the wave. The latter satisfies the equation

$$u = FV \tag{21.10}$$

if F exists.

The counterpart of the three-dimensional kinematic compatibility condition (6.7) for the function $f(\cdot\,,\ \cdot)$, for which the curve $\mathscr{C}$ is a curve of discontinuity and the functions $f||_{\mathscr{N}^+\cup\mathscr{C}}$ and $f||_{\mathscr{N}^-\cup\mathscr{C}}$ are differentiable in the sense of Definition 4.1, is

$$[\![\dot{f}]\!] = \frac{\delta_1}{\delta t}[\![f]\!] - V\left[\!\!\left[\frac{\partial f}{\partial X}\right]\!\!\right]. \tag{21.11}$$

The symbol $\delta_1/\delta t$ is the one-dimensional counterpart of the displacement derivative (2.6). The sets $\mathscr{N}^+\cup\mathscr{C}$ and $\mathscr{N}^-\cup\mathscr{C}$ are defined by

$$\begin{aligned} \mathscr{N}^+\cup\mathscr{C} &\equiv \{(X, t) \in B\times I\colon\ X \geqslant Y(t)\}, \\ \mathscr{N}^-\cup\mathscr{C} &\equiv \{(X, t) \in B\times I\colon\ X \leqslant Y(t)\}. \end{aligned} \tag{21.12}$$

21.3 Shock Waves

A shock wave is a curve of discontinuity of the particle velocity. In discussing shock waves we shall make use of the deformation E. The kinematic compatibility condition (21.11) leads to

$$\begin{aligned} [\![v]\!] &= -V[\![E]\!], \\ [\![\dot{E}]\!] &= -V[\![\partial_X E]\!] + \frac{\delta_1}{\delta t}[\![E]\!], \\ [\![\dot{v}]\!] &= -V[\![\dot{E}]\!] + \frac{\delta_1}{\delta t}[\![v]\!], \end{aligned} \tag{21.13}$$

where ∂_X denotes $\partial/\partial X$.

The conservation of mass equation (21.2) for a shock wave assumes the form

$$[\![\varrho]\!] = -\frac{(\varrho)^-(\varrho)^+}{\varrho_0}[\![E]\!]. \tag{21.14}$$

Hence, a shock wave is regarded as compressive if

$$[\![E]\!] < 0 \tag{21.15}$$

and expansive if

$$[\![E]\!] > 0. \tag{21.16}$$

At a shock wave the global equation (21.4) leads to

$$[\![T]\!] + \varrho_0 V[\![v]\!] = 0 \quad \text{when } [\![b]\!] = 0 \tag{21.17}$$

(recall the formulae in Section 10.2).

From (21.13) and (21.17) we deduce the well-known expression for the velocity of a shock

$$\varrho_0 V^2 = \frac{[\![T]\!]}{[\![E]\!]}, \tag{21.18}$$

and the amplitude equation

$$2V\frac{\delta_1}{\delta t}[\![E]\!]+[\![E]\!]\frac{\delta_1}{\delta t}V = V^2[\![\partial_X E]\!]-[\![\dot{v}]\!].$$

From the equation of motion (21.3) we obtain

$$[\![\partial_X T]\!] = \varrho_0[\![\dot{v}]\!]. \tag{21.19}$$

Substitution of this into the previous equation yields

$$2\sqrt{V}\frac{\delta_1}{\delta t}\left(\sqrt{V}[\![E]\!]\right) = V^2[\![\partial_X E]\!]-\frac{1}{\varrho_0}[\![\partial_X T]\!], \tag{21.20}$$

which is satisfied for any material medium in which a shock wave propagates.

The discontinuity of the quantities e, η and q in a thermodynamic process with a shock wave means that the energy balance equation and the entropy production inequality on a shock wave are reduced to

$$\begin{aligned} &-V[\![e+\tfrac{1}{2}\varrho_0 v^2]\!] = [\![Tv]\!]-[\![q]\!], \\ &\varrho_0 V[\![\eta]\!]-\left[\!\!\left[\frac{q}{\vartheta}\right]\!\!\right] \geqslant 0. \end{aligned} \tag{21.21}$$

On use of equations $(21.13)_1$ and (21.18) the energy conservation equation $(20.21)_1$ may be expressed as the familiar Rankine–Hugoniot equation for shock waves, namely

$$[\![e]\!] = \tfrac{1}{2}\{(T)^- + (T)^+\}[\![E]\!]+\frac{1}{V}[\![q]\!]. \tag{21.22}$$

21.4 Acceleration Waves

For acceleration waves the curve $\mathscr{C}$ is a curve of discontinuity of the second derivatives of χ. Using (21.11) we obtain

$$[\![\dot{v}]\!] = -V[\![\dot{F}]\!] = V^2[\![\partial_X F]\!]. \tag{21.23}$$

Application of (21.11) to $\dot{v}$ and $\dot{F}$ leads to the amplitude equation

$$2\sqrt{V}\frac{\delta_1}{\delta t}\left(\frac{[\![\dot{v}]\!]}{\sqrt{V}}\right) = [\![\ddot{v}]\!]-V^2[\![\partial_X \dot{F}]\!] \tag{21.24}$$

if $V > 0$; in the general case it may be written in the equivalent form

$$2\frac{\delta_1}{\delta t}[\![\dot{v}]\!]-\frac{[\![\dot{v}]\!]}{V}\frac{\delta_1 V}{\delta t} = [\![\ddot{v}]\!]-V^2[\![\partial_X \dot{F}]\!].$$

In deriving these equations we have used only the kinematics of the medium. From the equation of motion (21.3) we obtain

$$[\![\partial_X T]\!] = \varrho_0[\![\dot{v}]\!]$$

and, on differentiation,

$$[\![\partial_X \dot{T}]\!] = \varrho_0[\![\ddot{v}]\!]-\varrho_0[\![\dot{b}]\!].$$

Equation (21.24) may now be rewritten in the form

$$2\sqrt{V}\frac{\delta_1}{\delta t}\left(\frac{[\![\dot{v}]\!]}{\sqrt{V}}\right) = \frac{1}{\varrho_0}[\![\partial_X \dot{T}]\!]-V^2[\![\partial_X \dot{F}]\!]+[\![\dot{b}]\!]. \quad (21.25)$$

Equation (21.25) may be replaced by an equation which does not involve derivatives with respect to X. Since v is continuous then from (21.17) we have $[\![T]\!] = 0$ and hence, from (21.11),

$$[\![\dot{T}]\!] = -V[\![\partial_X T]\!].$$

Therefore, we have

$$[\![\dot{T}]\!]+\varrho_0 V[\![\dot{v}]\!] = 0. \quad (21.26)$$

Differentiation of (21.3) with respect to time followed by use of (21.26) and (21.11) for $\dot{T}$ we obtain

$$\frac{\delta_1}{\delta t}[\![\dot{T}]\!]+V[\![\dot{v}]\!]\frac{\delta_1\varrho_0}{\delta t}+\varrho_0[\![\dot{v}]\!]\frac{\delta_1 V}{\delta t}+\varrho_0 V\frac{\delta_1}{\delta t}[\![\dot{v}]\!] = 0,$$

$$\frac{\delta_1}{\delta t}[\![\dot{T}]\!] = V[\![\partial_X \dot{T}]\!]+[\![\ddot{T}]\!].$$

Further substitutions lead to

$$\varrho_0 V[\![\ddot{v}]\!]-\varrho_0 V[\![\dot{b}]\!]+[\![\ddot{T}]\!]+V[\![\dot{v}]\!]\frac{\delta_1\varrho_0}{\delta t}+\varrho_0[\![\dot{v}]\!]\frac{\delta_1 V}{\delta t}$$
$$+\varrho_0 V\frac{\delta_1}{\delta t}[\![\dot{v}]\!] = 0. \quad (21.27)$$

Finally, using (21.11) with $f = \dot{v}$ and making appropriate reductions we obtain the amplitude equation in the form

$$2\varrho_0 V \frac{\delta_1}{\delta t}[\![\dot{v}]\!] = \varrho_0 V[\![b]\!] - [\![\dot{T}]\!] + \varrho_0 V^2[\![\ddot{F}]\!]$$

$$- \left(V \frac{\delta_1 \varrho_0}{\delta t} + \varrho_0 \frac{\delta_1 V}{\delta t} \right)[\![\dot{v}]\!]. \tag{21.28}$$

This equation is a simple consequence of the momentum conservation equation and of the kinematic compatibility condition and should be satisfied for every material medium. For many problems it is assumed that $[\![b]\!] = [\![\dot{b}]\!] = 0$. If the medium is homogeneous in the reference placement then ϱ_0 does not depend on X and $\delta_1 \varrho_0 / \delta t = V \partial \varrho_0 / \partial X = 0$.

The continuity of F at an acceleration wave implies that mass is continuous. The differentiated form of the mass conservation equation (21.2) leads to

$$[\![\dot{\varrho}]\!] = \frac{\varrho^2}{\varrho_0 V}[\![\dot{v}]\!]. \tag{21.29}$$

If V is positive then the requirement that the density increases after the passage of the wave, that is $(\dot{\varrho})^- > (\dot{\varrho})^+$, is satisfied if

$$[\![\dot{v}]\!] > 0. \tag{21.30}$$

A wave satisfying (21.30) is called a **compressive acceleration wave.** The inequality

$$[\![\dot{v}]\!] < 0 \tag{21.31}$$

defines an **expansive wave.**

For an acceleration wave (21.5) and (21.7) yield respectively

$$[\![\dot{e}]\!] = T[\![\dot{F}]\!] - [\![\partial_X q]\!], \quad \varrho_0[\![\dot{\eta}]\!] \geqslant [\![\partial_X (q/\vartheta)]\!]. \tag{21.32}$$

21.5 Bibliographical Notes

Much has been written on the theory of one-dimensional waves. A synthetic treatment may be found in Chen (1973, 1976), Nunziato *et al.* (1974) and Kosiński (1976).

The literature on particular problems in one-dimensional wave propagation covers both shock waves and acceleration waves for fluid and solids, but we do not attempt to give an exhaustive list of references here.

The existence of the one-dimensional kinematic compatibility condition (21.11) has been established by Chen and Wicke (1971); see also Chen (1973, 1976). The equation (21.20) for the amplitude of a shock wave was first derived by Achenbach and Herrmann (1968).

Slightly different and particular forms of the amplitude equations (21.24) and (21.28) for acceleration waves were given by Coleman *et al.* (1965), Coleman, Greenberg and Gurtin (1966) and Chen (1973, 1976). Equation (21.28) is the most general form of the amplitude equation since it takes account of not only the heterogeneity of the medium but also any discontinuity in the derivative of the body force.

22 ONE-DIMENSIONAL THERMAL AND MECHANICAL WAVES IN A THERMO-VISCOPLASTIC MEDIUM*

The outline of the theory of one-dimensional waves presented in Section 21 requires illustration. It is easier to analyse acceleration waves than shock waves and it is worth stressing that the amplitude equation for acceleration waves leads to an equation of Bernoulli type for most known models of simple materials. Depending on the state ahead of the wave the equation will have constant or variable coefficients and in certain cases it can be integrated as indicated at the end of Section 17.3.

Shock waves are more difficult to deal with apart from in the linearized theory since the amplitude equation is of functional-differential type and therefore not directly integrable. Here we discuss thermomechanical waves of order one in an inelastic dissipative medium.

22.1 The Governing Equations

In the one-dimensional theory with internal state variables a reaction of the material is described by the set of functions $\{\psi, T, \eta, q\}$ where ψ is the free energy defined by

$$\psi = e - \vartheta\eta,$$

T is the stress, η the entropy and q the heat flux; see Section 20.1 for a parallel discussion in the three-dimensional case. We assume that each of the functions depends on the deformation E, the temperature ϑ, the gradient $\partial_X\vartheta$, and the mechanical and thermal internal state variables α and β. Thus we write

$$\psi = \Psi(E, \vartheta, \partial_X\vartheta, \alpha, \beta), \qquad \eta = \mathsf{N}(E, \vartheta, \partial_X\vartheta, \alpha, \beta),$$
$$T = \mathsf{T}(E, \vartheta, \partial_X\vartheta, \alpha, \beta), \qquad q = \mathsf{Q}(E, \vartheta, \partial_X\vartheta, \alpha, \beta).$$

For α and β we assume the evolution equations

$$\begin{aligned} \dot{\alpha} &= \mathsf{A}(E, \vartheta, \alpha, \beta), \\ \dot{\beta} &= \mathsf{B}_1(E, \vartheta, \alpha, \beta)\,\partial_X\vartheta + \mathsf{B}_2(E, \vartheta, \alpha, \beta). \end{aligned} \qquad (22.1)$$

---

* This section is contributed by Dr. K. Wołoszyńska-Saxton.

The dissipation inequality

$$-\dot{\psi}-\eta\dot{\vartheta}+\frac{1}{\varrho_0}T\dot{E}-\frac{1}{\varrho_0\vartheta}q\partial_X\vartheta \geqslant 0$$

implies that

$$\psi = \Psi(E,\vartheta,\alpha,\beta), \qquad \eta = -\partial_\vartheta\Psi(E,\vartheta,\alpha,\beta),$$
$$T = \varrho_0\,\partial_E\Psi(E,\vartheta,\alpha,\beta), \qquad q = -\varrho_0\vartheta \mathsf{B}_1\partial_\beta\Psi(E,\vartheta,a,\beta).$$

Using this model we shall analyse shock waves for the case of small deformations and small changes in temperature. We assume that the energy has the quadratic form

$$\Psi(E,\vartheta,\alpha,\beta) = \frac{1}{\varrho_0}\left[\frac{a_1}{2}(E-\alpha)^2+a_2(E-\alpha)(\vartheta-\vartheta^\#)\right.$$
$$\left.+\frac{a_3}{2}(\vartheta-\vartheta^\#)^2+\frac{a_4}{2}\beta^2\right]$$

and then

$$T = a_1(E-\alpha)+a_2(\vartheta-\vartheta^\#),$$
$$\eta = -\frac{a_2}{\varrho_0}(E-\alpha)-\frac{a_3}{\varrho_0}(\vartheta-\vartheta^\#), \tag{22.2}$$
$$q = -ba\vartheta^\#\beta,$$

with

$$a_1 > 0, \quad a_2 < 0, \quad a_3 < 0,$$

where

$$a_4 = a\tau, \qquad b = \mathsf{B}_1\tau.$$

The heat flux q in $(22.2)_3$ is obtained from a linearized form of the evolution equation $(22.1)_2$, namely

$$\dot{\beta} = \frac{b}{\tau}\,\partial_X\vartheta-\frac{1}{\tau}\beta, \tag{22.3}$$

which leads to the Maxwell–Cattaneo relation $\tau\dot{q} = -k\partial_X\vartheta-q$, where $k = b^2a\vartheta^\#$. When $\tau = 0$ this reduces to Fourier's law $q = -k\partial_X\vartheta$.

The constant τ is called the relaxation time and the parameter α is treated as an inelastic deformation here. Note that the difference $E-\alpha$ satisfies Hooke's law $(22.2)_1$ with temperature effects.

On using (22.2) and (22.3) together with the equation of motion, the compatibility conditions, the energy balance equation and the evolution equations we obtain the following non-homogeneous system of partial differential equations with constant coefficients:

$$\begin{aligned}
&\dot{v}-\frac{1}{\varrho_0}a_1\,\partial_X E-\frac{1}{\varrho_0}a_2\,\partial_X\vartheta+\frac{1}{\varrho_0}a_1\,\partial_X\alpha = 0,\\
&\dot{E}-\partial_X v = 0,\\
&\dot{\vartheta}+g_1\,\partial_X v+g_5\,\partial_X\beta+\mathsf{G} = 0,\\
&\dot{\alpha}-\mathsf{A}(E,\vartheta,\alpha,\beta) = 0,\\
&\dot{\beta}-\frac{b}{\tau}\,\partial_X\vartheta+\frac{1}{\tau}\beta = 0.
\end{aligned} \tag{22.4}$$

The notations

$$g_1 \equiv \frac{a_2}{a_3},\qquad g_5 \equiv \frac{ba}{a_3},\qquad \mathsf{G} = \mathsf{G}(E,\vartheta,\alpha,\beta) \equiv -\frac{a_2}{a_3}\mathsf{A}(E,\vartheta,\alpha,\beta)$$

have been introduced in (22.4).

Equation $(22.4)_3$ is valid under the assumption that the coefficients of $\dot{E}$, $\dot{\vartheta}$, $\dot{\alpha}$ and $\dot{\beta}$ in the energy conservation equation are the same as in the equilibrium state. By a state of equilibrium we understand a set of values $(E^{\#}, \vartheta^{\#}, \alpha^{\#}, \beta^{\#})$ such that the right-hand sides of $(22.1)_1$ and (22.3) vanish, that is

$$\mathsf{A}(E^{\#},\vartheta^{\#},\alpha^{\#},\beta^{\#}) = 0,\qquad \beta^{\#} = 0.$$

One can prove that in an asymptotically stable state of equilibrium

$$\partial_\alpha\Psi(E^{\#},\vartheta^{\#},\alpha^{\#},\beta^{\#}) = \partial_\beta\Psi(E^{\#},\vartheta^{\#},\alpha^{\#},\beta^{\#}) = 0.$$

22.2 The Velocity of a Shock Wave

The system (22.4) may be written in the form

$$\partial_t\mathfrak{U}+\partial_X\mathfrak{F}(\mathfrak{U})+\mathfrak{B}(\mathfrak{U}) = \mathbf{0} \tag{22.5}$$

where $\mathfrak{U} = (v, E, \vartheta, \alpha, \beta)$, $\partial_X\mathfrak{F} = \mathfrak{A}\partial_X\mathfrak{U}$, and

$$\mathfrak{A} = \begin{bmatrix} 0 & -\frac{1}{\varrho_0}a_1 & -\frac{1}{\varrho_0}a_2 & \frac{1}{\varrho_0}a_1 & 0\\ -1 & 0 & 0 & 0 & 0\\ g_1 & 0 & 0 & 0 & g_5\\ 0 & 0 & 0 & 0 & 0\\ 0 & 0 & -\frac{b}{\tau} & 0 & 0 \end{bmatrix}.$$

As a result we obtain the generalized Rankine–Hugoniot relation

$$V[\![\mathfrak{U}]\!] = [\![\mathfrak{F}]\!] \tag{22.6}$$

for the shock wave $X = Y(t)$ under the condition that the vector function $\mathfrak{B}(\mathfrak{U})$ is a continuous function of its arguments. The linearity of (22.5) implies that the velocity $V(t)$ is a constant, λ say. Equation (22.6) allows us to express the reciprocal relations between the jumps of the functions v, E, ϑ, α and β on the curve $\mathscr{C}$ in the form

$$\begin{aligned}
\varrho_0 \lambda [\![v]\!] &= -a_1 [\![E]\!] - a_2 [\![\vartheta]\!], \\
\lambda [\![E]\!] &= -[\![v]\!], \\
\lambda [\![\vartheta]\!] &= g_1 [\![v]\!] + g_5 [\![\beta]\!], \\
\lambda [\![\alpha]\!] &= 0, \\
\lambda [\![\beta]\!] &= -b_1 [\![\vartheta]\!], \qquad b_1 \equiv \frac{b}{\tau},
\end{aligned} \tag{22.7}$$

together with

$$[\![T]\!] = \varrho_0 \lambda^2 [\![E]\!].$$

The internal state variable α is clearly continuous on $\mathscr{C}$, from $(22.7)_4$.

If the jump in one of the functions is known we may compute the jumps in the remaining ones from the formulae

$$[\![\vartheta]\!] = -\frac{g_1 \lambda^2}{\lambda^2 + b_1 g_5} [\![E]\!], \qquad [\![\beta]\!] = \frac{b_1 g_1 \lambda}{\lambda^2 + b_1 g_5} [\![E]\!],$$

$$[\![v]\!] = -\lambda [\![E]\!], \qquad [\![\beta]\!] = -\frac{b_1}{\lambda} [\![\vartheta]\!].$$

The speed λ of the shock wave satisfies the equation

$$\lambda^4 + \lambda^2 \left(b_1 g_5 - \frac{1}{\varrho_0} a_1 + \frac{1}{\varrho_0} g_1 a_2 \right) - \frac{1}{\varrho_0} b_1 g_5 a_1 = 0, \tag{22.8}$$

obtained from (22.5) under the assumption that $[\![\mathfrak{U}]\!] \neq \mathbf{0}$. Clearly, there are two symmetric waves, with speeds $\pm \lambda_1$ and $\pm \lambda_2$ say, propagating in the material.

We consider two particular cases:

Case 1

If there is no thermomechanical coupling, that is

$$\partial_\vartheta T = a_2 = 0 \qquad (g_1 = 0),$$

then there are two symmetric thermal waves and two mechanical waves with speeds given by

$$\lambda_T^2 = -b_1 g_5, \qquad \lambda_m^2 = \frac{1}{\varrho_0} a_1.$$

Case 2

If the material is a non-conductor of heat so that $q = 0$ ($g_5 = 0$) we obtain only two symmetric waves, propagating with the so-called adiabatic speed given by

$$\lambda_a^2 = \frac{1}{\varrho_0} a_1 - \frac{1}{\varrho_0} g_1 a_2. \tag{22.9}$$

It can be proved that the wave speeds satisfy the inequalities

$$\lambda_1^2 > \lambda_T^2 > \lambda_2^2, \quad \lambda_1^2 > \lambda_m^2 > \lambda_2^2, \quad \lambda_1^2 > \lambda_a^2 > \lambda_m^2 > \lambda_2^2. \tag{22.10}$$

It is convenient to write the polynomial (22.8) in the dimensionless form

$$\left(\frac{\lambda}{\lambda_m}\right)^4 - \left(\frac{\lambda}{\lambda_m}\right)^2 (d^2+1+\delta)+d^2 = 0,$$

where

$$d^2 = \frac{\lambda_T^2}{\lambda_m^2}, \quad \delta = -\frac{g_1 a_2}{\varrho_0 \lambda_m^2}.$$

Its roots are then given by

$$\frac{\lambda_{1,2}^2}{\lambda_m^2} = \frac{d^2+1+\delta \pm \sqrt{(d^2+1+\delta)^2-4d^2}}{2}.$$

The constant δ is a positive quantity depending on the degree of coupling between the mechanical and thermal effects, and vanishes if there is no coupling. The condition $d = 0$ ($\lambda_T = 0$) arises for a non-heat-conducting material. For a material governed by Fourier's law of heat conduction d approaches infinity, as does λ_T.

22.3 The Amplitude Equation

The amplitude equation (21.20) is considerably simplified in the linear theory and assumes the form

$$\frac{\delta_1}{\delta t}[\![E]\!] = -c_1[\![\mathsf{A}]\!] - c_2[\![E]\!], \tag{22.11}$$

where

$$\begin{aligned} c_1 &\equiv \frac{1}{2}\left(\frac{\lambda}{\lambda_m}\right)^2 \left\{1-\left(\frac{\lambda}{\lambda_T}\right)^2\right\}\left\{1-\frac{\lambda^4}{\lambda_T^2\lambda_m^2}\right\}^{-1} > 0, \\ c_2 &\equiv \frac{1}{2\tau}\left\{1-\left(\frac{\lambda}{\lambda_m}\right)^2\right\}\left\{1-\frac{\lambda^4}{\lambda_T^2\lambda_m^2}\right\}^{-1} > 0. \end{aligned} \tag{22.12}$$

On substituting λ_1 or λ_2 for λ in (22.12) we obtain the coefficients in (22.11) corresponding to the fast and slow waves. We denote these by c_1^1, c_1^2 and c_2^1, c_2^2 respectively. Taking into account the inequalities (22.10) we find that

$$\begin{aligned} c_1^1 &= c_1^2, \quad c_2^1 = c_2^2 \quad \text{if } d^2 = 1+\delta, \\ c_1^1 &< c_1^2, \quad c_2^1 > c_2^2 \quad \text{if } d^2 > 1+\delta, \\ c_1^1 &> c_1^2, \quad c_2^1 < c_2^2 \quad \text{if } d^2 < 1+\delta. \end{aligned}$$

The constants c_1 and c_2 influence the damping of the amplitude $[\![E]\!]$ along a given wave. We distinguish three basic cases.

Case (a)

The material is elastic-viscoplastic without thermal effects; $c_1 = \frac{1}{2}$. Then

$$\frac{\delta_1}{\delta t}[\![E]\!] = -\tfrac{1}{2}[\![\mathsf{A}]\!],$$

and the viscosity of the material causes the damping of the amplitude.

Case (b)

Thermoelasticity; $\mathsf{A} = \mathbf{0}$. Equation (22.11) becomes linear and the amplitude $[\![E]\!]$ decreases in time:

$$[\![E]\!](t) = Ce^{-c_2 t}.$$

It can be seen that the damping is stronger for the fast wave ($c_2^1 > c_2^2$) if $d^2 > 1+\delta$. Then $d^2 > 1$ and so $\lambda_T^2 > \lambda_m^2$. If $d^2 < 1+\delta$ then stronger damping occurs with the slower wave; note that for $d^2 < 1$ ($\lambda_T^2 < \lambda_m^2$) the slower wave decays more rapidly regardless of the quantity δ.

Case (c)

The material is elastic. Then

$$[\![E]\!](t) = \text{const},$$

that is the amplitude is constant along a strong discontinuity wave and is equal to the value of the jump applied to the boundary at the initial instant.

For a visco-plastic material the rate of damping of the amplitude is influenced by both thermal and viscous properties of the material, as equation (22.11) shows.

22.4 Discussion of Initial Conditions

To solve equation (22.11) we need the initial conditions which are obtained from the boundary conditions for the system (22.4).

We assume that pressure $p(t)$ and temperature $\theta(t)$ are applied to the boundary $x = 0$, so that

$$T(0, t) = p(t), \quad \vartheta(0, t) = \theta(t), \quad t \geqslant 0, \tag{22.13}$$

and $p(0) \equiv p_0 \neq 0$, $\theta(0) \equiv \theta_0 \neq 0$.

At $t = 0$ two wave fronts of strong discontinuity are generated so that part of the applied quantities (22.13) will propagate with speed λ_1 and another part with speed λ_2. We write

$$[\![T]\!](0) = [\![T]\!]_1(0)+[\![T]\!]_2(0) \equiv T^p = p_0 - T^*,$$
$$[\![\vartheta]\!](0) = [\![\vartheta]\!]_1(0)+[\![\vartheta]\!]_2(0) \equiv \vartheta^p = \theta_0 - \vartheta^*,$$

where the indices "1" and "2" correspond to the first and second waves respectively. We assume that the material is in a state of equilibrium at $t = 0$.

The compatibility condition

$$[\![\vartheta]\!] = -\frac{g_1}{\varrho_0(\lambda^2+b_1 g_5)}[\![T]\!]$$

is satisfied for each wave. This leads to

$$[\![T]\!]_1(0) = \frac{\lambda_1^2+b_1 g_5}{\lambda_1^2-\lambda_2^2}\left\{\varrho_0\frac{\lambda_2^2+b_1 g_5}{g_1}\vartheta^p + T^p\right\},$$

$$[\![T]\!]_2(0) = -\frac{\lambda_2^2+b_1 g_5}{\lambda_1^2-\lambda_2^2}\left\{\varrho_0\frac{\lambda_1^2+b_1 g_5}{g_1}\vartheta^p + T^p\right\},$$

$$[\![\vartheta]\!]_1(0) = -\frac{g_1}{\varrho_0(\lambda_1^2-\lambda_2^2)}\left\{\varrho_0\frac{\lambda_2^2+b_1 g_5}{g_1}\vartheta^p + T^p\right\},$$

$$[\![\vartheta]\!]_2(0) = \frac{g_1}{\varrho_0(\lambda_1^2-\lambda_2^2)}\left\{\varrho_0\frac{\lambda_1^2+b_1 g_5}{g_1}\vartheta^p + T^p\right\}.$$

As a result of the thermomechanical coupling each wave carries disturbances in the temperature and stress fields which depend simultaneously on T^p and ϑ^p.

We consider two possibilities:

Case 1

If $[\![\vartheta]\!](0) = 0$ then the application of the initial condition $T^p \neq 0$ causes initial discontinuities in temperature and stress given by

$$[\![\vartheta]\!]_1(0) = \frac{\delta\lambda_m^2}{a_2(\lambda_1^2-\lambda_2^2)}T^p,$$
$$[\![\vartheta]\!]_2(0) = -[\![\vartheta]\!]_1(0), \tag{22.14}$$

$$
\begin{aligned}
[\![T]\!]_1(0) &= \frac{\lambda_1^2 - d^2\lambda_m^2}{\lambda_1^2 - \lambda_2^2}\, T^p, \\
[\![T]\!]_2(0) &= \frac{d^2\lambda_m^2 - \lambda_2^2}{\lambda_1^2 - \lambda_2^2}\, T^p.
\end{aligned}
\tag{22.15}
$$

From (22.10) we see that the signs of the coefficients appearing in (22.14) and (22.15) imply that $[\![T]\!]_l(0) = \operatorname{sign} T^p$, $l = 1, 2$. For $T^p < 0$ (compression) we have $[\![\vartheta]\!]_1(0) < 0$ (cooling), while $[\![\vartheta]\!]_2(0) > 0$. If $T^p > 0$ then $[\![\vartheta]\!]_1(0) > 0$ and $[\![\vartheta]\!]_2(0) < 0$.

Which wave carries the greater initial mechanical disturbance (in absolute value) depends on the values of d^2 and δ. Thus

$$
\begin{aligned}
|[\![T]\!]_1(0)| &> |[\![T]\!]_2(0)| \Leftrightarrow d^2 < 1+\delta, \\
|[\![T]\!]_1(0)| &< |[\![T]\!]_2(0)| \Leftrightarrow d^2 > 1+\delta, \\
[\![T]\!]_1(0) &= [\![T]\!]_2(0) \Leftrightarrow d^2 = 1+\delta.
\end{aligned}
\tag{22.16}
$$

This condition $(22.16)_3$ implies that it is possible for a mechanical disturbance to be equally distributed between both waves at $t = 0$. The magnitude of the initial strain jump, by contrast, has the property

$$
\begin{aligned}
|[\![E]\!]_1(0)| &> |[\![E]\!]_2(0)| \Leftrightarrow d^2 < 1-\delta, \\
|[\![E]\!]_1(0)| &< |[\![E]\!]_2(0)| \Leftrightarrow d^2 > 1-\delta, \\
[\![E]\!]_1(0) &= [\![E]\!]_2(0) \Leftrightarrow d^2 = 1-\delta.
\end{aligned}
$$

Case 2

If $[\![T]\!](0) = 0$ and $[\![\vartheta]\!](0) = \vartheta^p \neq 0$ on the boundary $x = 0$ then

$$
\begin{aligned}
[\![T]\!]_1(0) &= \varrho_0 \frac{(\lambda_1^2 - d^2\lambda_m^2)(\lambda_2^2 - d^2\lambda_m^2)}{g_1(\lambda_1^2 - \lambda_2^2)}\, \vartheta^p, \\
[\![T]\!]_2(0) &= -[\![T]\!]_1(0), \\
[\![\vartheta]\!]_1(0) &= \frac{d^2\lambda_m^2 - \lambda_2^2}{\lambda_1^2 - \lambda_2^2}\, \vartheta^p, \\
[\![\vartheta]\!]_2(0) &= \frac{\lambda_1^2 - d^2\lambda_m^2}{\lambda_1^2 - \lambda_2^2}\, \vartheta^p.
\end{aligned}
$$

Here, $\operatorname{sign}[\![\vartheta]\!]_l(0) = \operatorname{sign}\vartheta^p$, $l = 1, 2$, and $\operatorname{sign}[\![T]\!]_1(0) = -\operatorname{sign}\vartheta^p$. Clearly

$$
\begin{aligned}
|[\![\vartheta]\!]_1(0)| &> |[\![\vartheta]\!]_2(0)| \Leftrightarrow d^2 > 1+\delta, \\
|[\![\vartheta]\!]_1(0)| &< |[\![\vartheta]\!]_2(0)| \Leftrightarrow d^2 < 1+\delta, \\
[\![\vartheta]\!]_1(0) &= [\![\vartheta]\!]_2(0) \Leftrightarrow d^2 = 1+\delta.
\end{aligned}
\tag{22.17}
$$

The inequality

$$
|[\![E]\!]_1(0)| \leqslant |[\![E]\!]_2(0)|
\tag{22.18}
$$

holds independently of the relative magnitudes of d^2 and δ. The case $\lambda_1^2 = \lambda_2^2$ corresponds to equality in (22.18).

22.5 Bibliographical notes

There are many papers in which Fourier's law is not used in the description of the thermal properties of a material. The first papers to modify this law were by Maxwell (1867), Cattaneo (1948), Vernotte (1958) and Kaliski (1965). There are different ideas for deriving the wave equation for heat conduction; Gurtin and Pipkin (1968) make use of the idea of a material with memory, Kosiński (1975) and Suliciu (1975) the idea of a material with internal parameters, Bogy and Naghdi (1970) that of a rate-type material, while Müller (1971) defines the universal functions of coldness.

Thermal waves have been analysed in a number of papers for both elastic and non-elastic solids. The Maxwell–Cattaneo equation has been used by, for example, Achenbach (1968) and Beevers (1973) in the case of an elastic material and by Tokuoka (1973a, b, 1974) for a plastic material. A different approach to the analysis of thermal waves has been provided by Lord and Shulman (1967), Lord and Lopez (1970), Norwood and Warren (1969), McCarthy (1970, 1971), Chen (1969a, b, 1976), Baltov (1973), Kukudžanov (1977), Lindsay and Straughan (1979) and Ignaczak (1980).

Coupled thermo-mechanical waves in a visco-plastic solid have been considered by Szmit-Wołoszyńska (1978). Her results have been published as Wołoszyńska (1981a, b) together with the initial boundary-value problem presented here in Section 22.

References

Achenbach, J. D., (1968). The influence of heat conduction on propagating stress jumps, *J. Mech. Phys. Solids*, **16** (2), 273–282.

Achenbach, J. D. (1973). *Wave Propagation in Elastic Solids*, North-Holland Publ. Co., American Elsevier, Amsterdam–New York.

Achenbach, J. D. and Herrmann, G. (1970). Propagation of second-order thermomechanical disturbances in viscoelastic solids, in: *Proc. IUTAM Symp., East Kilbride, June 1968*, Boley, B. A. (ed.), Springer-Verlag, Vienna–New York.

Adamson, A. W. (1967). *Physical Chemistry of Surfaces*, Interscience Publishers, New York–London.

Anile, A. M. 1984. Propagation of weak shock waves, *Wave Motion*, 571–578.

Anile, A. M. (1985). A geometric characterization of the compatibility relations for regularly discontinuous tensor fields, submitted to Italian journal.

Aris, R. (1962). *Vectors, Tensors and the Basic Equations of Fluid Mechanics*, Prentice-Hall, Englewood Cliffs, New York.

Bachman, R. C. and Cohen, H. (1979). Wave propagation in elastic rods with multiple wave speeds, *Math. Proc. Cambridge Philos. Soc.*, **86**, 179–191.

Balaban, M. M., Green, A. E. and Naghdi, P. M. (1970). Acceleration waves in elastic-plastic materials, *Int. J. Engng. Sci.* **8** (4), 315–335.

Baltov, A. (1973). Dynamical problems in infinitesimal thermoplasticity, in: *Proc. Summer School, Jabłonna*, P. Perzyna (ed.), Zakład im. Ossolińskich, Wrocław–Warszawa–Kraków–Gdańsk, 3–31.

Bampi, F. and Morro, A. (1980). Viscous fluids with hidden variables and hyperbolic systems, *Wave Motion* **2**, 153–157.

Banfi, C. and Fabrizio, M. (1979). Sul concetto di sottocorpo nella meccanica dei continui, *Accademia Nazionale dei Lincei* **LXVI** (2), 136–142.

Banfi, C. and Fabrizio, M. (1981). Global theory for thermodynamic behaviour of a continuous medium, *Ann. Univ. Ferrara, Sez. VII-Sc. Mat.* **27**, 181–199.

Barański, W. (1974). *Interactions in a Continuous Medium*, Habilitation Thesis, IFTR-Reports, **36** (in Polish).

Becker, E. and Schmitt, H. (1968). Die Entstehung von ebenen, zylinder- und kugelsymmetrischen Verdichtungsstössen in relaxierenden Gasen, *Ing. Arch.*, **36** (5), 335–347.

Beevers, C. E. (1973). Stability of waves and shock structure in generalized thermoelasticity at low temperatures, *Acta Mech.* **17** (1–2), 55–68.

Beevers, C. E. (1974). Evolutionary dilatational shock waves in a generalized theory of thermoelasticity, *Acta Mech.*, **20**, 67–79.

Bejda, J. (1971a). Dynamics of a plane deformation, in: *Applications of Viscoplasticity*, P. Perzyna, J. Klepaczko, J. Bejda, W. K. Nowacki, T. Wierzbicki (eds.), Zakład Narodowy im. Ossolińskich, Wrocław–Warszawa–Kraków–Gdańsk, 117–171 (in Polish).

Bejda, J. (1971b). On travelling multidimensional non-elastic waves, in: *Problèmes de la rheologie*, W. K. Nowacki (ed.), PWN, Warszawa, 29–45.

Bejda, J. (1972). Propagation of two-dimensional strong discontinuity waves in an elastic-viscoplastic medium, *Arch. Mech.*, **24** (3), 337–344.

Bejda, J. (1977). Propagation of non-linear dispersive and dissipative waves, *Arch. Mech.*, **29** (3), 477–490.

Biot, M. A. (1954). Theory of stress-strain relations in anisotropic visco-elasticity and relaxation phenomena, *J. Appl. Phys.*, **25** (11), 1385–1391.

Bland, D. R. (1964). Dilatation waves and shocks in large displacement isentropic elasticity, *J. Mech. Phys. Solids*, **12** (4), 245–267.

Bland, D. R. (1969). *Nonlinear Dynamic Elasticity*, Ginn (Blaisdell), Boston, Massachusetts.

Blinowski, A. (1970). A new approach to problems of surface phenomena in an elastic solid, *Proc. Vibr. Probl.*, **11** (4), 383–397.

Blinowski, A. (1971). On curvature dependent surface tension, *Arch. Mech.* **23** (2), 213–222.

Blinowski, A. and Trzęsowski, A. (1981). Surface energy in liquids and the Hadwiger integral theorem, *Arch. Mech.*, **33** (1), 133–146.

Bogy, D. B. and Naghdi, P. M. (1970). On heat conduction and wave propagation in rigid solids, *J. Math. Phys.*, **11** (3), 917–923.

Boillat, G. (1965). *La propagation des ondes*, Traité de Physique Théoretique et de Physique Mathématique XXIII, Gauthier-Villars, Paris.

Boillat, G. and Ruggeri, T. (1980). Symmetric form of nonlinear mechanics equations and entropy growth across a shock, *Acta Mech.*, **35**, 271–274.

Borejko, P. (1979). Reflection and refraction of an acceleration wave at a boundary between two non-linear elastic materials, *Arch. Mech.*, **31** (3), 373–384.

Born, M. and Wolf, E. (1959). *Principles of Optics*, Pergamon Press, London–New York–Paris–Los Angeles.

Bowen R. M. (1968). Thermochemistry of reacting materials, *J. Chem. Phys.*, **49**, 1625–1637.

Bowen, R. M. and Chen, P. J. (1972). Acceleration waves in anisotropic thermoelastic materials with internal state variables, *Acta Mech.*, **15** (1–2), 95–104.

Bowen, R. M., Chen, P. J. and McCarthy, M. F. (1976). Thermodynamic influences on the behaviour of curved shock waves in elastic fluids and the vorticity jumps, *J. Elasticity*, **6** (4), 369–382.

Bowen, R. M. and Rankin, R. L. (1973). Acceleration waves in ideal fluid mixtures with several temperatures, *Arch. Rat. Mech. Anal.*, **51** (4), 261–277.

Bowen, R. M. and Wang, C.-C. (1970). Acceleration waves in inhomogeneous isotropic elastic bodies, *Arch. Rat. Mech. Anal.*, **38** (1), 13–45. Corrigendum, **40** (5) 1971, 403.

Bowen, R. M. and Wang, C.-C. (1971a). On displacement derivatives, *Quart. Appl. Math.*, **29** (1), 29–39.

Bowen, R. M. and Wang, C.-C. (1971b). Thermodynamic influences on acceleration waves in inhomogeneous isotropic bodies with internal state variables, *Arch. Rat. Mech. Anal.*, **41** (4), 287–318.

Bowen, R. M. and Wang, C.-C. (1972). Acceleration waves in orthotropic elastic materials, *Arch. Rat. Mech. Anal.*, **47** (2), 149–170.

Braun, M. (1974). *Zur Ausbreitung von Unstetigkeitsflächen in thermo-elastischen Stoffen*, Doctoral Thesis, Universität Stuttgart.

Braun, M. (1977). Nonlinear progressive waves in elastic materials, *Rheol. Acta*, **16** (2), 146–154.

Braun, M. (1983). Wave propagation in elastic membranes, in: *Nonlinear Deformation Waves, Proc. IUTAM Symp., Tallinn 1982*, U. Nigul and T. Engelbrecht (eds.), Springer, 379–384.

Braun, M. (1985). Geometry of wave fronts and rays in homogeneous media, Euromech. 166, Aachen 1983, to appear in *Wave Motion.*

Bree, J. and Beevers, C. E. (1979). A thermodynamic theory of isotropic elastic-plastic materials, *Arch. Mech.*, **31**, 385–395.

Brillouin, L. (1960). *Wave Propagation and Group Velocity*, Academic Press, New York and London.

Bruns, H. (1895). Das Eikonal, *Abhandl. Math.-Phys. Kl. Sächs. Akad. Wiss.*, **21**, 323–436.

Buff, F. B. (1960). The theory of capillarity, in: *Handbuch der Physik*, Vol. X, Berlin–Göttingen–Heidelberg, 281–304.

Bürger, W. (1966). Zur Entstehung von Verdichtungsstössen beim „Kolbenversuch" in Gasen mit thermodynamischer Relaxation, *ZAMM*, **46** (2), 149–151, 187–189.

Bürger, W. (1968). Schwache Unstetigkeiten in räumlichen stationären Strömungen von Gasen mit thermodynamischer Relaxation, *ZAMM*, **48** (5), 185–187.

Carathéodory, C. (1909). Untersuchungen über die Grundlagen der Thermodynamik, *Math. Ann.*, **67**, 355–386.

Carrassi, M. and Morro, A. (1972). A modified Navier–Stokes equation and its consequences on sound dispersion, *Nuovo Cimento*, **9B**, 321–343.

Cattaneo, C. (1948). Sulla conduzione del calore, *Atti del. Seminario Matem. e Fisico, Univ. di Modena*, **3**, 83–101.

Cattaneo, C. (1958). Sur une forme d'equation de la chaleur éliminant le paradoxe d'une propagation instantanée, *Compt. Rend. Sci.*, **247**, 431–433.

Cauchy, A.-L. (1827). Sur les relations qui existent dans l'etat d'équilibre d'un corps solide ou fluide, entre les pressions ou tensions et les forces accélératrices, *Ex. de math.*, **2**, 108–111.

Chadwick, P. and Currie, P. K. (1972). The propagation and growth of acceleration waves in heat-conducting elastic materials, *Arch. Rat. Mech. Anal.*, **49** (2), 137–158.

Chadwick, P. and Ogden, R. W. (1971). On the definition of elastic moduli, *Arch. Rat. Mech. Anal.*, **44** (1), 41–53.

Chadwick, P. and Powdrill, B. (1965). Singular surfaces in linear thermoelasticity, *Int. J. Engng. Sci.*, **3**, 561–595.

Chen, P. J. (1968a). Thermodynamic influences on the propagation and growth of acceleration waves in elastic materials, *Arch. Rat. Mech. Anal.*, **31** (3), 228–254.

Chen, P. J. (1968b). The growth of acceleration waves of arbitrary form in homogeneously deformed elastic materials, *Arch. Rat. Mech. Anal.*, **30** (1), 81–89.

Chen, P. J. (1968c). Growth of acceleration waves in isotropic elastic materials, *J. Acoust. Soc. Amer.*, **43** (5), 982–987.

Chen, P. J. (1969a). On the growth and decay of one-dimensional temperature rate waves, *Arch. Rat. Mech. Anal.*, **35** (1), 1–15.

Chen, P. J. (1969b). On the growth and decay of one-dimensional temperature rate waves of arbitrary form, *J. Appl. Math. Phys.* (*ZAMP*), **20**, 448–453.

Chen, P. J. (1973). Growth and decay of waves in solids, in: *Handbuch der Physik*, Vol. VIa/3, Springer-Verlag, Berlin–Heidelberg–New York, 303–402.

Chen, P. J. (1976). *Selected Topics in Wave Propagation*, Noordhoff International Publishing, Leyden.

Chen, P.J. and Gurtin, M. E. (1971). Growth and decay of one-dimensional shock waves in fluids with internal state variables, *Phys. Fluids*, **14** (6), 1091–1094.

Chen, P. J. and McCarthy, M. F. (1976). The electrical responses of a dynamically loaded deformable dielectric material with memory, *Arch. Rat. Mech. Anal.*, **62** (4), 353–366.

Chen, P. J. and Wicke, H. H. (1971). Existence of the one-dimensional kinematical conditions of compatibility, *Istit. Lombardo di Sci., Rend. A*, **105**, 322–328.

Chen, P. J. and Wright, T. W. (1975). Three-dimensional shock waves and their behaviour in elastic fluids, *Meccanica, J. Italian Assoc. Theor. Appl. Mech.* **10** (4), 232–238.

Christoffel, E. B. (1877). Untersuchungen über die mit dem Fortbestehen linearer partieller Differentialgleichungen verträglichen Unstetigkeiten, *Ann. Mat.*, **8** (2), 18–113.

Chu, B. T. (1964). Finite amplitude waves in incompressible perfectly elastic materials, *J. Mech. Phys. Solids*, **12** (1), 45–57.

Chu, B. T. (1967). Transverse shock waves in incompressible elastic solids, *J. Mech. Phys. Solids*, **15** (1), 1–14.

Cohen, H. (1978a). Decay of acceleration waves in elastic rods, *Arch. Rat. Mech. Anal.*, **67** (2), 151–163.

Cohen, H. (1978b). Shock wave decay in elastic rods, *Iranian J. Sci. Tech.*, **7**, 83–91.

Cohen, H. and DeSilva, C. N. (1966). Nonlinear theory of elastic surfaces, *J. Math. Phys.* **7** (2), 246–253.

Cohen, H. and Epstein, M. (1979). Acceleration waves in constrained elastic rods, *Arch. Rat. Mech. Anal.*, **72** (2), 141–154.

Cohen, H. and Suh, S. L. (1970). Wave propagation in elastic surfaces, *J. Math. Mech.*, **19** (12), 1117–1129.

Cohen, H. and Tallin, A. G. (1982). Waves in thermo-viscoelastic rods, *Acta Mech.*, **42**, 85–97.

Cohen, H. and Thomas, R. S. D. (1983). Plane wave propagation and evolution for quasilinear hyperbolic systems, *Utilitas Mathematica*, **24** 199–223.

Cohen, H. and Wang, C.-C. (1982). On compatibility conditions for singular surface, *Arch. Rat. Mech. Anal.*, **80** (3), 205–261.

Cohen, H. and Wang, C.-C. (1983). Principal waves in elastic subfluids, *Arch. Rat. Mech. Anal.*, **83** (2), 139–168.

Coleman, B. D., Greenberg, J. M. and Gurtin, M. E. (1966). Waves in materials with memory, Part V. On the amplitude of acceleration waves and mild discontinuities, *Arch. Rat. Mech. Anal.*, **22**, 333–354.

Coleman, B. D. and Gurtin, M. E. (1965). Waves in materials with memory, Part IV, Thermodynamics and the velocity of general acceleration waves, *Arch. Rat. Mech. Anal.*, **19** (5), 317–330.

Coleman, B. D. and Gurtin, M. E. (1967a). Growth and decay of discontinuities in fluids with internal state variables, *Phys. Fluids*, **10** (7), 1454–1458.

Coleman, B. D. and Gurtin, M. E. (1967b). Thermodynamics with internal state variables, *J. Chem. Phys.*, **47**, 597–613.

Coleman, B. D., Gurtin, M. E., Herrera, I. R. and Truesdell, C. A. (1965). *Wave Propagation in Dissipative Materials*, Springer-Verlag, Berlin–Heidelberg–New York.

Coleman, B. D. and Noll, W. (1963). The thermodynamics of elastic materials with heat conduction and viscosity, *Arch. Rat. Mech. Anal.*, **13** (3), 167–178.

Coleman, B. D. and Owen, D. R. (1974). A mathematical foundation for thermodynamics, *Arch. Rat. Mech. Anal.*, **54** (1), 1–104.

Coleman, B. D. and Owen, D. R. (1977). On the thermodynamics of semi-systems with restriction on the accessibility of states, *Arch. Rat. Mech. Anal.*, **66** (3), 173–181.

Coleman, B. D., Owen, D. R. and Serrin, J. (1981). The second law of thermodynamics for systems with approximate cycles, *Arch. Rat. Mech. Anal.*, **77**, 103–142.

Courant, R. (1962). Partial differential equations, in: *Methods of Mathematical Physics*, Vol. II, R. Courant and D. Hilbert (eds.), Interscience Publishers, New York.

Courant, R. and Friedrichs, K. O. (1948). *Supersonic Flow and Shock Waves*, Interscience Publishers, New York.

Currie, P. K. (1972). Shock waves in homogeneously-strained incompressible elastic materials: The Money–Rivlin material, *Acta Mech.*, **14** (1–2), 53–58.

Currie, P. K. and Hayes, M. A. (1969). Longitudinal and transverse waves in finite elastic strain. Hadamard and Green materials, *J. Inst. Math. Appl.*, **5**, 140–161.

Currie, P. K. and O'Leary, P. M. (1978). Viscoelastic Rayleigh waves II, *Quart. Appl. Math.*, **35**, 445–454.

Dafermos, C. and Kosiński, W. (1984). Continuous dependence result for quasi-linear hyperbolic systems, Preprint No. 666, Sonderforschungsbereich 72, Universität Bonn.

Day, W. A. (1969). A theory of thermodynamics for materials with memory, *Arch. Rat. Mech., Anal.*, **34**, 85–96.

Day, W. A. (1972a). *The Thermodynamics of Simple Materials with Fading Memory*, Springer-Verlag, Berlin–New York.

Day, W. A. (1972b). A condition equivalent to the existence of entropy in classical thermodynamics, *Arch. Rat. Mech. Anal.*, **49** (2), 159–171.

Day, W. A. (1975). Continuum thermodynamics based on a notion of rate of loss of information, *Arch. Rat. Mech. Anal.*, **59** (1), 53–62.

Day, W. A. (1976), Entropy and hidden variables in continuum thermodynamics, *Arch. Rat. Mech. Anal.*, **62** (4), 367–389.

Day, W. A. and Šilhavy, M. (1977). Efficiency and the existence of entropy in classical thermodynamics, *Arch. Rat. Mech. Anal.*, **66** (1), 73–81.

De Grott, S. R. and Mazur, P. (1962). *Non-equilibrium Thermodynamics*, North-Holland, Amsterdam.

De Silva, S. N. and Whitman, A. B. (1971). Thermodynamical theory of directed curves, *J. Math. Phys.*, **12** (8), 1603–1609.

Doria, M. L. and Bowen, R. M. (1970). Growth and decay of curved acceleration waves in chemically reacting fluids, *Phys. Fluids*, **13** (4), 867–876.

Drugan, W. J. and Rice, J. R. (1983). Restrictions on quasi-statically moving surfaces of strong discontinuity in elastic-plastic solids, in: *Mechanics of Material Behaviour*, Elsevier, Amsterdam.

Dunwoody, J. (1972). One-dimensional shock waves in heat-conducting materials with memory, *Arch. Rat. Mech. Anal.*, **47** (2), 117–148; **48** (3), 192–204.

Duvall, G. E. (1961). Shock wave stability in solids, in: *Des Ondes de Détonation, Proc. Symp. Gif-Sur-Yvette, 28 Aug.–2 Sept.*, 337–353.

Eckart, C. (1940). The thermodynamics of irreversible processes, I. The simple fluid, *Phys. Rev.*, **58** (2), 267–269.

Elcrat, A. R. (1977). On the propagation of sonic discontinuities in the unsteady flow of a perfect gas, *Int. J. Engng. Sci.*, **15** (1), 29–34.

Elżanowski, M. (1975). *Geometrical Properties of Space of States in Neo-classical Thermodynamics*, Doctoral Thesis, IPPT PAN, Warszawa (in Polish).

Engelking, R. (1979). *General Topology*, PWN, Warsaw.

Ericksen, J. L. (1953). On the propagation of waves in isotropic incompressible perfectly elastic materials, *J. Rat. Mech. Anal.*, **2** (2), 329–337.

Eringen, A. C. (1975). Basic principles, in: *Continuum Physics*, A. C. Eringen (ed.), Vol. II, 1–127, Academic Press, New York–San Francisco–London.

Eringen, A. C. and Şuhubi, E. S. (1974). *Elastodynamics*, Vol. I, *Finite Motions*, Academic Press, New York–London.

Federer, H. (1966). *Geometric Measure Theory*, Springer-Verlag, Berlin.

Fergola, P. and Romano A. (1983). On the thermodynamics of fluid and solid phases, *Ricerche di Matematica, Napoli*, **XXXII** (2), 221–235.

Fisher, G. M. C. and Leitman, M. J. (1968). On continuum thermodynamics with surfaces, *Arch. Rat. Mech. Anal.*, **30** (3), 225–262.

Fisher, G. M. C. and Leitman, M. J. (1970). Continuum thermodynamics with surfaces: restrictions on constitutive equations, *Quart. Appl. Math.*, **28** (3), 303–311.

Fiszdon, W., Herczyński, R. and Walenta, Z. (1974). The structure of a plane shock wave of a monotonic gas. Theory and experiment, in: *Rarefied Gas Dynamics, Proc. IX Inter. Symp. Göttingen*, M. Becker and M. Fiebig (eds.), DFVLR-Press, Porz-Wahn.

Fiszdon, W., Herczyński, R. and Walenta, Z. (1976). Plane shock waves in monotonic gas. Comparison between experimental and theoretical results, *Rozpr. Inżyn.* (Engin. Trans.), **24** (3), 629–650.

Fiszdon, W., Herczyński, R. and Walenta, R. (1983). Shock wave structures, shock reflections—gaps between experiment and theory, Bericht 110/1983, Max-Planck-Institute für Strömungsforschung, Göttingen.

Fox, N. (1969). Generalized thermoelasticity, *Int. J. Engng. Sci.*, **7**, 437–445.

Freudenthal, A. M. and Geiringer, H. (1958). The mathematical theories of the inelastic continuum, in: *Handbuch der Physik*, Vol. VI, 229–343, Springer-Verlag, Berlin–Göttingen–Heidelberg.

Friedlander, F. G., (1958). *Sound Pulses*, Cambridge University Press, Cambridge.

Frischmuth, K. and Perzyna, P., (1983). Thermodynamics of the modified material structure with internal state variables, *Arch. Mech.*, **35**, 279–286.

Germain, P. (1972). Shock waves, jump relations, and structure, in: *Advances in Applied Mechanics*, **12**, 131–194.

Germain, P. and Lee, E. H. (1973). On shock waves in elastic-plastic solids, *J. Mech. Phys. Solids*, **21**, 359–382.

Ghez, R. (1966). A generalized Gibbsian surface, *Surface Science*, **4**, 125–140.

Gibbs, J. W. (1928). *The scientific Papers of J. Willard Gibbs*, Vol. I, *Thermodynamics*, Yale University Press, New Haven.

Goetz, A. (1965). *Differential Geometry*, PWN, Warszawa (in Polish).

Green, W. A. (1964). The growth of plane discontinuities propagating into a homogeneously deformed elastic material, *Arch. Rat. Mech. Anal.*, **16** (1), 79–88.

Green, W. A. (1965). The growth of plane discontinuties propagating into a homogeneously deformed elastic material, corrections and additional results, *Arch. Rat. Mech. Anal.*, **19** (1), 20–23.

Green, A. E. and Laws, N. (1972). On a global entropy production inequality, *Quart. J. Mech. Appl. Math.*, **25**, 1–11.

Green, A. E. and Naghdi, P. M. (1977). On thermodynamics and the nature of the second law, *Proc. Roy. Soc. Lond.*, **A357**, 253–270.

Green, A. E. and Rivlin, R. S. (1964). On Cauchy's equations of motion, *ZAMP*, **15**, 290–292.

Greguss, P. (1960). On the relation between the ultrasonic velocity and the surface properties of metals, *Proc. Vibr. Probl.*, **1** (5), 3–10.

Guggenheim, E. A. (1957). *Thermodynamics, an Advanced Treatment for Chemists and Physicists*, North-Holland, Amsterdam.

Gumiński, K. (1962). *The Thermodynamics of Irreversible Processes*, PWN, Warszawa (in Polish).

Gurtin, M. E. (1972). The Linear Theory of Elasticity, in: *Handbuch der Physik*, Vol. VIa/2, 1–295, Springer-Verlag, Berlin–Heidelberg–New York.

Gurtin, M. E. (1973). Modern Continuum Thermodynamics, in: *Mechanics Today*, S. Nemat-Nasser (ed.), Vol. I, 168–213, Pergamon Press, New York–Toronto–Oxford–Sydney–Braunschweig.

Gurtin, M. E. (1976). On the existence of a single temperature in continuum thermodynamics, *J. Appl. Math. Phys.* (*ZAMP*), **27** (6), 775–779.

Gurtin, M. E., Markenscoff, X. and Thurston, R. N. (1976). Effect of surface stress on the natural frequency of thin crystals, *Appl. Phys. Letters*, **29** (9), 529–530.

Gurtin, M. E. and Martins, L. C. (1976). Cauchy's theorem in classical physics, *Arch. Rat. Mech. Anal.*, **60** (4), 305–324.

Gurtin, M. E., Mizel, V. J. and Williams, W. O. (1968). A note on Cauchy's stress theorem, *J. Math. Anal. Appl.*, **22**, 398–401.

Gurtin, M. E. and Murdoch, A. I. (1975). A continuum theory of elastic material surfaces, *Arch. Rat. Mech. Anal.*, **57** (4), 291–323.

Gurtin, M. E. and Murdoch, A. I. (1976). Effect of surface stress on wave propagation in solids, *J. Appl. Phys.*, **47** (10), 4414–4421.

Gurtin, M. E. and Pipkin, A. C. (1968). A general theory of heat conduction with finite wave speed, *Arch. Rat. Mech. Anal.*, **31** (2), 113–126.

Gurtin, M. E. and Walsh, E. K. (1967). Extrinsically induced acceleration waves in elastic bodies, *J. Acoust. Soc. Am.*, **41**, 1320–1324.

Gurtin, M. E. and Williams, W. O. (1967). An axiomatic foundation for continuum thermodynamics, *Arch. Rat. Mech. Anal.*, **26** (2), 83–117.

Gyarmati, I., (1977). On the wave approach of thermodynamics and some problems of non-linear theories, *J. Non-Equilib. Thermodyn.*, **2** (4), 233–260.

Hadamard, J. (1903). *Leçons sur la propagation des ondes et les equations de l'hydrodynamique*, Hermann, Paris; reprinted in 1949 by Chelsea Publishing Company, New York.

Hadamard, J. (1927). *Course d'analyse*, Hermann and Company, Paris, Vol. I.

Hanyga, A. (1986). *Modern Thermodynamics of Continuous Media*, PWN, Warszawa (in Polish).

Hanyga, A. (ed.) (1984). *Seismic Wave Propagation in the Earth*, in: *Physics and Evolution of the Earth's Interior*, R. Teisseyre (ed.), 1–377, PWN, Warszawa, and Elsevier, Amsterdam–Oxford–New York.

Hanyga, A. (1985). *Mathematical Theory of Non-Linear Elasticity*, PWN, Warszawa, Ellis Horwood, Chichester.

Hayes, M. A. (1968). On pure extension, *Arch. Rat. Mech. Anal.*, **31** (5), 155–164.

Hayes, M. A. (1984). Inhomogeneous plane waves, *Arch. Rat. Mech. Anal.*, **85** (1), 41–79.

Hayes, M. A. and Rivlin, R. S. (1972). Energy propagation in a deformed material, *Arch. Rat. Mech. Anal.*, **45** (1), 54–62.

Hayes, W. D. (1957). The vorticity jump across a gasdynamic discontinuity, *J. Fluid Mech.*, **2** (6), 595–600.

Hayes, W. D. (1970). Kinematic wave theory, *Proc. Roy. Soc. Lond.* **A 320**, 209–226.

Hayes, W. D. (1974). Introduction to wave propagation, in: *Non-linear Waves*, S. Leibowich, A.R. Seebass (eds.), 1–43, Cornell University Press, Ithaca–London.

Helmholtz, H. (1858, 1867). Über Integrale der hydrodynamischen Gleichungen, welche den Wirbelbewegungen entsprechen, *J. Reine Angew. Math.*, **55**, 25–55; see also English translation: On integrals of the hydrodynamical equations, which express vortex motion, *Phil. Mag.*, **33** (4), 485–512.

Herring, C. (1953). The use of classical macroscopic concepts in surface energy problems, in: *Structure and Properties of Solid Surfaces*, R. Gomer, C. S. Smith (eds.), University of Chicago Press, Chicago.

Hill, R. (1961). Discontinuity relations in the mechanics of solids, in: *Progress in Solid Mechanics*, Vol. II, I. N. Sneddon, R. Hill (eds.), North-Holland, Amsterdam.

Hill, R. (1962). Acceleration waves in solids, *J. Mech. Phys. Solids*, **10** (1), 1–16.

Hill, R. (1967). Eigenmodal deformations in elastic-plastic continua, *J. Mech. Phys. Solids*, **15** (6), 371–386.

Hugoniot, H. (1885). Sur la propagation du mouvement dans un fluide indéfini, *C.R. Acad. Sci.*, **101**, 1118–1120, 1229–1232, Paris.

Hugoniot, H. (1887, 1889). Mémoire sur la propagation du mouvement dans les corps et spécialement dans les gaz parfaits, *J. École polytech.*, **57**, 3–97, **58**, 1–125.

Huilgol, R. R. (1973). Growth of plane shock waves in materials with memory, *Int. J. Engng. Sci.*, **11** (1), 75–86.

Hutter, K. (1977). The foundations of thermodynamics, its basic postulates and implications; A review of modern thermodynamics, *Acta Mech.*, **27**, 1–54.

Ignaczak, J. (1980). Thermoelasticity with finite wave speeds—a survey, in: *Proc. Conf. Thermal Stresses, March, 1980, Virginia*, Polyt. Inst. State Univ. Blackburg.

Israel, W. (1976). Non-stationary irreversible thermodynamics: a causal relativistic theory, *Ann. Phys.*, **100**, 310–331.

Jeffrey, A. (1964). A note on the derivation of the discontinuity conditions across contact discontinuities, shocks and phase fronts, *ZAMP*, **15**, 68–71.

Jeffrey, A. (1965). A note on the integral form of the fluid dynamic conservation equations relative to an arbitrary moving volume, *ZAMP*, **16**, 835–837.

Jeffrey, A. (1976). *Quasilinear Hyperbolic Systems and Waves*, Research Notes in Mathematics, **5**, Pitman, London–San Francisco–Melbourne.

Jeffrey, A. and Taniuti, T. (1964). *Non-linear Wave Propagation with Application to Physics and Magnetohydrodynamics*, Academic Press, New York–London.

John, F. (1978). *Partial Differential Equations* (3rd edition), Springer-Verlag.

Kaliski, S. (1965). Wave equation of heat conduction, *Bull. Acad. Polon, Sci., Série Sci. Tech.*, **13** (4), 211–219.

Karpman, V. U. (1973). *Nelineynye volny v dispergiruyushchikh sredakh*, Izd. Nauka, Moskva.

Keller, J. B. (1962). Geometrical theory of diffraction, *J. Opt. Soc. Amer.*, **52**, 116–130.

Kennedy, J. E. and Nunziato, J. W. (1976). Shock wave evolution in a chemically reacting solid, *J. Mech. Phys. Solids*, **24** (2/3), 107–124.

Kestin, J. (1966, 1968). *A Course in Thermodynamics*, Blaisdell, Vols. I, II, London.

Kline, M. and Kay, I. W. (1965). *Electromagnetic Theory and Geometrical Optics*, Interscience, New York–London–Sydney.

Kotchine, N. E. (1926). Sur la théorie des ondes de choc dans un fluide, *Rend.Circ. Mat. Palermo*, **50**, 305–344.

Kočin, N. E. (1965). *Vektornoe ischislenie i nachal atenzornogo ischisleniya*, Izd. Nauka, Moskva.

Kosiński, W. (1974). Behaviour of acceleration and shock waves in materials with internal state variables, *Int. J. Non-Linear Mech.*, **9** (6), 481–499.

Kosiński, W. (1975). Thermal waves in inelastic bodies, *Arch. Mech.*, **27** (5–6), 733–748.

Kosiński, W. (1976). Analysis of one-dimensional shock and acceleration waves in an inelastic medium, *Theor. Appl. Mech.*, **14** (1), 95–126 (in Polish).

Kosiński, W. (1980). Fracture effects in the propagation of shock waves through a bulk solid, *Arch. Mech.*, **32** (3), 421–430.

Kosiński, W. (1982). A note on stability of dissipative bodies, *Arch. Mech.*, **34** (3), 401–407.

Kosiński, W. (1983a). *Evolution Equation of Dissipative Bodies*, Habilitation Thesis, IFTR-Reports, 9/83 (in Polish).

Kosiński, W. (1983b). On global evolution of states of deformable bodies, in: *Global Analysis—Analysis on Manifolds*, T. M. Rassias (ed.), Teubner Texte zur Mathematik, Band **57**, 180–208, Leipzig.

Kosiński, W. (1985). Thermodynamics of singular surfaces and phase transitions, in: *Free Boundary Problems: Applications and Theory*, Vol. III, A. Bossavit, A. Damlanian, A. M. Fremond (eds.), 140–151, Pitman, Boston–London–Melbourne.

Kosiński, W. (1986). *A note on balance laws*, in preparation.

Kosiński, W. and Perzyna, P. (1972). Analysis of acceleration waves in material with internal parameters, *Arch. Mech.*, **24** (4), 629–643.

Kosiński, W. and Perzyna, P. (1973). The unique material structures, *Bull. Acad. Polon. Sci., Série Sci. Tech.*, **21** (12), 655–662.

Kosiński, W. and Szmit, K. (1977). On waves in elastic materials at low temperatures, Part I. Hyperbolicity in thermoelasticity, Part II. Principal waves, *Bull. Acad. Polon. Sci., Série Sci. Tech.*, **25** (1), 17–32.

Kranyš, M. (1972). Kinetic derivation of non-stationary general relativistic thermodynamics, *Nuovo Cimento*, **8B**, 417–441,

Kukudžanov, V. (1974). On the propagation of strong discontinuity waves in an elastic-viscoplastic medium, in: *Dynamics of Inelastic Media*, 255–271, Zakład Narodowy im. Ossolińskich, Wrocław–Warszawa–Kraków–Gdańsk.

Kukudžanov, V. (1977). On wave propagation in a coupled thermoelastic-plastic medium, *Arch. Mech.*, **29**, 325–338.

Landau, L. and Lifshitz, E. (1959). *Fluid Mechanics of Continuous Media*, Pergamon Press and Addison–Wesley, Reading, Mass.

Lawenda, B. H. (1979). *Thermodynamics of Irreversible Processes*, MacMillan, London.

Lewis, R. M. (1965). Asymptotic theory of wave propagation, *Arch. Rat. Mech. Anal.*, **20** (3), 191–250.

Lighthill, M. J. (1957). Dynamics of a dissociating gas, Part I. Equilibrium flow, *J. Fluid Mech.*, **2** (1), 1–32.

Lighthill, M. J. and Whitham, G. B. (1955). On kinematic waves, I. Flood movement in long rivers, *Proc. Roy. Soc. Lond.*, **A229**, 281–316.

Lindsay, K. A. and Straughan, B. (1979a). A thermodynamic viscous interface theory and associated stability problems, *Arch. Rat. Mech. Anal.*, **71** (4), 307–326.

Lindsay, K. A. and Straughan, B. (1979b). Propagation of mechanical and temperature acceleration waves in thermoelastic materials, *ZAMP*, **30**, 477–490.

Liu, I-Shuh (1972). Method of Lagrange multipliers for exploitation of entropy principle, *Arch. Rat. Mech. Anal.*, **46**, 131–148.

Lord, H. W. and Lopez, A. A. (1970). Wave propagation in thermoelastic solids at very low temperature, *Acta Mech.*, **10**, 85–98.

Lord, H. W. and Shulman, Y. (1967). A generalized dynamical theory of thermoelasticity, *J. Mech. Phys. Solids*, **15**, 299.

Łojasiewicz, S. (1973). *Introduction to the Theory of Real Functions*, PWN, Warszawa (in Polish).

Malecki, I. (1969). *Physical Foundations of Technical Acoustics*, PWN–Pergamon Press, Warszawa - Oxford - London - Edinburgh - New York - Toronto - Sydney - Paris - Braunschweig.

Mandel, J. (1962). Ondes plastiques dans un milieu indéfini à trois dimensions, *J. Méc.*, **1** (1), 3–30.

Mandel, J. (1976). Notions générales sur les ondes, in: *Mechanical Waves in Solids*, J. Mandel, L. Brun (eds.), 1–62, CISM Courses and Lectures, No. 222, Springer-Verlag.

(1972). Manual of Definitions, Terminology and Symbols in Colloid and Surface Chemistry, *Pure Appl. Chem.*, **31**, 578.

Maurin, K. (1976). *Analysis*, Part I, PWN, Warszawa, D. Reidel, Dordrecht–Boston.

Maxwell, J. C. (1867). On the dynamical theory of gases, *Phil. Trans. Roy. Soc. Lond.*, **157**, 49–101.

McCarthy, M. F. (1966). The growth of magneto-elastic waves in a Cauchy elastic material of finite electrical conductivity, *Arch. Rat. Mech. Anal.*, **23** (3), 191–227.

McCarthy, M. F. (1969). The growth of discontinuities in relativistic gas dynamics, *Int. J. Eng. Sci.*, **7**, 209–216.

McCarthy, M. F. (1970). Wave propagation in thermomechanical materials with memory, I. The speed of propagation and amplitude of first-order waves, *Proc. Vibr. Probs.*, **11** (2), 123–133.

McCarthy, M. F. (1971). The growth of thermal waves, *Int. J. Engng. Sci.*, **9**, 163–174.

McCarthy, M. F. (1973). Thermodynamic influences on the propagation of waves in electro-elastic materials, *Int. J. Engng. Sci.*, **11**, 1301–1316.

McCarthy, M. F. (1975). Singular surfaces and waves, in: *Continuum Physics*, A. C. Eringen (ed.), Vol. II, 449–521, Academic Press, New York–San Francisco–London.

McCarthy, M. F. and O'Leary, P. M. (1975). Thermodynamic influences on the propagation of electromagnetic shock waves, *Royal Irish Academy*, **74**, Sec. A, (9), 85–96.

McCarthy, M. F. and O'Leary, P. M. (1978). The growth and decay of a hydraulic jump, *J. Hydrol.*, **39**, 272–285.

McCarthy, M. F. and Tiersten, H. F. (1978). On integral forms of the balance laws for deformable semiconductors, *Arch. Rat. Mech. Anal.*, **68** (1), 27–36.

McCarthy, M. F. and Tiersten, H. F. (1980). Wave propagation in composite media viewed as interpenetrating continua, in: *Continuum Models of Discrete Systems*, 531–549, University of Waterloo Press.

McCarthy, M. F. and Tiersten, H. F. (1982). On one-dimensional acceleration waves in composite materials modelled as interpenetrating solid continua, *J. Elasticity*, **12** (4), 345–366.

Meixner, J. (1966). Consequences of an inequality in non-equilibrium thermodynamics, *J. Appl. Mech.*, **33** (3), 481–488.

Meixner, J. (1968). TIP has many faces, *IUTAM Symp. on Irreversible Aspects of Continuum Mechanics, Vienna, June, 1966*, Springer-Verlag, Wien.

Meixner, J. and Reik, H. B. (1959). Thermodynamik der irreversiblen Prozesse, in: *Handbuch der Physik*, Vol. III/2, 413–523, Springer-Verlag, Berlin–Göttingen–Heidelberg.

Mihăilescu, M. and Suliciu, I. (1976). Finite and symmetric thermomechanical waves in materials with internal state variables, *Int. J. Solids Structures*, **12** (8), 559–575.

Mindlin, R. D. (1963). High frequency vibrations of plated crystal plates, in: *Progress in Applied Mechanics, The Prager Anniversary Volume*, 73–84, Macmillan, New York.

Mindlin, R. D. (1965). Second gradient of strain and surface tension in linear elasticity, *Int. J. Solids Structures*, **1** (4), 417–438.

Moeckel, G. P. (1974). Thermodynamics of an interface, *Arch. Rat. Mech. Anal.*, **57** (3), 255–280.

Morro, A. (1980a). Acceleration waves in thermo-viscous fluids, *Rend. Sem. Mat. Univ. Padova*, **63**, 169–184.

Morro, A. (1980b). Wave propagation in thermo-viscous materials with hidden variables, *Arch. Mech.*, **32**, 145–161.

Morro, A. (1981). Oblique interaction of waves with shocks, *Acta Mech.*, **38**, 241–248.

Müller, I. (1967a). On the entropy inequality, *Arch. Rat. Mech. Anal.*, **26** (2), 118–141.

Müller, I. (1967b). Zum Paradoxen der Wärmeleitungstheorie, *Zeits. für Phys.*, **198**, 329–344.

Müller, I. (1971). Die Kältefunktion, eine universelle Funktion in der Thermodynamik viskoser wärmeleitender Flüssigkeiten, *Arch. Rat. Mech. Anal.*, **40** (1), 1–36.

Müller, I. (1973). Entropy, absolute temperature and coldness in thermodynamics, *CISM Lectures*, 1971, No. 76, Springer-Verlag, Vienna–New York.

Müller, I. (1979). Entropy in non-equilibrium thermodynamics. A challenge to mathematicians, in: *Trends in the Application of Pure Mathematics to Mechanics*, Vol. II, H. Zorski (ed.), 281–295, Pitman, Boston–London–Melbourne.

Murdoch, A. I. (1976a). A thermodynamical theory of elastic material interfaces, *Quart. J. Mech. Appl. Math.*, **29** (3), 245–275.

Murdoch, A. I. (1976b). The propagation of surface waves in bodies with material boundaries, *J. Mech. Phys. Solids*, **24** (2/3), 137–146.

Murdoch, A. I. (1977). The effect of interfacial stress on the propagation of Stoneley waves, *J. Sound Vibration*, **50** (1), 1–11.

Murdoch, A. I. (1979). Symmetry considerations for materials of second grade, *J. Elasticity*, **9** (1), 43–50.

Murdoch, A. I. (1983). On material frame-indifference and constitutive relations motivated by the kinematic theory of gases, *Arch. Rat. Mech. Anal.*, **83** (2), 185–194.

Musgrave, M. J. P. (1970). *Crystal Acoustics*, Holden Day, San Francisco–Cambridge–London–Amsterdam.

Naghdi, P. M. (1972). The theory of shells and plates, in: *Handbuch der Physik*, Vol. VIa/2, 425–640, Springer-Verlag, Berlin–Heidelberg–New York.

Napolitano, L. G., (1977). Decomposition of fluid dynamics balance equations, *L'Aerotecnica Missili e Spazio*, **56** (4), 183–194.

Napolitano, L. G. (1978). Thermodynamics and dynamics of pure interfaces, *Acta Astronautica*, **5**, 655–670.

Napolitano, L. G. (1979). Thermodynamics and dynamics of surface phase, *Acta Astronautica*, **6**, 1093–1112.

Napolitano, L. G. (1982). Properties of parallel-surface coordinate systems, *Meccanica*, **17**, 107–118.

Nariboli, G. A. (1966). Wave propagation in anisotropic elasticity, *J. Math. Anal.*, **16**, 108–122.

Nemat-Nasser, S. (1973). On non-linear thermoelasticity and non-equilibrium thermodynamics, in: *Non-linear Elasticity*, R. W. Dickey (ed.), Academic Press, New York–London.

Nickerson, H. K., Spencer, D. C. and Steenrod, N. E. (1959). *Advanced Calculus*, Van Nostrand, Princeton–New York.

Noll, W. (1955). On the continuity of the solid and fluid states, *J. Rat. Mech. Anal.*, **4** (1), 3–81.

Noll, W. (1972). A new mathematical theory of simple materials, *Arch. Rat. Mech. Anal.*, **48** (1) 1–50.

Noll, W. (1973). Lectures on the foundations of continuum mechanics and thermodynamics, *Arch. Rat. Mech. Anal.*, **52** (1), 62–92.

Noll, W. (1978). A general framework for problems in the statics of finite elasticity, in: *Contemporary Developments in Continuum Mechanics and Partial Differential Equations*, 363–387, G. M. de la Penha, L. A. Medeiros (eds.), North-Holland, Amsterdam.

Norwood, F. R. and Warren W. E. (1969). Wave propagation in the generalized dynamical theory of thermoelasticity, *Quart. J. Mech. Appl. Math.*, **22**, 283–290.

Nowacki, W. (1975a). *Dynamic Problems in Thermoelasticity*, PWN, Warszawa.

Nowacki, W. (1975b). Ondes dans les milieux continus généralisés, in: *Mechanical Waves in Solids*, J. Mandel, L. Brun (eds.), 255–292, CISM Courses and Lectures, No. 222, Springer-Verlag, Wien–New York.

Nowacki, W. K. (1978). *Stress Waves in Non-Elastic Solids*, Pergamon Press, Oxford–New York–Toronto–Sydney–Paris–Frankfurt.

Nowacki, W. K. (1980). Propagation des ondes dans un sol elasto-plastique, in: *Problèmes non-linéaires de mécanique*, W. K. Nowacki (ed.), PWN, Warszawa, 423–432.

Nunziato, J. W. (1973). One-dimensional shock waves in a chemically reacting mixture of elastic materials, *J. Chem. Phys.*, **58** (3), 961–965.

Nunziato, J. W. and Walsh, E. K. (1975). Acceleration waves and mild dicontinuities in a non-simple mixture of chemically reacting elastic materials, *Arch. Rat. Mech. Anal.*, **58** (2), 115–126.

Nunziato, J. W., Walsh, E. K., Schuler, K. W. and Barker, L. M. (1974). Wave propagation in nonlinear viscoelastic solids, in: *Handbuch der Physik*, Vol. VIa/4, 1–108, Springer-Verlag, Berlin–Heidelberg–New York.

Ogden, R. W. (1970). Waves in isotropic elastic materials of Hadamard, Green or harmonic type, *J. Mech. Phys. Solids*, **18**, 149–163.

Ogden, R. W. (1974). Growth and decay of acceleration waves in incompressible elastic solids, *Quart. J. Mech. Appl. Math.*, **27**, 451–464.

Ogden, R. W. (1984). *Non-linear Elastic Deformations*, Ellis Horwood, Chichester.

Olszak, W., Perzyna, P. and Sawczuk, A. (1965). *The Theory of Plasticity*, PWN, Warszawa (in Polish).

Orowan, E. (1970). Surface energy and surface tension in solids and fluids, *Proc. Roy. Soc. Lond.*, **A316**, 473–491.

Ościk, J. (1982). *Adsorption*, PWN, Warszawa, and Ellis Horwood, Chichester.

Parker, D. F. and Seymour, B. R. (1980). Finite amplitude one-dimensional pulses in an inhomogeneous granular material, *Arch. Rat. Mech. Anal.*, **72**, 265–284.

Perzyna, P. (1974). Thermodynamic theory of rheological materials with internal structure changes, in: *Problèmes de la rhéologie*, W. K. Nowacki (ed.), 277–306, PWN, Varsovie.

Perzyna, P. (1978). *Thermodynamics of Inelastic Materials*, PWN, Warszawa (in Polish).

Perzyna, P. (1982). Discussion of strain rate effects on necking phenomenon, in: *Mechanics of Inelastic Media and Structures, Proc. Int. Symp., Warszawa, 1978*, O. Mahrenholtz, A. Sawczuk (eds.), PWN, Warszawa–Poznań.

Perzyna, P. and Kosiński, W. (1973). A mathematical theory of materials, *Bull. Acad. Polon. Sci. Série Sci. Tech.*, **21** (12), 647–654.

Piau M. (1975). Ondes d'accélération dans les milieux élastoplastiques, viscoplastiques, *J. Méc.*, **14** (1), 1–38.

Pierotti, R. A. and Thomas, H. E. (1971). *Physical Adsorption: The Interaction of Gases and Solids*, John Wiley, New York.

Płatkowski, T. (1981). Structure of a shock wave in mixtures of monatomic and diatomic gases with rotational degrees of freedom, IFTR-Reports, 7/81 (in Polish).

Rademacher, H. (1919). Über partielle und totale Differenzierbarkeit von Funktionen mehrerer Variabeln und über die Transformationen der Doppelintegral, *Math. Anal.*, **79**, 340–359.

Ram, R. and Pandev, B. D. (1979). Local and global behavior of acceleration waves in radiating gases, *Indian J. Pure Appl. Math.*, **10** (8), 950–958.

Raniecki, B. (1976). Ordinary waves in inviscid media, in: *Mechanical Waves in Solids*, J. Mandel, L. Brun (eds.), 157–219, CISM Courses and Lectures, No. 222, Springer-Verlag, Wien–New York.

Reynolds, O., (1903). *The Sub-Mechanics of the Universe, Collected Papers*, Vol. 3.

Rice, J. R. (1976). The localization of plastic deformation, in: *Theoretical and Applied Mechanics*, W. T. Koiter (ed.), 207–220, North-Holland, Amsterdam.

Romano, A. (1982). Thermodynamics of a continuum with an interface and Gibbs' rule, *Ricerche di Matematica*. Napoli, **XXXI** (2), 277–294.

Ruggeri, T. (1983). Symmetric-hyperbolic system of conservation equations for a viscous heat-conducting fluid, *Acta Mech.*, **47** (34), 167–183.

Rychlewski, J. (1970–71). *Nonlinear Mechanics of the Continuum* (unpublished lectures), IPPT PAN (in Polish).

Sawczuk, A. (1973). *Foundations of Plasticity*, Noordhoff, Leyden.

Sawczuk, A. (1974). *Problems of Plasticity*, Noordhoff, Leyden.

Schapery, R. A. (1964). Application of thermodynamics to thermomechanical, fracture and birefringent phenomena in viscoelastic media, *J. Appl. Phys.*, **35** (5), 1451–1465.

Schmitt, H. (1972). Entstehnung von Verdichtungsstössen in strahlenden Gasen, *ZAMM*, **52** (11), 529–537.

Scholte, J. G. (1947). The range and existence of Rayleigh and Stoneley waves, *Monthly Notices Roy. Astron. Soc. Geophys. Supp.*, **5**, 120–126.

Schuler, K. W., Nunziato, J. W. and Walsh, E. K. (1973). Recent results in non-linear viscoelastic wave propagation, *Int. J. Solids Structures*, **9** (10), 1237–1281.

Scott, N. H. (1975a). Acceleration waves in constrained elastic materials, *Arch. Rat. Mech. Anal.*, **58** (1), 57–75.

Scott, N. H. (1975b). Acceleration waves in incompressible elastic solids, *Quart. J. Mech. Appl. Math.*, **29** (3), 295–310.

Scriven, L. E. (1960). Dynamics of a fluid interface, *Chem. Engng. Sci.*, **12**, 98–108.

Semenchenko, V. K. (1957). *Poverkhnostnye yavleniya v metalakh i splavakh*, Gosud. Izdat. Tekh.-Teoret. Lit., Moskva.

Seymour, B. R. and Mortell, M. P. (1975). Nonlinear geometrical acoustics, *Mechanics Today*, S. Nemat-Nasser (ed.), Vol. II, 251–312, Pergamon Press., New York–Toronto.

Sikorski, R. (1969). Advanced Calculus. Functions of Several Variables, PWN, Warszawa.

Šilhavy, M. (1977). A note on the existence of entropy in classical thermodynamics, *Arch. Mech.*, **29** (2), 289–298.

Šilhavy, M. (1978). A condition equivalent to the existence of non-equilibrium entropy and temperature for materials with internal variables, *Arch. Rat. Mech. Anal.*, **6**, 299–332.

Šilhavy, M. (1980). On measures, convex cones and foundations of thermodynamics, I. Systems with vector-valued actions, II. Thermodynamic systems, *Czech. J. Phys.*, **B30**, 841–861, 961–991.

Šilhavy, M. (1982). On the second law of thermodynamics, Parts I and II, *Czech. J. Phys.*, **B32**, 987–1010; 1073–1099.

Singh, R. S. and Sharma, V. D. (1979). The growth of discontinuities in a non-equilibrium flow of a relaxing gas, *Quart. J. Mech. Appl. Math.*, **32** (4), 331–338.

Slattery, J. C. (1964). Surfaces, I. Momentum and moment momentum balance for moving surfaces, II. Kinematics of diffusion in a heterogeneous surface, *Chem. Engng. Sci.*, **19**, 379–385, 435–455.

Slattery, J. C. (1967). General balance equation for a phase interface, *Ind. Engng. Chem. Fundamentals*, **6** (1), 108–115.

Sobczyk, K. (1984). *Stochastic Wave Propagation*, PWN, Warszawa, Elsevier, Amsterdam.

Sommerfeld, A. and Runge, J. (1911). Anwendung der Vektorrechnung auf die Grundlagen der Geometrischen Optik, *Ann. Phys.*, **35**, 277–298.

Spence, D. A. (1973). Nonlinear wave propagation in viscoelastic materials, in: *Nonlinear Elasticity*, R. W. Dickey (ed.), 365–396, Academic Press, New York–London.

Spielrein, J. (1916). *Lehrbuch der Vectorrechnung nach den Bedürfnissen in der technischen Mechanik und Elektrizitätslehre*, Stuttgart.

Stokes, G. G., (1848). On a difficulty in the theory of sound, *Phil. Mag.*, **23**, 349–356.

Stoneley, R. (1924). Elastic waves at the surface of separation of two solids, *Proc. Roy. Soc.*, **A106**, 416–428.

Şuhubi, E. S. (1970). The growth of acceleration waves of arbitrary form in deformed hyperelastic materials, *Int. J. Engng. Sci.*, **8** (8), 699–710.

Suliciu, I. (1975). Symmetric waves in materials with internal state variables, *Arch. Mech.*, **27** (5–6), 841–856.

Synge, J. L. (1960). Classical dynamics, in: *Handbuch der Physik*, Vol. III/1, 1–225, Springer-Verlag, Berlin–Göttingen–Heidelberg.

Synge, J. L. and Conway, W. (1931). *The Mathematical Papers of W. R. Hamilton*, Vol. I, Cambridge University Press, London.

Szmit-Wołoszyńska, K. (1978). *Wave Problems in Coupled Thermoviscoplasticity*, Doctoral Thesis, IPPT PAN, Warszawa (in Polish).

Taylor, A. E. (1965), *General Theory of Functions and Integration*, Blaisdell, New York–Toronto–London.

Thomas, T. Y. (1949). The fundamental hydrodynamical equations and shock conditions for gases, *Math. Mag.*, **22**, 169–189.

Thomas, T. Y. (1957a). Extended compatibility conditions for the study of surfaces of discontinuity in continuum mechanics, *J. Math. Mech.*, **6**, 311–322, 907–908.

Thomas, T. Y. (1957b). The growth and decay of sonic discontinuities in ideal gases, *J. Math. Mech.*, **6** (4), 455–469.

Thomas, T. Y. (1961). *Plastic Flow and Fracture in Solids*, Academic Press, New York–London.

Thomas, T. Y. (1965). *Concepts from Tensor Analysis and Differential Geometry*, 2nd Edition, Academic Press, New York–London.

Thurston, R. N. (1974). Waves in solids, in: *Handbuch der Physik*, Vol. VIa/4, 109–308, Springer-Verlag, Berlin–Heidelberg–New York.

Tiersten, H. F. (1969). Elastic surface waves guided by thin films, *J. Appl. Phys.*, **40**, 770–789.

Tiersten, H. F., Sinha, B. K. and Meeker, T. R. (1981). Intrinsic stress in thin films deposited on anisotropic substrates and its influence on the natural frequencies of piezoelectric resonators, *J. Appl. Phys.*, **52** (9), 5614–5624.

Ting, T. C. T. (1976). Shock waves and weak discontinuities in anisotropic elastic-plastic media, *ASME—AMD*, **17**, 41–64.

Ting, T. C. T. (1977). Plastic wave speeds in isotropically work-hardening materials, *J. Appl. Mech.*, **44**, 68–72.

Tokuoka, T. (1973a). Thermo-acoustical waves in linear thermo-elastic materials, *J. Engng. Math.*, **7** (2), 115–122.

Tokuoka, T. (1973b). Growth and decay of thermo-acoustical waves in thermo-plastic materials, *Trans. Jap. Soc. Aeron. Space Sci.*, **16** (33), 143–159.

Tokuoka, T. (1974). Thermo-acoustical waves in thermo-plastic materials, *J. Engng. Math.*, **8** (1), 9–22.

Tokuoka, T. (1981a). Distortion of wave-form of weak and short waves in Eulerian fluid, *Memoirs of the Faculty of Engineering, Kyoto University*, **63** (1), 1–9.

Tokuoka, T. (1981b). Waves in elastic materials with internal constraints, *Int. J. Engng. Sci.* **19** (12), 1695–1704.

Tokuoka, T. and Kusunoki, H. (1982). Waves in elastic subfluids, *Arch. Rat. Mech. Anal.*, **79** (2), 175–187.

Tolman, R. C. and Fine, P. C. (1948). On the irreversible production of entropy, *Rev. Mod. Phys.*, **20** (1), 51–77.

Truesdell, C. A. (1952). On curved shocks in steady plane flow of an ideal fluid, *J. Aeron. Sci.*, **19**, 826–828.

Truesdell, C. A. (1961). General and exact theory of waves in finite elastic strain, *Arch. Rat. Mech. Anal.*, **8** (4), 263–296.

Truesdell, C. A. (1969). *Rational Thermodynamics*, McGraw Hill, New York.

Truesdell, C. A. (1977). *A First Course in Rational Continuum Mechanics*, Vol. I, Academic Press, New York.

Truesdell, C. A. and Noll, W. (1965). The non-linear field theories of mechanics, in: *Handbuch der Physik*, Vol. III/3, 1–602, Springer-Verlag, Berlin–Heidelberg–New York.

Truesdell, C. A. and Toupin, R. A. (1960). The classical field theories, in: *Handbuch der Physik*, Vol. III/1, 226–793, Springer-Verlag, Berlin–Göttingen–Heidelberg.

Turhan, D. and Mengi, Y. (1977). The propagation of initially plane waves in nonhomogeneous viscoelastic media, *Int. J. Solids Structures*, **13**, 79–92.

Uziembło, B. (1974). *Fundamentals of the Axiomatic Thermodynamics of Multicomponent Continuous Media*, Doctoral Thesis, IPPT PAN, Warszawa (in Polish).

Valanis, K. C. (1968). Unified theory of thermomechanical behaviour of viscoelastic materials, in: *Proc. Symp. Mech. Behav. Mater. Dyn. Loads, San Antonio, 1967*, 343–364, Springer-Verlag, New York.

Valanis, K. C. (1971). Irreversibility and existence of entropy, *Int. J. Non-Linear Mech.*, **6** (3), 337–360.

Varley, E. (1965a). Acceleration fronts in viscoelastic materials, *Arch. Rat. Mech. Anal.*, **19** (3), 215–225.

Varley, E. (1965b). Simple waves in general elastic materials, *Arch. Rat. Mech. Anal.*, **20** (5), 309–328.

Varley, E. and Cumberbatch, E. (1965). Non-linear theory of wave-front propagation, *J. Inst. Maths. Applics.*, **1**, 101–112.

Varley, E. and Dunwoody, J. (1965). The effect of non-linearity at an acceleration wave, *J. Mech. Phys. Solids*, **13** (1), 17–28.

Vernotte, M. P. (1958). Les paradoxes de la théorie continue de l'équation de la chaleur, *Compt. Rend.*, **246**, 3154–3155.

Vol'pert, A. I. (1967). The space BV and quasi-linear equations, *Math. USSR-Sbornik*, **2**, 225–267; translation from Russian: *Mat. Sb. AN*, **73** (115), 225–303.

Wang, C.-C. and Truesdell, C. A. (1973). *Introduction to Rational Elasticity*, Noordhoff, Leyden.

Weinberg, S. (1962). Eikonal method in magnetohydrodynamics, *Phys. Rev.*, **2**, **126** (6), 1899–1909.

Weingarten, G. (1901). Sulle superficie di discontinuità nella teoria della elasticità dei corpi solidi, *Rend. Accad. Lincei* (5), **10**, 57–60.

Wesołowski, Z. (1968). On surface tension and edge tension, *Arch. Mech.*, **20** (2), 243–258.

Wesołowski, Z. (1974). *Dynamic Problems of the Non-Linear Theory of Elasticity*, PWN, Warszawa (in Polish).

Wesołowski, Z. (1976). On the propagation of an acoustic wave in a non-linear elastic material, in: *Dynamics of Elastic Systems*, Cz. Woźniak (ed.), Zakład Narodowy im. Ossolińskich, Wrocław–Warszawa–Kraków–Gdańsk (in Polish).

Wesołowski, Z. (1977). Brittle fracture as a wave, *Arch. Mech.*, **29** (3), 491–496.

Wesołowski, Z. (1982). Shock wave in piecewise linear elastic material, *Arch. Mech.*, **34** (3), 351–358.

Wesołowski, Z. and Bürger, W. (1977). Shock waves in incompressible elastic solids, *Rheol. Acta*, **16** (2), 155–160.

Weyl, H. (1949). Shock waves in arbitrary fluids, *Comm. Pure Appl. Math.*, **2**, 103–122.

Whitham, G. B. (1960). A note on group velocity, *J. Fluid Mech.*, **9** (3), 347–352.

Whitham, G. B. (1961). Group velocity and energy propagation for three-dimensional waves, *Comm. Pure Appl. Math.*, **14** (3), 675–691.

Whitham, G. B. (1974). *Linear and Nonlinear Waves*, Wiley-Interscience, New York–London–Sydney–Toronto.

Whitham, G. B. (1979). *Lectures on Wave Propagation*, Tata Institute of Fundamental Research, Bombay, Springer-Verlag, Berlin–Heidelberg–New York.

Willems, J. C. (1972). Dissipative dynamical systems, Part I. General theory, *Arch. Rat. Mech. Anal.*, **45** (5), 321–351.

Williams, W. O. (1971). Axioms for work and energy in general continua, I. Smooth velocity fields, *Arch. Rat. Mech. Anal.*, **42** (2), 93–114.

Williams, W. O. (1972). Axioms for work and energy in general continua, II. Surfaces of discontinuity, *Arch. Rat. Mech. Anal.*, **49** (3), 225–240.

Wilmański, K. (1974a). *Fundamentals of Phenomenological Thermodynamics*, PWN. Warszawa (in Polish).

Wilmański, K. (1974b). Note on Clausius–Duhem inequality for a singular surface, *Bull. Acad. Polon, Sci., Série Sci. Tech.*, **20** (10), 493–500.

Wilmański, K. (1975a). Thermodynamic properties of singular surfaces in continuous media, *Arch. Mech.*, **27** (3), 517–529.

Wilmański, K. (1975b). *Outlines of the Thermodynamics of Continuous Media*, IFTR-Report, 6 (in Polish).

Wilmański, K. (1977). On the Galilean invariance of balance equations for a singular surface in a continuum, *Arch. Mech.*, **29** (3), 459–475.

Wiśniewski, S., Staniszewski, B. and Szymanik, R. (1976). *Thermodynamics of Nonequilibrium Processes*, PWN, Warszawa, and D. Reidel, Dordrecht–Boston.

Włodarczyk, E. (1977). *Shock Waves in Continuous Media, I. Phenomenological Theory of Shock Waves*, Publ. WAT, Warszawa (in Polish).

Wołoszyńska, K. (1981a). On coupling acceleration and shock waves in a thermoviscoplastic medium, I. Symmetry and hyperbolicity conditions, *Arch. Mech.*, **33**, 261–272.

Wołoszyńska, K. (1981b). On coupling acceleration and shock waves in a thermoviscoplastic medium, II. One-dimensional waves, *Arch. Mech.*, **33**, 451–468.

Woźniak, Cz. (1966) *Nonlinear Theory of Shells*, PWN, Warszawa (in Polish).

Wright, T. W. (1973). Acceleration waves in simple elastic materials, *Arch. Rat. Mech. Anal.*, **50** (4), 237–277.

Wright, T. W. (1976). An intrinsic description of unsteady shock waves, *Quart. J. Mech. Appl. Math.*, **29** (3), 311–325.

Zielke, W. (1977). *Reduction of Nonlinear Equations to KdV and Periodic Solutions*, Habilitation Thesis, IFTR-Reports, 58, Warszawa (in Polish).

Index

(The symbol ↗ denotes that the notion appears in the title of a section

Mathematics and its Applications

Series Editor: G. M. BELL, Professor of Mathematics, King's College (KQC), University of London

Artmann, B. — **The Concept of Number***
Balcerzyk, S. & Józefiak, T. — **Commutative Rings***
Balcerzyk, S. & Józefiak, T. — **Commutative Noetherian and Krull Rings***
Baldock, G. R. & Bridgeman, T. — **Mathematical Theory of Wave Motion**
Ball, M. A. — **Mathematics in the Social and Life Sciences: Theories, Models and Methods**
de Barra, G. — **Measure Theory and Integration**
Bell, G. M. and Lairs, D. A. — **Co-operative Phenomena in Lattice Models Vols. I & II**
Berkshire, F. H. — **Mountain and Lee Waves**
Berry, J. S., Burghes, D. N., Huntley, I. D., James, D. J. G. & Moscardini, A. O. — **Teaching and Applying Mathematical Modelling**
Burghes, D. N. & Borrie, M. — **Modelling with Differential Equations**
Burghes, D. N. & Downs, A. M. — **Modern Introduction to Classical Mechanics and Control**
Burghes, D. N. & Graham, A. — **Introduction to Control Theory, including Optimal Control**
Burghes, D. N., Huntley, I. & Mc Donald, J. — **Applying Mathematics**
Burghes, D. N. & Wood, A. D. — **Mathematical Models in the Social, Management and Life Sciences**
Butkovskiy, A. G. — **Green's Functions and Transfer Functions Handbook**
Butkovskiy, A. G. — **Structural Theory of Distributed Systems**
Cao, Z-Q., Kim, K. H. & Roush, F. W. — **Incline Algebra and Applications**
Chorlton, F. — **Textbook of Dynamics, 2nd Edition**
Chorlton, F. — **Vector and Tensor Methods**
Crapper, G. D. — **Introduction to Water Waves**
Cross, M. & Moscardini, A. O. — **Learning the Art of Mathematical Modelling**
Cullen, M. R. — **Linear Models in Biology**
Dunning-Davies, J. — **Mathematical Methods for Mathematicians, Physical Scientists and Engineers**
Eason, G., Coles, C. W. & Gettinby, G. — **Mathematics and Statistics for the Bio-sciences**
Exton, H. — **Handbook of Hypergeometric Integrals**
Exton, H. — **Multiple Hypergeometric Functions and Applications**
Exton, H. — ***q*-Hypergeometric Functions and Applications**
Faux, I. D. & Pratt, M. J. — **Computational Geometry for Design and Manufacture**
Firby, P. A. & Gardiner, C. F. — **Surface Topology**
Gardiner, C. F. — **Modern Algebra**
Gardiner, C. F. — **Algebraic Structures: with Applications**
Gasson, P. C. — **Geometry of Spatial Forms**
Goodbody, A. M. — **Cartesian Tensors**
Goult, R. J. — **Applied Linear Algebra**

Graham, A.	**Kronecker Products and Matrix Calculus: with Applications**
Graham, A.	**Matrix Theory and Applications for Engineers and Mathematicians**
Griffel, D. H.	**Applied Functional Analysis**
Griffel, D. H.	**Linear Algebra***
Hanyga, A.	**Mathematical Theory of Non-linear Elasticity**
Harris, D. J.	**Mathematics for Business, Management and Economics**
Hoskins, R. F.	**Generalised Functions**
Hoskins, R. F.	**Standard and Non-standard Analysis***
Hunter, S. C.	**Mechanics of Continuous Media, 2nd (Revised) Edition**
Huntley, I. & Johnson, R. M.	**Linear and Nonlinear Differential Equations**
Jaswon, M. A. & Rose, M. A.	**Crystal Symmetry: The Theory of Colour Crystallography**
Johnson, R. M.	**Theory and Applications of Linear Differential and Difference Equations**
Kim, K. H. & Roush, F. W.	**Applied Abstract Algebra**
Kosiński, W.	**Field Singularities and Wave Analysis in Continuum Mechanics**
Lindfield, G. & Penny, J. E. T.	**Microcomputers in Numerical Analysis**
Lord, E. A. & Wilson, C. B.	**The Mathematical Description of Shape and Form**
Marichev, O. I.	**Integral Transforms of Higher Transcendental Functions**
Massey, B. S.	**Measures in Science and Engineering**
Meek, B. L. & Fairthorne, S.	**Using Computers**
Mikolas, M.	**Real Function and Orthogonal Series**
Moore, R.	**Computational Functional Analysis**
Müller-Pfeiffer, E.	**Spectral Theory of Ordinary Differential Operators**
Murphy, J. A. & McShane, B.	**Compution in Numerical Analysis***
Nonweiller, T. R. F.	**Computational Mathematics: An Introduction to Numerical Approximation**
Ogden, R. W.	**Non-linear Elastic Deformations**
Oldknow, A. & Smith, D.	**Learning Mathematics with Micros**
O'Neill, M. E. & Chorlton, F.	**Ideal and Incompressible Fluid Dynamics**
O'Neill, M. E. & Chorlton, F.	**Viscous and Compressible Fluid Dynamics***
Rankin, R. A.	**Modular Forms**
Ratschek, H. & Rokne, J.	**Computer Methods for the Range of Functions**
Scorer, R. S.	**Environmental Aerodynamics**
Smith, D. K.	**Network Optimisation Practice: A Computational Guide**
Srivastava, H. M. & Karlsson, P. W.	**Multiple Gaussian Hypergeometric Series**
Srivastava, H. M. & Manocha, H. L.	**A Treatise on Generating Functions**
Shivamoggi, B. K.	**Stability of Parallel Gas Flows***
Stirling, D. S. G.	**Mathematical Analysis***
Sweet, M. V.	**Algebra, Geometry and Trigonometry in Science, Engineering and Mathematics**
Temperley, H. N. V. & Trevena, D. H.	**Liquids and Their Properties**
Temperley, H. N. V.	**Graph Theory and Applications**
Thom, R.	**Mathematical Models of Morphogenesis**
Toth, G.	**Harmonic and Minimal Maps**
Townend, M. S.	**Mathematics in Sport**
Twizell, E. H.	**Computational Methods for Partial Differential Equations**
Wheeler, R. F.	**Rethinking Mathematical Concepts**
Willmore, T. J.	**Total Curvature in Riemannian Geometry**
Willmore, T. J. & Hitchin, N.	**Global Riemannian Geometry**
Wojtyński, W.	**Lie Groups and Lie Algebras***

Statistics and Operational Research

Editor: **B. W. CONOLLY**, Professor of Operational Research, Queen Mary College, University of London

Beaumont, G. P.	**Introductory Applied Probability**
Beaumont, G. P.	**Probability and Random Variables***
Conolly, B. W.	**Techniques in Operational Research: Vol. 1, Queueing Systems***
Conolly, B. W.	**Techniques in Operational Research: Vol. 2, Models, Search, Randomization**
Conolly, B. W.	**Lecture Notes in Queueing Systems**
French, S.	**Sequencing and Scheduling: Mathematics of the Job Shop**
French, S.	**Decision Theory**
Griffiths, P. & Hill, I. D.	**Applied Statistics Algorithms**
Hartley, R.	**Linear and Non-linear Programming**
Jolliffe, F. R.	**Survey Design and Analysis**
Jones, A. J.	**Game Theory**
Kemp, K. W.	**Dice, Data and Decisions: Introductory Statistics**
Oliveira-Pinto, F.	**Simulation Concepts in Mathematical Modelling***
Oliveira-Pinto, F. & Conolly, B. W.	**Applicable Mathematics of Non-physical Phenomena**
Schendel, U.	**Introduction to Numerical Methods for Parallel Computers**
Stoodley, K. D. C.	**Applied and Computational Statistics: A First Course**
Stoodley, K. D. C., Lewis, T. & Stainton, C. L. S.	**Applied Statistical Techniques**
Thomas, L. C.	**Games, Theory and Applications**
Whitehead, J. R.	**The Design and Analysis of Sequential Clinical Trials**

**In preparation*